Introduction to Reactive Gas Dynamics

Introduction to Reactive Gas Dynamics

Raymond Brun

OXFORD
UNIVERSITY PRESS

Great Clarendon Street, Oxford OX2 6DP

Oxford University Press is a department of the University of Oxford.
It furthers the University's objective of excellence in research, scholarship,
and education by publishing worldwide in

Oxford New York

Auckland Cape Town Dar es Salaam Hong Kong Karachi
Kuala Lumpur Madrid Melbourne Mexico City Nairobi
New Delhi Shanghai Taipei Toronto

With offices in

Argentina Austria Brazil Chile Czech Republic France Greece
Guatemala Hungary Italy Japan Poland Portugal Singapore
South Korea Switzerland Thailand Turkey Ukraine Vietnam

Oxford is a registered trade mark of Oxford University Press
in the UK and in certain other countries

Published in the United States
by Oxford University Press Inc., New York

© Raymond Brun 2009

The moral rights of the author have been asserted
Database right Oxford University Press (maker)

First Published 2009

All rights reserved. No part of this publication may be reproduced,
stored in a retrieval system, or transmitted, in any form or by any means,
without the prior permission in writing of Oxford University Press,
or as expressly permitted by law, or under terms agreed with the appropriate
reprographics rights organization. Enquiries concerning reproduction
outside the scope of the above should be sent to the Rights Department,
Oxford University Press, at the address above

You must not circulate this book in any other binding or cover
and you must impose the same condition on any acquirer

British Library Cataloguing in Publication Data

Data available

Library of Congress Cataloging in Publication Data

Brun, R. (Raymond), 1932–
  Introduction to reactive gas dynamics / Raymond Brun.
    p. cm.
  ISBN 978–0–19–955268–9
 1. Relaxation (Gas dynamics)   2. Gas dynamics.   3. Nonequilibrium statistical
mechanics. I. Title.
QC168.85.R45B78   2009
533′.21—dc22   2008054980

Typeset by Newgen Imaging Systems (P) Ltd., Chennai, India
Printed in the UK on acid-free paper
by CPI Antony Rowe, Chippenham, Wiltshire

ISBN 978–0–19–955268–9

10 9 8 7 6 5 4 3 2 1

Contents

Introduction		xiii
General Notations		xvii
Part I Fundamental Statistical Aspects		1
Notations to Part I		3

1 Statistical Description and Evolution of Reactive Gas Systems — 5

- 1.1 Introduction — 5
- 1.2 Statistical description — 6
  - 1.2.1 State parameters — 7
  - 1.2.2 Transport parameters — 9
- 1.3 Evolution of gas systems — 11
  - 1.3.1 Boltzmann equation — 11
  - 1.3.2 General properties — 12
  - 1.3.3 Macroscopic balance equations — 12
- 1.4 General properties of collisions — 14
  - 1.4.1 Elastic collisions — 14
  - 1.4.2 Inelastic collisions — 17
  - 1.4.3 Reactive collisions — 18
- 1.5 Properties of collisional terms — 18
  - 1.5.1 Collisional term expressions — 18
  - 1.5.2 Characteristic times: collision frequencies — 21
- Appendix 1.1 Elements of tensorial algebra — 22
- Appendix 1.2 Elements of molecular physics — 25
- Appendix 1.3 Mechanics of collisions — 31

2 Equilibrium and Non-Equilibrium Collisional Regimes — 36

- 2.1 Introduction — 36
- 2.2 Collisional regimes: generalities — 37
- 2.3 Pure gases: equilibrium regimes — 38
  - 2.3.1 Monatomic gases — 39
  - 2.3.2 Diatomic gases — 41

2.4	Pure diatomic gases: general non-equilibrium regime		43
2.5	Pure diatomic gases: specific non-equilibrium regimes		46
	2.5.1	Dominant TV collisions	47
	2.5.2	Dominant VV collisions	47
	2.5.3	Dominant resonant collisions	49
	2.5.4	Physical applications of the results	50
2.6	Gas mixtures: equilibrium regimes		50
	2.6.1	Mixtures of monatomic gases	50
	2.6.2	Mixtures of diatomic gases	51
2.7	Mixtures of diatomic gases in vibrational non-equilibrium		52
2.8	Mixtures of reactive gases		53
	2.8.1	Reactive gases without internal modes	53
	2.8.2	Reactive gases with internal modes	55

Appendix 2.1 The H theorem — 56
Appendix 2.2 Properties of the Maxwellian distribution — 57
Appendix 2.3 Models for internal modes — 59
Appendix 2.4 General vibrational relaxation equation — 60
Appendix 2.5 Specific vibrational relaxation equations — 62
Appendix 2.6 Properties of the Eulerian integrals — 65

3 Transport and Relaxation in Quasi-Equilibrium Regimes: Pure Gases — 66

3.1	Introduction		66
3.2	Expansion of the distribution function		66
	3.2.1	Definition of flow regimes	66
	3.2.2	Classification of flow regimes	68
3.3	First-order solutions		69
	3.3.1	Pure gases with elastic collisions: monatomic gases	70
	3.3.2	Pure diatomic gases with one internal mode	75
	3.3.3	Pure diatomic gases with two internal modes	82

Appendix 3.1 Orthogonal bases — 87
Appendix 3.2 Systems of equations for a, b, d coefficients — 91
Appendix 3.3 Expressions of the collisional integrals — 92
Appendix 3.4 Influence of the collisional model on the transport terms — 95
Appendix 3.5 Linearization of the relaxation equation — 96
Appendix 3.6 Vibrational non-equilibrium distribution — 98

4 Transport and Relaxation in Quasi-Equilibrium Regimes: Gas Mixtures — 100

4.1	Introduction	100

4.2	Gas mixtures with elastic collisions		100
	4.2.1	Chapman–Enskog method	100
	4.2.2	Transport terms: Navier–Stokes equations	103
4.3	Binary mixtures of diatomic gases		106
	4.3.1	One internal mode	106
	4.3.2	Two internal modes	109
4.4	Mixtures of reactive gases		112

Appendix 4.1 Systems of equations for a, b, l, d coefficients — 113
Appendix 4.2 Collisional integrals and simplifications — 117
Appendix 4.3 Simplified transport coefficients — 122
Appendix 4.4 Alternative technique: Gross–Jackson method — 124
Appendix 4.5 Alternative technique: method of moments — 128

5 Transport and Relaxation in Non-Equilibrium Regimes — 131

5.1	Introduction		131
5.2	Vibrational non-equilibrium gases: SNE case		131
	5.2.1	Pure diatomic gases	131
	5.2.2	Mixtures of diatomic gases	135
	5.2.3	Usual approximations: SNE case	137
5.3	Mixtures of reactive gases: $(SNE)_C$ case		138
	5.3.1	$(SNE)_C + (WNE)_V$ case	138
	5.3.2	$(SNE)_C + (SNE)_V$ case	144

Appendix 5.1 Pure gases in vibrational non-equilibrium — 147
Appendix 5.2 First-order expression of the vibrational relaxation equation — 149
Appendix 5.3 Gas mixtures in vibrational non-equilibrium — 150
Appendix 5.4 Expressions of g coefficients and relaxation pressure — 154
Appendix 5.5 Vibration–dissociation–recombination interaction — 156

6 Generalized Chapman–Enskog Method — 160

6.1	Introduction		160
6.2	General method		160
6.3	Vibrationally excited pure gases		162
	6.3.1	Transport terms	164
	6.3.2	Approximate expressions of heat fluxes	165
6.4	Extension to mixtures of vibrational non-equilibrium gases		166
6.5	Reactive gases		167
6.6	Conclusions on non-equilibrium flows		169

Appendix 6.1 Vibrationally excited pure gases — 169
Appendix 6.2 Transport terms in non-dissociated media — 171

Appendix 6.3 Example of gases with dominant VV collisions — 173
Appendix 6.4 A simplified technique: BGK method — 175
Appendix 6.5 Boundary conditions for the Boltzmann equation — 178
Appendix 6.6 Free molecular regime — 181
Appendix 6.7 Direct simulation Monte Carlo methods — 183
Appendix 6.8 Hypersonic flow regimes — 186

Part II Macroscopic Aspects and Applications — 189

Notations to Part II — 191

7 General Aspects of Gas Flows — 195
- 7.1 Introduction — 195
- 7.2 General equations: macroscopic aspects and review — 195
  - 7.2.1 Comments on the transport terms — 196
  - 7.2.2 Particular forms of balance equations — 197
  - 7.2.3 Entropy balance — 199
  - 7.2.4 Boundary conditions — 200
- 7.3 Physical aspects of the general equations — 201
  - 7.3.1 Characteristic quantities — 201
  - 7.3.2 Dimensionless conservation equations — 202
  - 7.3.3 Dimensionless numbers: flow classification — 204
- 7.4 Characteristic general flows — 207
  - 7.4.1 Steady flows — 207
  - 7.4.2 Unsteady flows — 209
  - 7.4.3 Simplified flow models — 210
  - 7.4.4 Stability of the flows: turbulent flows — 211
- Appendix 7.1 General equations: review — 212
- Appendix 7.2 Unsteady heat flux at a gas–solid interface — 216
- Appendix 7.3 Gas–liquid interfaces — 217
- Appendix 7.4 Dimensional analysis — 219
- Appendix 7.5 Generalities on total balances — 220
- Appendix 7.6 Elements of magnetohydrodynamics — 221

8 Elements of Gas Dynamics — 224
- 8.1 Introduction — 224
- 8.2 Ideal gas model: consequences — 224
- 8.3 Isentropic flows — 226
  - 8.3.1 One-dimensional steady flows — 226

		8.3.2	Multidimensional steady flows	226

		8.3.2	Multidimensional steady flows	226
		8.3.3	One-dimensional unsteady flows	227
	8.4	Shock waves and flow discontinuities		229
		8.4.1	Straight shock wave: Rankine–Hugoniot relations	229
		8.4.2	Ideal gas model	230
	8.5	Dissipative flows		231
		8.5.1	Domain of influence: boundary layer	231
		8.5.2	General equations: two-dimensional flows	233

Appendix 8.1 Method of characteristics 236
Appendix 8.2 Fundamentals of supersonic nozzles 237
Appendix 8.3 Shock waves: configuration and kinematics 239
Appendix 8.4 Generalities on the boundary layer 242
Appendix 8.5 Simple boundary layers: typical cases 247
Appendix 8.6 The turbulent boundary layer 252
Appendix 8.7 Flow separation and drag in MHD 255

9 Reactive Flows 259

9.1	Introduction		259
9.2	Generalities on chemical reactions		259
9.3	Equilibrium flows		260
	9.3.1	Law of mass action: chemical equilibrium constant	260
	9.3.2	Examples of reactions	261
	9.3.3	Examples of equilibrium flows	264
9.4	Non-equilibrium flows		266
	9.4.1	Chemical kinetics	266
	9.4.2	Vibrational kinetics	268
	9.4.3	General kinetics	271
9.5	Typical cases of Eulerian non-equilibrium flows		271
	9.5.1	Flow behind a straight shock wave	271
	9.5.2	Flow in a supersonic nozzle	278
	9.5.3	Flow around a body	282

Appendix 9.1 Evolution of vibrational populations behind a shock wave 283
    9.1.1 Evolution without dissociation 284
    9.1.2 Evolution with dissociation 285
Appendix 9.2 Air chemistry at high temperature 286
    9.2.1 Air chemistry in equilibrium conditions 286
    9.2.2 Ionization phenomena 287
Appendix 9.3 Reaction-rate constants 290
Appendix 9.4 Nozzle flows 292

10 Reactive Flows in the Dissipative Regime — 294
- 10.1 Introduction — 294
- 10.2 Boundary layers in chemical equilibrium — 295
  - 10.2.1 The flat plate — 295
  - 10.2.2 The stagnation point — 296
  - 10.2.3 Reactive boundary layer and wall catalycity — 298
  - 10.2.4 Boundary layer along a body — 300
- 10.3 Boundary layers in vibrational non-equilibrium — 300
  - 10.3.1 Example 1: boundary layer behind a moving shock wave — 300
  - 10.3.2 Example 2: boundary layer in a supersonic nozzle — 301
  - 10.3.3 Example 3: boundary layer behind a reflected shock wave — 303
- 10.4 Two-dimensional flows — 305
  - 10.4.1 Hypersonic flow in a nozzle — 305
  - 10.4.2 Hypersonic flow around a body — 308
  - 10.4.3 Mixtures of supersonic reactive jets — 311
- Appendix 10.1 Catalycity in the vibrational non-equilibrium regime — 313
- Appendix 10.2 Generalized Rankine–Hugoniot relations — 315
- Appendix 10.3 Unsteady boundary layers — 316
- Appendix 10.4 CO_2/N_2 gas-dynamic lasers — 317
- Appendix 10.5 Transport terms in the non-equilibrium regime — 320
- Appendix 10.6 Numerical method for solving the Navier–Stokes equations — 323

11 Facilities and Experimental Methods — 326
- 11.1 Introduction — 326
- 11.2 The shock tube — 327
  - 11.2.1 Simple shock tube theory — 327
  - 11.2.2 Disturbing effects — 330
  - 11.2.3 Reflected shock waves — 335
  - 11.2.4 General techniques: configurations and operation — 337
  - 11.2.5 General methods of measurement — 341
- 11.3 The hypersonic tunnel — 347
  - 11.3.1 Generalities — 347
  - 11.3.2 The hypersonic shock tunnel — 347
- Appendix 11.1 Experiments in real flight — 350
- Appendix 11.2 Optimum flow duration in a shock tube — 352
- Appendix 11.3 Heat flux measurements in a shock tube — 353
- Appendix 11.4 Shock–interface interactions — 355
- Appendix 11.5 Operation of a free-piston shock tunnel — 356
- Appendix 11.6 Source flow in hypersonic nozzles — 358

12 Relaxation and Kinetics in Shock Tubes and Shock Tunnels — 360

- 12.1 Introduction — 360
- 12.2 Vibrational relaxation — 361
  - 12.2.1 Relaxation times: general methods — 361
  - 12.2.2 Vibrational populations — 366
  - 12.2.3 Vibrational catalycity — 372
- 12.3 Chemical kinetics — 374
  - 12.3.1 Dissociation-rate constants — 374
  - 12.3.2 Time-resolved spectroscopic methods — 376
  - 12.3.3 Chemical catalycity — 382
  - 12.3.4 Hypersonic flow around bodies — 383
- Appendix 12.1 Generalities on IR emission — 385
- Appendix 12.2 Models for vibration relaxation times — 386
- Appendix 12.3 Simulation of emission spectra — 387
- Appendix 12.4 Precursor radiation in shock tubes — 391
- Appendix 12.5 Examples of kinetic models — 394

References — 397

Index — 405

Introduction

For more than a century, the properties of gaseous flows have been systematically analysed, both for the basic knowledge itself and for practical applications. This endeavour can be viewed from two aspects: firstly, the analysis of the elementary or microscopic phenomena of gaseous media, belonging to 'atomic and molecular physics'; and secondly, the study of macroscopic processes, incorporating 'fluid mechanics'. These fields have developed separately, with connections made with only the 'kinetic theory of gases'. As for applications, impressive strides have been made, especially in the domain of aeronautics and astronautics.

These applications are indeed at the origin of the increased interest in high-enthalpy gas flows, related to supersonic and hypersonic flight as well as to laser and plasma flows. In these flows, the important energies involved give rise to high temperatures and then to chemical processes such as the vibrational excitation of molecules, dissociation, ionization, and various reactions. As a consequence, the connection between microscopic and macroscopic aspects, mentioned above, has been considerably reinforced.

Analysis of the coupling and interaction between chemical phenomena and aerodynamic processes is the subject of this book. This subject has previously been dealt with in several relatively old general textbooks[1] and also more extensively in several others.[2] The present book is not intended to replace the previous ones, nor to present an exhaustive study of this field, but to analyse the essential features of non-equilibrium phenomena which generally result from the interaction between processes often possessing characteristic times of the same order of magnitude. Thus, the properties of gaseous flows at high velocity and/or at high temperature cannot be described using local 'state' quantities, and depend on their 'history', thus constituting typical non-equilibrium media.

The book is divided into two parts. Part I includes the statistical description of a gaseous reactive medium, starting (Chapter 1) with the elementary interactions between the particles of the medium, and the evolution equations, either at a semi-microscopic level (Boltzmann equation) or at the macroscopic level (fluid mechanics equations). Particular solutions of these equations are

---

[1] For a general and detailed understanding of subjects and methods exposed in the first part, the reader may refer to Refs. 1–8, and for the second part, Refs. 80–84 and 99–100.

[2] An insight into the themes of the first part may be found in the "Proceedings of the Rarefied Gas Dynamics Symposiums, RGD", organized and published every two years since 1958. In the same way, the topics treated in the second part are detailed in the "Proceedings of the International Symposiums of Shock Waves, ISSW", also biennial since 1967.

developed in Chapter 2, especially those corresponding to an 'equilibrium state' (Maxwell–Boltzmann distribution, for example) and also 'non-equilibrium solutions' essentially related to the excitation of the vibrational levels of the molecules or chemical reaction processes. These solutions are called 'zero-order solutions' and correspond to 'closed' gaseous media, i.e. they are 'dominated' by the intermolecular collisions or by only a few of them. Then, in Chapters 3 and 4, first-order solutions are developed, and the resulting transport properties are determined for pure gases as well as for mixtures, taking into account external influences; these solutions correspond to a linearized non-equilibrium of zero-order solutions. Chapter 5 is also devoted to properties of the first-order solutions (transport and relaxation) in media considered in non-equilibrium at zero order, taking into account also the possible interaction between chemical processes, such as vibration–dissociation coupling. Finally, in Chapter 6, a general method of modelling the reactive gas flows is proposed (generalized Chapman–Enskog method), whatever the degree and type of non-equilibrium may be.

In Part II, also composed of six chapters, the macroscopic properties of the reactive flows are analysed, mainly by way of typical examples. In Chapter 7, the general equations governing the reactive flows are thus presented, as well as the main dimensionless characteristic numbers and various typical flows. Some of these flows, such as shock waves, unsteady flows, and boundary layers, are thoroughly examined in Chapter 8. Chapters 9 and 10 are entirely devoted to inviscid and dissipative reactive flows, exemplified by flows behind strong shock waves, expansion flows in supersonic nozzles, and hypersonic flows along bodies. The non-equilibrium character of these flows is emphasized and its influence on aerodynamic and physical parameters is examined, as well as the exchanges with adjacent media. Chapter 11 is reserved for the description and operation of experimental facilities generating non-equilibrium flows, shock tubes, and shock tunnels and for the corresponding measurement techniques. Finally, in Chapter 12, the experimental results concerning the relaxation times, vibrational populations, reaction rates, and so on are interpreted and compared to results given by various models. Concrete examples of non-equilibrium flows in simulated planetary atmospheres are also presented.

No detailed quantitative result is given in the book insofar as many data can be found in the numerous references cited in the text. There is also no exhaustive development of various processes such as ionization and plasma flows requiring significant developments. In the same way, topics that are omitted include the physics of the gas–wall interaction as well as the interaction between the radiation and the flow. Use is made of the results of the quantum analysis of molecular and atomic processes without derivation. Moreover, no detailed numerical analysis of the equations is described, and must be found in the references. From a general point of view, and as mentioned above, this book is essentially devoted to a general analysis of non-equilibrium phenomena and processes, illustrated by examples and supported by the Appendices, which develop and highlight particular points in detail.

A portion of this book is an outgrowth of several graduate and undergraduate courses and is directed towards students possessing a basic knowledge of thermodynamics, statistical physics, and fluid mechanics. Other more specialized topics constitute the result of studies led by the author and his coworkers in the analysis, modelling, and experimental simulation of non-equilibrium flows, often in the framework of particular applications to space science. Thus, this book may also be of interest for scientists and engineers engaged in research or industry related to these applications and, of course, for people wishing to gain knowledge in the domain of reactive flows.

The author is grateful to his coworkers, essentially students, who, while preparing their theses, have contributed to the progress and/or the investigation of numerous topics presented herein. All cannot be mentioned here, but their contribution can be appreciated in the extensive citations to their work in the bibliographic references. The author is particularly grateful to J.G. Meolans for his direct contribution to various theoretical subjects exposed here, to D. Zeitoun for the numerical processing of various problems, and also to L.Z. Dumitrescu for his participation in many experiments. Thanks are also owed to N. Belouaggadia for her contribution to the editing of Chapters 5 and 6.

The suggestions and corrections brought to the initial text by G. Duffa and J.C. Lengrand have been quite pertinent, and these contributors have to be thanked for the significant improvements brought to this text; furthermore, without the (friendly) insistence of G. Duffa, this book would probably never have been written. Many thanks are also due to G. de Terlikowska for having read the complete manuscript and bringing substantial improvements to it.

Finally, the author expresses his deep gratitude to B. Shizgal for reading the English adaptation of the French edition and for his many helpful comments.

General Notations

Only the more commonly used symbols are defined here. A few symbols listed below may have more than one or two meanings; other very specific symbols are defined in the text where they are used.

Scalar symbols are in *italic*, vectorial symbols in ***bold italic***, and tensorial ones in **BOLD BLOCK CAPITAL**.

a	ideal speed of sound
$a_{i,j}^{k,l}$, $a_{k,l}^{i,j}$	collision rates for transitions $i, j \to k, l$, and $k, l \to i, j$
c_p, c_q	mass concentration of component p, of component q
C	total effective cross section, specific heat per molecule
C_T, C_R, C_V	translation, rotation, vibration specific heats
C_{TR}	$C_T + C_R$
C_{TRV}	$C_T + C_R + C_V$
D	binary diffusion coefficient
e	average energy per mass
E	average energy per molecule
E_T, E_R, E_V	average translation, rotation, vibration energies
f	distribution function
F_i	incident energy flux (normal to a wall)
g_i	statistic weight of level i
h	Planck constant (6.63×10^{-34} J·s), enthalpy per mass unit
i, j, k, l	internal energy levels
$i_r, i_v, \ldots$	rotation, vibration energy levels
I	average quantum number
I	unit tensor
J	rotation quantum number
j_p, j_q	mass flux of component p, component q
k	Boltzmann constant (1.38×10^{-23} J·K^{-1}), reaction-rate constant
k_D, k_R	dissociation, recombination-rate constants
K_C	equilibrium constant
m_p m_q	mass of particle p, particle q
m, m_r	average mass of a particle, reduced mass of two particles
M	molar mass, Mach number
n	particle density
N	unit vector normal to a surface S

N_i	incident particle flux (normal to a wall)
p	static pressure
p_r	relaxation pressure
p, q	component p, component q
P	Prandtl number
$\mathbf{P}$	stress tensor
q	heat flux
Q_R, Q_V	rotation, vibration partition functions
$\mathbf{r}$	generalized spatial coordinate
r, θ	semi-polar coordinates
R	gas constant ($\mathcal{R}/M$)
$\mathcal{R}$	universal gas constant ($8.32 \text{ J} \cdot \text{K}^{-1}$)
S	surface area, cross section
t	time
T	temperature
T_T, T_R, T_V	translation, rotation, vibration temperatures
T_{TR}, T_{TRV}	translation–rotation, translation–rotation–vibration temperatures
$\mathbf{U}_p$	diffusion velocity of species p
$\mathbf{V}$	macroscopic velocity
$\dot{w}_p$	mass production rate of species p
x, y, z	Cartesian coordinates
X_p	molar concentration of component p
$\overline{X}$	quantity X in equilibrium, mean value of quantity X
α	accommodation coefficient
γ	intermode exchange coefficient, wall recombination coefficient, specific-heat ratio
ε	'small parameter' ($\varepsilon \ll 1$)
η	bulk viscosity coefficient
$\theta_R, \theta_V, \theta_D$	rotation, vibration, dissociation characteristic temperatures
λ	mean free path, conductivity coefficient
$\lambda_T, \lambda_R, \lambda_V$	translation, rotation, vibration conductivity coefficients
λ_{TR}	$\lambda_T + \lambda_R$
λ_{TRV}	$\lambda_T + \lambda_R + \lambda_V$
μ	viscosity coefficient
ν	characteristic frequency
ρ	mass density
τ	characteristic time, relaxation time
$\boldsymbol{\tau}$	shear stress
ξ_p, ξ_q	concentration of component p, (n_p/n), of component q, (n_q/n)

Subscripts and Superscripts

C	chemical reaction
D	dissociation
e	electronic, equilibrium conditions
el	elastic collisions
f	forward reaction, frozen conditions
g	gas (at a wall)
$i, i_r, i_v, \ldots$	internal, rotational, vibrational level
in	inelastic collisions
p, q	component p, component q
r, R	backward reaction, rotation
R	recombination
T	translation
v, V	vibration
TR, TV	translation–rotation, translation–vibration exchanges
VV, Vr	vibration–vibration, resonant exchanges
w	wall
w_r	adiabatic wall
$*$	dimensionless quantity

Abbreviations

BGK	Bathnagar–Gross–Krook
CE	Chapman–Enskog
DSMC	direct simulation Monte Carlo
GCE	generalized Chapman–Enskog
LT	Landau–Teller
MBE	Maxwell–Boltzmann–Euler
MS	mixed solution
NS	Navier–Stokes
SNE	strong non-equilibrium
SSH	Schwarz–Slavsky–Herzfeld
STS	state-to-state
T, TR, TRV	translation, translation–rotation, translation–rotation–vibration
TV, VV, Vr	translation–vibration, vibration–vibration, resonant
WCU	Wang–Chang–Uhlenbeck
WNE	weak non-equilibrium

PART I
Fundamental Statistical Aspects

Notations to Part I

a, b, d, f, g, l, x	expansion terms of the corresponding coefficients A, B, D, F, G, L, X of the perturbation of the distribution function
b	impact parameter (binary collision)
d	molecule diameter (hard sphere model)
E_{VD}, E_{VR}	vibration energy loss due to dissociation, recombination, or reaction (per molecule)
F	energy flux
$\mathbf{g}$	relative velocity of two particles
$\mathbf{G}$	centre of mass velocity of two particles
$H_{i...n}^{(n)}$	Hermite polynomials
I	differential effective cross section
J	collisional term (Boltzmann equation)
K	parameter of the Treanor distribution, collisional integral (Gross–Jackson method)
N	number of vibrational levels, molecule flux
$P_{i,j}^{k,l}$	probability of the transition $i, j \to k, l$
P	Wang-Chang–Uhlenbeck polynomials
$Q_{i,j}^{k,l}$	average probability of the transition $i, j \to k, l$
S	Sonine–Laguerre polynomials
$\mathbf{u}_p$	peculiar velocity of particles p
$\mathbf{v}_p$	velocity of particles p
V	intramolecular potential, vibration–dissociation coupling factor
$\mathbf{w}$	reduced peculiar velocity
$\mathbf{W}$	root-mean-square velocity
Z	collision frequency (for one particle)
Z_0	collision number per unit time
α	azimuthal angle of deviation
$\alpha, \beta, \gamma, \delta, \lambda$	collisional integrals
$\varepsilon_i, \varepsilon_j, \ldots$	internal energy of a molecule on the level i, on the level $j, \ldots$
γ	non-dimensional peculiar velocity
$\Delta \varepsilon$	reduced internal energy balance

4 NOTATIONS TO PART I

Φ	perturbation of the distribution function, intermolecular potential
θ	reference time
Ψ_p	quantity related to a particle p
Ψ	eigenfunctions of the collisional operator
$[\Psi_p]$	collisional balance of the quantity Ψ_p
Ω	solid angle of deviation
χ	angle of deviation

Subscripts

c	continuum regime
fm	free molecular regime
m, n, q, r, s, t	expansion orders for translation, rotation, and vibration modes (0 or 1)

Superscripts

$'$	relative to a quantity after collision
$0, 1$	expansion orders
m, n, q, r, s, t	expansion orders for translation, rotation, and vibration modes (0 or 1)

ONE

Statistical Description and Evolution of Reactive Gas Systems

1.1 Introduction

The macroscopic representation of gaseous media is based on their discrete structure and is deduced from the behaviour of individual particles such as molecules, atoms, and so on[1-7]. A statistical description is therefore necessary in order to explain the properties and the evolution of these media, particularly when reactions are included.

This description is essentially based on two general principles:

- The first arises from the large number of particles in these gaseous systems for a large pressure range including rarefied as well as compressed gases (Table 1). A statistical description is therefore used whereby the macroscopic quantities are determined from appropriate local 'averages' over a large number of particles.

- The second observation, valid for about the same pressure range, is that the particles themselves experience only infrequent 'collisions'. Thus, they may be considered independent except as regards collisions which have a characteristic duration τ_C much smaller than the mean time between collisions τ_{el} (Table 1).

These observations enable us to define a local ensemble of particles possessing a definite 'state' which may be modified by particle collisions spreading information in the medium.

Table 1. General parameters for air[4].

Air	τ_{el} (s)	$\tau_C = \sqrt{C}/g$ (s)	λ (cm)	n (cm^{-3})
1	10^{-9}	10^{-13}	10^{-5}	10^{19}
2	10^{-11}	10^{-13}	10^{-7}	10^{21}
3	10^{-5}	10^{-13}	10^{-1}	10^{13}

1. Normal conditions (10^5 Pa, 300 K)
2. Compressed air (10^7 Pa, 300 K)
3. Atmosphere (100 km altitude).

1.2 Statistical description

Let us consider a gaseous medium consisting of various particles (molecules, atoms, ions, and so on) of different species p having a velocity v and an internal energy ε. In a semi-classical formulation, the velocity variable is continuous ($-\infty < v < +\infty$) whereas the internal energy defined in Appendix 1.2 is quantized with discrete levels i, each corresponding to a rotational state i_r and a vibrational state i_v, denoted collectively as

$$i = (i_r, i_v) \tag{1.1}$$

If we take the independence of particles into account, we may define a probability density for the particles of level i and of species p having the velocity v_p and located at the coordinate r at the instant t. This probability density f_{ip} is called the *distribution function*, with

$$f_{ip} = f_{ip}(r, v_p, t) \tag{1.2}$$

The probable number of these particles in the differential volume $dr\, dv_p\, dt$ is equal to

$$f_{ip} dv_p\, dr\, dt \tag{1.3}$$

Knowledge of the distribution function is therefore key to the statistical description of a gaseous system. A deterministic macroscopic description may also be deduced from this knowledge, since the moments of the distribution function are macroscopic variables. Thus, if $\Psi_{ip}(v_p, r, t)$ is a particular property of an individual particle, the corresponding macroscopic quantity $\Psi(r, t)$ is given by

$$n\Psi(r, t) = \sum_p \sum_i \int_{v_p} f_{ip} \Psi_{ip} dv_p \tag{1.4}$$

where n represents the total number density of the particles.

1.2 STATISTICAL DESCRIPTION

Other moments obtained by integrating over the velocity space, without summing over the species or levels, provide the properties of particles p in the level i. The sum over the levels gives quantities specific to the molecules p. Analogously, the sum over the rotational levels gives properties dependent on the particles p in the vibrational level i_v and so on.

1.2.1 State parameters

Thus, the 'state' quantities, which include mass, momentum, and energy, are defined as follows:

Mass density ρ:

$$\rho = \sum_i \sum_p \int_{v_p} f_{ip} m_p d\bm{v}_p = \sum_p \rho_p \tag{1.5}$$

where ρ_p represents the mass density of each component p. Here, $\rho_p = n_p m_p$, with

$$n_p = \sum_i n_{ip} = \sum_i \int_{v_p} f_{ip} d\bm{v}_p \tag{1.6}$$

Thus, n_p and n_{ip} respectively represent the 'population' (number density) of the particles p and, among them, those in the level i.

$$n = \sum_p n_p \quad \text{(total number density)}$$

and

$$\rho = nm$$

where m represents the mean mass of the particles, i.e. $m = \frac{1}{n} \sum_p n_p m_p$.

Average or macroscopic velocity V:

$$\rho \bm{V} = \sum_p \sum_i \int_{v_p} f_{ip} m_p \bm{v}_p d\bm{v}_p = \sum_p \rho_p \bm{V}_p \tag{1.7}$$

Thus, $\bm{V}$ represents the mass barycentric velocity of the flow and $\bm{V}_p$ the average velocity of the species p, with

$$\rho_p \bm{V}_p = \sum_i \int_{v_p} f_{ip} m_p \bm{v}_p d\bm{v}_p \tag{1.8}$$

The 'thermal' or 'peculiar' velocity of each particle, independent of the average velocity, is represented by $\boldsymbol{u}_p = \boldsymbol{v}_p - \boldsymbol{V}$ so that $\boldsymbol{U}_p = \boldsymbol{V}_p - \boldsymbol{V}$ represents the diffusion velocity of the species p at the macroscopic level, with

$$\sum_p \rho_p \boldsymbol{U}_p = 0 \tag{1.9}$$

Average energy of the particles, E:

$$nE = \sum_p \sum_i \int_{\boldsymbol{v}_p} f_{ip}\left(\frac{1}{2}m_p u_p^2 + \varepsilon_{ip}\right) d\boldsymbol{v}_p \tag{1.10}$$

This energy is independent of the mean velocity and is composed of a translational energy connected to the peculiar velocity, E_T, and an internal energy. This energy is the sum of the rotational energy E_R and of the vibrational energy E_V, with

$$nE_T = \sum_p \sum_i \int_{\boldsymbol{v}_p} f_{ip}\frac{1}{2}m_p u_p^2 d\boldsymbol{v}_p \tag{1.11}$$

The internal energies of each species may also be defined as

$$n_p E_{Rp} = \sum_{i_r} \int_{\boldsymbol{v}_p} f_{ip}\varepsilon_{i_rp} d\boldsymbol{v}_p = \sum_{i_r} n_{i_rp}\varepsilon_{i_rp} \tag{1.12}$$

$$n_p E_{Vp} = \sum_{i_v} \int_{\boldsymbol{v}_p} f_{ip}\varepsilon_{i_vp} d\boldsymbol{v}_p = \sum_{i_v} n_{i_vp}\varepsilon_{i_vp} \tag{1.13}$$

and

$$E = E_T + E_R + E_V = E_T + \sum_p \xi_p(E_{Rp} + E_{Vp}) \tag{1.14}$$

where ξ_p represents the number concentration of the component p.

General comments on state properties

The definition of one single mean quantity for the translation energy independent of the type of particle is generally possible if these particles are not too 'different'. The case of an electron–ion–atom mixture, for example, requires a definition of the translation energy for each species or group of species (heavy and light particles, for instance). Such mixtures are considered here only exceptionally (for example, partially ionized plasmas).

From the definition of E_T, the translational temperature T is defined as

$$E_T = \frac{3}{2}kT \tag{1.15}$$

This definition implies that there is no preferential direction, which is generally the case; exceptions, however, may exist, such as in supersonic expansions of rarefied gases or when polar molecules are affected by magnetic fields.

For each internal mode of the molecules, the corresponding energy E_{rp}, E_{vp} may, under specific conditions, give rise to the definition of a particular temperature associated with either the rotational or vibrational degrees of freedom; these situations are examined in Chapters 2 and 3.

1.2.2 Transport parameters

Owing to the constant movement of the particles, local fluxes take place. At the macroscopic level, they correspond to possible exchanges of various quantities which characterize 'transport phenomena'. Just as the state quantities give the description of a system at each point r and at each instant t, the transport quantities characterize local and instantaneous exchanges; thus, they represent local flux densities, independent of the mean velocity V. If the exchanges of fundamental quantities only—mass, momentum, and energy—are taken into account, the following transport quantities may be defined:

Mass flux:

$$\boldsymbol{j}_p = \rho_p \boldsymbol{U}_p = \sum_i \int_{\boldsymbol{v}_p} f_{ip} m_p \boldsymbol{u}_p d\boldsymbol{v}_p \tag{1.16}$$

where $\boldsymbol{j}_p$ represents the mass diffusion flux of species p. The total flux, of course, is zero, i.e.

$$\sum_p \rho_p \boldsymbol{U}_p = 0 \tag{1.17}$$

The same applies for a pure gas.

Momentum flux:

$$\mathbf{P} = \sum_p \sum_i \int_{\boldsymbol{v}_p} f_{ip} m_p \boldsymbol{u}_p \boldsymbol{u}_p d\boldsymbol{v}_p \tag{1.18}$$

This quantity (1.18) represents the total momentum flux: it is a symmetrical second-order tensor corresponding to the 'internal' forces of the medium

(Newton's law), due to the momentum exchanges between the mean streamlines of the flow. As is well known, the diagonal terms represent the stresses normal to the considered surface element (normal N), and the others the tangential stresses. Thus, the force acting on this element, τ, is such that

$$\tau = N \cdot P \qquad (1.19)$$

Energy flux (or 'heat flux'):

$$q = \sum_p \sum_i \int_{v_p} f_{ip} \left(\frac{1}{2} m_p u_p^2 + \varepsilon_{ip}\right) u_p dv_p \qquad (1.20)$$

Particular energy fluxes can also be defined. Thus, the kinetic energy flux (or translation energy flux) is equal to

$$q_T = \sum_p \sum_i \int_{v_p} f_{ip} \frac{1}{2} m_p u_p^2 u_p dv_p \qquad (1.21)$$

In the same way, the rotational energy flux specific to the species p is

$$q_R = \sum_i \int_{v_p} f_{ip} \varepsilon_{i_r p} u_p dv_p \qquad (1.22)$$

and the vibrational energy flux is such that

$$q_V = \sum_i \int_{v_p} f_{ip} \varepsilon_{i_v p} dv_p \qquad (1.23)$$

General comments on transport properties

The momentum flux proper to each species has no real interest in the framework of the general conditions indicated above, since the stresses are essentially due to the motion of particles. The same applies to the translational energy flux of each species and, as seen above, the temperature itself. In contrast, internal energy fluxes are very sensitive to the nature of the species (Chapters 3 and 4).

Particular hypotheses

The previous definitions, necessary but purely descriptive, mask a complex and changing reality arising from the importance of collisions, collisions between particles of the medium and between these particles with the outside medium through interfaces or 'walls'. Thus, these collisional processes contribute to mass, momentum, and energy exchanges between adjacent media, the information of which is transmitted by interparticle collisions.

Other external influences, such as gravity, electric, or magnetic fields, capable of modifying the trajectory of the particles are not taken into account here.

Furthermore, the long-range behaviour of the interparticle interaction potential is generally neglected in the statistical description of the medium in order to preserve the notion of a mean free path between collisions; as a consequence, a cut-off is generally used in the expression of these potentials (Appendix 1.3).

Finally, as a result of the previous observations, the collisions generally involve only two partners, and they are called binary collisions.

1.3 Evolution of gas systems

As previously discussed, the system under study evolves because of the collisions; the problem is therefore the determination of the evolution equation of the distribution function.

1.3.1 Boltzmann equation

The formal variation of the distribution function along the streamlines of the particles is simply $\frac{df_{ip}}{dt}$, corresponding to a variation of the probable number of the particles in the elementary generalized volume $\frac{df_{ip}}{dt} d\boldsymbol{v}_p \, d\boldsymbol{r} \, dt$, with $\frac{d}{dt} = \frac{\partial}{\partial t} + \boldsymbol{v}_p \cdot \frac{\partial}{\partial \boldsymbol{r}}$.

This variation is due only to the collisions; thus, if we call $J d\boldsymbol{v}_p \, d\boldsymbol{r} \, dt$ the collisional balance of these particles (species p, internal level i, and velocity $\boldsymbol{v}_p$, at the generalized coordinate $\boldsymbol{r}$ and at the instant t in the volume $d\boldsymbol{v}_p \, d\boldsymbol{r} \, dt$), we have

$$\frac{df_{ip}}{dt} = J \qquad (1.24)$$

In this formulation, the collisional term J characterizes the collisions between particles of the medium, whereas those with the outer medium constitute boundary conditions for the distribution function.

Equation (1.24) is the so-called Boltzmann equation, from which, in principle, it is possible to determine f_{ip} and, therefore, the macroscopic quantities previously defined and thus to know the evolution of the system. However, it is also possible to obtain equations of evolution of these macroscopic quantities from the Boltzmann equation without solving it, and even without knowing the details of the collisional term J.

1.3.2 General properties

The system may be considered isolated, so that the principle of conservation for mass, momentum, and energy leads to the following results:

$$\sum_p \sum_i \int_{v_p} J m_p d\boldsymbol{v}_p = 0$$

$$\sum_p \sum_i \int_{v_p} J m_p \boldsymbol{v}_p d\boldsymbol{v}_p = 0$$

$$\sum_p \sum_i \int_{v_p} J \left(\frac{1}{2} m_p v_p^2 + \varepsilon_{ip}\right) d\boldsymbol{v}_p = 0 \tag{1.25}$$

Thus, whatever transformations and exchanges may occur in the interactions, these total collisional balances are null.

The same does not apply, of course, to partial balances of mass, momentum, and energy, or for total or partial balances of other quantities. Thus, for example, the collisional balance of the number of particles p in level i, that is $\int_{v_p} J \, d\boldsymbol{v}_p$, is not generally zero. Similarly, the total balance of the number of particles p, that is $\sum_i \int_{v_p} J \, d\boldsymbol{v}_p$, is not zero in a reactive medium.

1.3.3 Macroscopic balance equations

After multiplying the Boltzmann equation (1.24) successively by m_p, $m_p \boldsymbol{v}_p$, and $\frac{1}{2} m_p v_p^2 + \varepsilon_{ip}$, integrating over the velocity space and summing over the levels and the species, we obtain the macroscopic balance equations for mass, momentum, and total energy. Taking the expressions (1.25) into account, we find the equations in the following classical forms:

$$\frac{\partial \rho}{\partial t} + \frac{\partial \cdot \rho \boldsymbol{V}}{\partial \boldsymbol{r}} = 0$$

$$\rho \frac{d\boldsymbol{V}}{dt} = -\frac{\partial \cdot \mathbf{P}}{\partial \boldsymbol{r}}$$

$$\rho \frac{de}{dt} = -\frac{\partial \cdot \boldsymbol{q}}{\partial \boldsymbol{r}} - \mathbf{P} : \frac{\partial \boldsymbol{V}}{\partial \boldsymbol{r}} \tag{1.26}$$

with the definitions $\frac{d}{dt} = \frac{\partial}{\partial t} + \boldsymbol{V} \cdot \frac{\partial}{\partial \boldsymbol{r}}$ and $e = \frac{E}{m}$.

Equations (1.26) should enable us to determine the state quantities ρ, $\boldsymbol{V}$, e, but at this stage, the transport quantities $\mathbf{P}$ and $\boldsymbol{q}$ are unknown. One solution might be to deduce the evolution equations of these last quantities from the

Boltzmann equation, but the corresponding collisional balances are not zero and are difficult to evaluate. Furthermore, higher-order moments of the distribution function appear in these equations and require other assumptions.[4,8,9] However, the so-called methods of n moments are widely used, requiring a larger number of macroscopic equations, ($n = 13, n = 20,\ldots$; Appendix 4.5).

From a macroscopic point of view, it seems simpler to restrict ourselves to the three equations (1.26) for well-defined physical situations and to try to obtain information on the distribution function so as to close the system of equations (1.26) (Chapters 2 and 3).

Moreover, it is often important to know the evolution of the 'intermediate' quantities, such as the species concentrations and/or the population of the internal levels, especially the vibrational states; these quantities may also be determined from macroscopic equations deduced from the Boltzmann equation. Thus, the population of the level i of the species p, n_{ip}, is given by the following equation obtained by integrating the Boltzmann equation over the velocity space:

$$\frac{\partial n_{ip}}{\partial t} + \frac{\partial \cdot n_{ip} V_p}{\partial r} = \int_{v_p} J\, d v_p \tag{1.27}$$

The evolution of the population of the vibrational levels is obtained by summing Eqn. (1.27) over the rotational levels. Thus, we have

$$\frac{\partial n_{i_v p}}{\partial t} + \frac{\partial \cdot n_{i_v p} V_p}{\partial r} = \sum_{i_r} \int_{v_p} J\, d v_p \tag{1.28}$$

Equivalently, the evolution of the population of the species p is obtained by summing Eqn. (1.28) over the vibrational levels i_v; the equation giving the mass density of the species p is then

$$\frac{\partial \rho_p}{\partial t} + \frac{\partial \cdot \rho_p V_p}{\partial r} = \sum_{i_v} \int_{v_p} J\, m_p d v_p \tag{1.29}$$

The collisional terms of Eqns (1.28) and (1.29) respectively represent the rate of change of the population of the level i_v (of the species p) and the mass production rate of the species p due to collisions; these terms will be developed later in Chapter 2.

Before examining the possible methods of solving the Boltzmann equation and the associated macroscopic conservation equations, it is necessary to develop the collisional term, at least partially, by analysing the various possible types of collision, their frequency, and the resulting consequences.

1.4 General properties of collisions

The media considered here generally consist of molecules, atoms, and occasionally ions and electrons. After collision between two particles (sometimes three), there may be transformation or creation of species, with change of internal state and velocity: then, the collision is called *reactive*; it is called *inelastic* if there is change of internal state and velocity only, and *elastic* if only the velocities of particles are modified. The elastic and inelastic collisions concern two particles only, p and q, identifiable before, during, and after the collision; they are typically binary collisions ($p' = p$, $q' = q$). The reactive collisions may involve several particles and intermediate components during the 'reaction' (Chapter 9).

1.4.1 Elastic collisions

In a large 'moderate' temperature range, most collisions are elastic: the relative velocities are relatively low, the collisions are not too violent, and only translational energy exchanges can occur, without modification of the internal state of the interacting molecules. The same applies to atoms under the ionization threshold and, more generally, in the case of monatomic gases.

The problem is the determination of the velocities of both particles after the interaction given the velocities before collision and the impact parameter b (Fig. 1). In the (isolated) system consisting of just two particles, the usual principles of conservation of mass, momentum, and energy apply; thus, we may

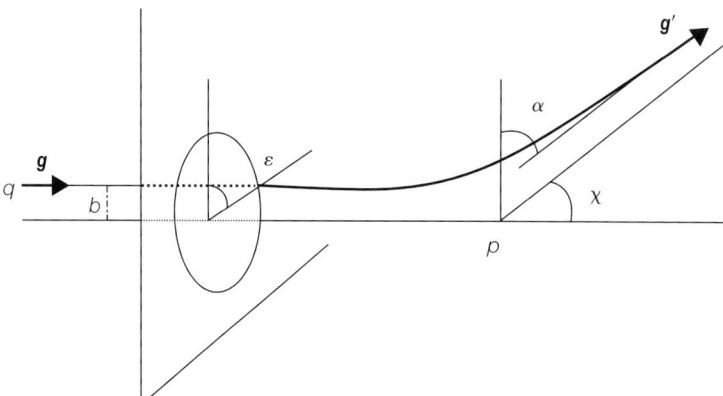

Figure 1. Collision parameters.

write the following relations between quantities before and after collision:

$$m_p + m_q = m_p + m_q$$
$$m_p v_p + m_q v_q = m'_p v'_p + m'_q v'_q$$
$$\frac{1}{2} m_p v_p^2 + \frac{1}{2} m_q v_q^2 = \frac{1}{2} m_p v'^2_p + \frac{1}{2} m_q v'^2_q \quad (1.30)$$

The quantities m_p, $m_p v_p$, $\frac{1}{2} m_p v_p^2$ are 'collisional invariants'.

From the momentum conservation equation in (1.30), also valid during the collision, we deduce that the centre of mass velocity of the two particles $G = \frac{m_p v_p + m_q v_q}{m_p + m_q}$ remains constant. Thus

$$G = G' \quad (1.31)$$

The motion of the centre of mass is therefore rectilinear and uniform during and after the collision. From the energy conservation equation (1.30), we can also deduce that the modulus of the relative velocity before collision, $g = v_p - v_q$, is preserved, that is

$$g = g' \quad (1.32)$$

Two unknowns still need to be determined in order to find the directions of the velocities after collision, for example the angles χ (deviation angle) and α (azimuth angle) (Fig. 1), which completely determine the relative directions of g and g'. To do this, it is necessary to analyse the trajectory of the particles in the interaction zone, and therefore to take into account the forces acting on the particles in this zone.

In the case of elastic collisions, it is possible to assume that the interaction force between particles p and q, F_{pq}, depends on their distance r only (Appendix 1.3), that is, a spherical potential φ is defined such that

$$F_{pq} = -F_{qp}(r) = -\frac{d\varphi}{dr} \quad (1.33)$$

Considering neutral particles only, this force is repulsive for short distances and attractive for long ones. Thus, in the interaction zone, only the repulsive force is important and governs the collision. Of course, in the case of complex molecules and inelastic collisions, the interaction potential is not spherical and generally depends on the relative orientation of the interacting particles.

In the present case, we have

$$F_{pq} = m_q \frac{dv_q}{dt} = -m_p \frac{dv_p}{dt} = -F_{qp} \quad (1.34)$$

Then
$$m_r \frac{d\mathbf{g}}{dt} = \mathbf{F}_{pq}$$

with
$$m_r = \frac{m_p m_q}{m_p + m_q} \quad \text{(reduced mass)}$$

Therefore
$$\frac{d}{dt}(\mathbf{r} \wedge \mathbf{g}) = 0 \tag{1.35}$$

Thus, the plane P including the relative velocity of particles p and q and their distance r remains normal to a constant vector during the interaction. This plane is moving parallel to itself (Fig. 2), so that the collision process may be described in this plane (Fig. 3).

Therefore, applying the energy conservation principle and Eqn. (1.35), we finally obtain the deviation χ, that is

$$\chi = \pi - 2 \int_{r_{min}}^{\infty} \left(\frac{r^4}{b^2} - 1 - \frac{2\phi r^2}{m_r b^2 g^2} \right)^{-1/2} dr \tag{1.36}$$

where b is the impact parameter (Fig. 3).

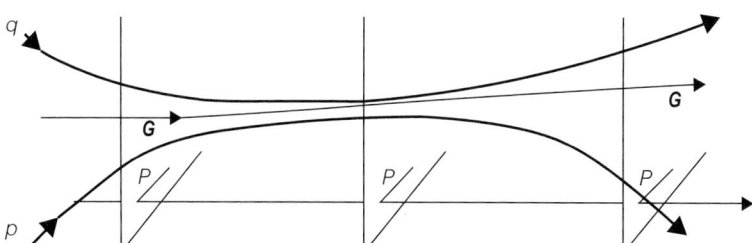

Figure 2. Plane collision.

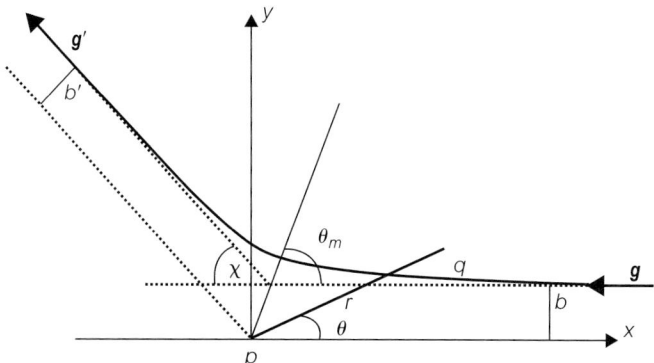

Figure 3. Relative trajectory for a plane collision.

The straight line $\theta = \theta_{min}$ is a symmetry axis for the trajectory, where $\frac{dr}{d\theta} = 0$; $r = r_{min}$ then represents the minimum distance between particles (Appendix 1.3). As the collision is planar, we have $\alpha = \varepsilon$ and $b' = b$ (Fig. 3).

The interaction is therefore completely determined if the potential φ (or its repulsive component) is known (Appendix 1.3). For the simplest model (rigid elastic sphere model), an explicit value for χ is obtained:

$$\chi = 2\operatorname{Arccos}\frac{b}{d} \quad \text{if} \quad b \leq d$$
$$\chi = 0 \quad \text{if} \quad b > d \tag{1.37}$$

with

$$d = \frac{d_p + d_q}{2} \tag{1.38}$$

In this case, the deviation is independent of the relative velocity of the particles. More generally, for more complex potentials, from Eqn. (1.36) we have

$$\chi = f(\varphi, b, g) \tag{1.39}$$

Finally, in the case of elastic collisions, a complete deterministic description of the collision is available.

1.4.2 Inelastic collisions

In this type of collision, the exchange between particles include not only translation energy but also internal energy (essentially, rotational and vibrational energy). These collisions therefore concern molecules. The peculiar velocities are higher than in the case of elastic collisions.

The conservation equations for mass and momentum (1.30) are still valid, but the energy conservation equation is written in the following form:

$$\frac{1}{2}m_p v_p^2 + \varepsilon_{ip} + \frac{1}{2}m_q v_q^2 + \varepsilon_{jq} = \frac{1}{2}m_p v_p'^2 + \varepsilon_{kp} + \frac{1}{2}m_q v_q'^2 + \varepsilon_{lq} \tag{1.40}$$

which includes the possible transition $(i \to k)$ for the molecule p and $(j \to l)$ for the molecule q. Then, the collisional invariants are

$$m_p, m_p v_p, \frac{1}{2}m_p v_p^2 + \varepsilon_{ip} \tag{1.41}$$

with

$$g_{pq}^2 = g_{pq}'^2 + 2\frac{\Delta\varepsilon}{m_r} \tag{1.42}$$

where

$$\Delta\varepsilon = \frac{\varepsilon_{kp} + \varepsilon_{lq} - \varepsilon_{ip} - \varepsilon_{jq}}{kT} \quad \text{(internal energy balance)} \quad (1.43)$$

The less energetic collisions involve translation–rotation exchanges only (TR collisions), since the rotational levels are closely spaced (Appendix 1.2). During these collisions, it is possible that only one interacting molecule changes its level ($k_r = i_r, l_r \neq j_r$), or both molecules ($k_r \neq i_r, l_r \neq j_r$), when the elastic collisions (TT collisions) involve translation energy exchange only.

The more intense collisions involve rotational and vibrational exchanges. There are translation–vibration (TV) collisions, in which only one molecule changes its vibrational level ($k_v = i_v, l_v \neq j_v$), and vibration–vibration (VV) collisions, in which both molecules change their vibrational level ($k_v \neq i_v, l_v \neq j_v$). Generally, in this last case, one molecule gets excited to an upper level, while the other goes to a lower state. The transitions may be monoquantum or polyquantum, depending on the intensity of the collision. One important class of collisions is that of resonant collisions (Vr collisions), in which the molecules seem to exchange their level ($k_v = j_v, l_v = i_v$).

1.4.3 Reactive collisions

These collisions are intense enough to create new species (dissociation, ionization, various reactions). They are of course more complex than the previous ones. The general conservation principles are still valid, but more than two particles and intermediate components may be involved; energy is also necessary to break chemical bonds and to create activated species. Collisional invariants, however, exist, such as the number of atoms or the global electrical neutrality: a number of examples are detailed in later chapters.

1.5 Properties of collisional terms

1.5.1 Collisional term expressions

The above classical and deterministic description of the collisions does not give indications of the probability of occurrence. Thus, a probability P must be assigned to each particular type of collision. As a general example, at low temperature, the probability of elastic collisions is practically equal to one, but it decreases with increasing temperature, when the probability of inelastic collisions, and then reactive ones, increases. It is then possible, at least formally,

1.5 PROPERTIES OF COLLISIONAL TERMS

to take this probability into account in the collisional term of the Boltzmann equation.

Elastic and inelastic collisions

A target particle p in level i and with the velocity $\boldsymbol{v}_p$ 'collides' with a probable number of particles q in level j and with the velocity $\boldsymbol{v}_q$, that is, per unit time and unit volume, $f_{jq}d\boldsymbol{v}_q$. These particles cross the elementary section $b\,db\,d\varepsilon$ with the relative velocity g (Fig. 1). Then, if $P_{ip,jq}^{kp,lq}$ is the probability for a particle p to pass from a level i to a level k while the particle q passes from a level j to a level l, the probable number of particles colliding with a molecule p on level i is

$$Z_{ip} = \sum_{kp,jq,lq} \sum_{q} \int_{b,\varepsilon,\boldsymbol{v}_q} P_{ip,jq}^{kp,lq} f_{jq} g b\, db\, d\varepsilon\, d\boldsymbol{v}_q \qquad (1.44)$$

where Z_{ip} is the collision frequency of a molecule i_p.

The total number of particles i_p 'lost' by collisions per volume and time unit is therefore equal to

$$J_P d\boldsymbol{v}_p = \sum_{kp,jq,lq} \sum_{q} \int_{b,\varepsilon,\boldsymbol{v}_q} P_{ip,jq}^{kp,lq} f_{ip} f_{jq} g b\, db\, d\varepsilon\, d\boldsymbol{v}_q\, d\boldsymbol{v}_p = Z_{ip} f_{ip} d\boldsymbol{v}_p \qquad (1.45)$$

Analogously, the probable number of particles i_p gained by collisions is

$$J_G d\boldsymbol{v}'_p = \sum_{kp,jq,lq} \sum_{q} \int_{b',\varepsilon',\boldsymbol{v}'_q} P_{kp,lq}^{ip,jq} f'_{kp} f'_{lq} g' b'\, db'\, d\varepsilon'\, d\boldsymbol{v}'_q\, d\boldsymbol{v}'_p \qquad (1.46)$$

The variation of the probable number of particles i_p with velocity $\boldsymbol{v}_p$ along the streamlines, per unit volume and unit time, is equal to the corresponding collisional balance of the Boltzmann equation and is given by

$$\frac{df_{ip}}{dt} d\boldsymbol{v}_p = J_G d\boldsymbol{v}'_p - J_P d\boldsymbol{v}_p \qquad (1.47)$$

In the case of elastic collisions, we have $b' = b$, $g' = g$, $\varepsilon' = \varepsilon$ and $d\boldsymbol{v}'_p\, d\boldsymbol{v}'_q = d\boldsymbol{v}_p\, d\boldsymbol{v}_q$ (Appendix 1.3).

Then, Eqn. (1.47) becomes

$$\frac{df_p}{dt} = J = \sum_{q} \int_{b,\varepsilon,\boldsymbol{v}_q} P_{el}\left(f'_p f'_q - f_p f_q\right) g b\, db\, d\varepsilon\, d\boldsymbol{v}_q \qquad (1.48)$$

and is independent of internal levels.

As mentioned above, when the temperature is not too high, it may be assumed that P_{el} is equal to 1.

For inelastic collisions, it is also commonly assumed that they are reversible, so that we can write

$$P_{ip,jq}^{kp,lq} gb \, db \, d\varepsilon \, d\bm{v}_p \, d\bm{v}_q = P_{kp,lq}^{ip,jq} g' b' \, db' \, d\varepsilon' \, d\bm{v}'_p \, d\bm{v}'_q \tag{1.49}$$

This hypothesis may indeed mask some properties that are dependent on the relative orientation of the interacting molecules. However, in this framework, we can write for the inelastic collisions

$$\frac{df_{ip}}{dt} = \sum_{kp,jq,lq} \sum_q \int_{b,\varepsilon,\bm{v}_q} P_{ip,jq}^{kp,lq} (f'_p f'_q - f_p f_q) gb \, db \, d\varepsilon \, d\bm{v}_q \tag{1.50}$$

This equation (Wang-Chang–Uhlenbeck equation) of course includes elastic collisions. Thus, the symmetry of the direct and inverse processes is assumed, which is not valid for the non-degenerate energy modes (rotation mode, for example). Nevertheless, when the asymmetry effects (magnetic field, wall vicinity, etc.) are negligible, this hypothesis may be considered valid.

Reactive collisions

The diversity of all possible types of reactive collisions does not allow us to write general expressions for the collisional terms in the Boltzmann equation. However, probabilities may also be defined, and in each case, the collisional term may be developed. Examples are presented in the next chapters.

Collision cross sections

The flux of particles crossing the elementary area $b \, db \, d\varepsilon$ in the region of influence of a particle i_p may be found again in the elementary solid angle $d\Omega = \sin \chi \, d\chi \, d\alpha$ (Fig. 1). Thus, taking into account the probability of various exchanges (velocity, level, species), we have

$$Pb \, db \, d\varepsilon = I \sin \chi \, d\chi \, d\alpha \tag{1.51}$$

Here, I, a proportionality factor, has the dimension of a surface and is called the differential cross section. It characterizes the type of collision as P. The collisional term of the Boltzmann equation (1.24) may then be written as

$$\sum_{kp,jq,lq} \sum_q \int_{\Omega,\bm{v}_q} I_{ip,jq}^{kp,lq} \left(f'_p f'_q - f_p f_q \right) g \, d\Omega \, d\bm{v}_q \tag{1.52}$$

For moderate temperatures, the elastic collisions are dominant, so that we have $P_{el} \sim 1$, $b \, db \, d\varepsilon = I_{el} d\Omega$, and

$$P = \frac{I}{I_{el}} \tag{1.53}$$

1.5.2 Characteristic times: collision frequencies

If the collisional term of the Boltzmann equation is decomposed into a sum of specific terms corresponding to the main collision groups mentioned above, we can write

$$\frac{df_{ip}}{dt} = J_{el} + J_{in} + J_c \qquad (1.54)$$

Every group, of course, may be also subdivided into other particular collision types (Chapter 2), but starting from Eqn. (1.54), we may assign a characteristic time (or frequency) to each group. Taking Eqn. (1.47) into account, Eqn. (1.54) may be written in the quasi-non-dimensional following form:

$$\frac{1}{\theta}\frac{df^*_{ip}}{dt^*} = \frac{1}{\tau_{el}}J^*_{el} + \frac{1}{\tau_{in}}J^*_{in} + \frac{1}{\tau_c}J^*_c \qquad (1.55)$$

where $\tau_{el}, \tau_{in}, \tau_c$ represent probable characteristic times between, respectively, elastic, inelastic, and reactive collisions, with $Z_{el} = 1/\tau_{el}$, $Z_{in} = 1/\tau_{in}$, $Z_c = \tau_c$.

These times depend on respective 'populations' and on collision processes (Appendices 1.2 and 1.3). Here, θ is a reference time, chosen according to the specific problem under consideration.

Thus, Eqn. (1.55) may give rise to a 'hierarchy' between the various types of collisions according to their respective characteristic times, that is, according to their average global probability. This property is used in the next chapters. As mentioned above, there are many particular types of collisions, with different probabilities. A typical example is represented by collisions populating a particular energy level and those depopulating this level. Thus, in an environment where the temperature increases, the probability of depopulating the fundamental vibrational energy level is much higher than the inverse process, and the corresponding characteristic time may be very short.

Of course, it is possible to evaluate the mean collision frequencies undergone by one particle, these quantities representing moments of the distribution function. Thus, the mean collision frequency undergone by a particle p, Z_p, is such that

$$n_p Z_p = \sum_i \int_{v_p} f_{ip} Z_{ip} d\boldsymbol{v}_p$$

or

$$n_p Z_p = \sum_{ip,kp,jq,lq} \int_{v_p,v_q} f_{ip}f_{jq}g\, d\boldsymbol{v}_p\, d\boldsymbol{v}_q \int_\Omega I^{kp,lq}_{ip,jq} d\Omega \qquad (1.56)$$

The second integral of Eqn. (1.56) depends on collisions only: it is called the total cross section:

$$C_{ip,jq}^{kp,lq} = \int_\Omega I_{ip,jq}^{kp,lq} d\Omega = C(g,\varphi) \qquad (1.57)$$

The mean change per second of a property Ψ_{ip} of the particle p, $\Delta\Psi_p$, due to the collisions is such that

$$n_p \Delta\Psi_p = \sum_{ip,kp,jq,lq} \sum_q \int_{v_p,v_q} f_{ip} f_{jq} g \, d\boldsymbol{v}_p \, d\boldsymbol{v}_q \int_\Omega (\Psi'_{kp} - \Psi_{ip}) I_{ip,jq}^{kp,lq} d\Omega \qquad (1.58)$$

The second integral of Eqn. (1.58) depends only on the collisions and not on the populations. It is the total cross section for the change in the property Ψ_{ip} that is caused by the collisions. This change is also equal to

$$\sum_{ip} \int_{v_p} J\Psi_{ip} d\boldsymbol{v}_p \qquad (1.59)$$

Similarly, the collision frequency Z undergone by any particle of the mixture is such that

$$nZ = \sum_p n_p Z_p = \sum_{i,p} \int_{v_p} f_{ip} Z_{ip} d\boldsymbol{v}_p \qquad (1.60)$$

and

$$n\Delta\Psi = \sum_p n_p \Delta\Psi_p \qquad (1.61)$$

Other expressions are defined in Appendix 2.3.

Appendix 1.1 Elements of tensorial algebra

Vectors

The tensorial product $\boldsymbol{ab}$ of the vectors $\boldsymbol{a}\,(a_x, a_y, a_z)$ and $\boldsymbol{b}(b_x, b_y, b_z)$ is a dyadic or second-order tensor $a_i b_k$, written in Cartesian coordinates as

$$\begin{pmatrix} a_x b_x & a_x b_y & a_x b_z \\ a_y b_x & a_y b_y & a_y b_z \\ a_z b_x & a_z b_y & a_z b_z \end{pmatrix}$$

with $\boldsymbol{ab} \neq \boldsymbol{ba}$, except when $\boldsymbol{a}$ is parallel to $\boldsymbol{b}$.

Example

The tensorial product of the symbolic vector $\frac{\partial}{\partial r}$ (components $\frac{\partial}{\partial x}, \frac{\partial}{\partial y}, \frac{\partial}{\partial z}$) and of the vector V, (u, v, w), that is, $\frac{\partial V}{\partial r}$, is written in Cartesian coordinates as

$$\begin{pmatrix} \frac{\partial u}{\partial x} & \frac{\partial v}{\partial x} & \frac{\partial w}{\partial x} \\ \frac{\partial u}{\partial y} & \frac{\partial v}{\partial y} & \frac{\partial w}{\partial y} \\ \frac{\partial u}{\partial z} & \frac{\partial v}{\partial z} & \frac{\partial w}{\partial z} \end{pmatrix}$$

The contracted tensorial product of ab is $a \cdot b$, obtained by setting the indices $i = k$, then summing over i:

$$a \cdot b = \sum_i a_i b_i$$

This product is a scalar quantity (scalar product).

Example

Divergence of V, $\frac{\partial \cdot V}{\partial r}$, $\left(\frac{\partial u}{\partial x} + \frac{\partial v}{\partial y} + \frac{\partial w}{\partial z} \right)$.

Of course, if b is a scalar quantity, ab is a vector.

Example

Gradient of b, $\frac{\partial b}{\partial r}$ (components $\frac{\partial b}{\partial x}, \frac{\partial b}{\partial y}, \frac{\partial b}{\partial z}$).

Second-order tensors

These are represented by a 3×3 matrix; they are from the type ab (dyadic) or from the type W, that is, in Cartesian coordinates:

$$\begin{pmatrix} W_{xx} & W_{xy} & W_{xz} \\ W_{yx} & W_{yy} & W_{yz} \\ W_{zx} & W_{zy} & W_{zz} \end{pmatrix}$$

This tensor is symmetrical if $W_{ij} = W_{ji}$.

Example

Stress tensor **P**.

A transposed tensor $\overline{W}$ is obtained by reversing rows and columns.

Example

$\overline{\frac{\partial V}{\partial r}}$ is the transposed tensor of $\frac{\partial V}{\partial r}$. A symmetrical tensor $\overline{\overline{\frac{\partial V}{\partial r}}}$ is obtained by putting

$$\overline{\overline{\frac{\partial V}{\partial r}}} = \frac{1}{2}\left(\frac{\partial V}{\partial r} + \overline{\frac{\partial V}{\partial r}}\right)$$

This tensor becomes non-divergent by subtracting one-third of its trace, that is

$$\overset{0}{\overline{\overline{\frac{\partial V}{\partial r}}}} = \overline{\overline{\frac{\partial V}{\partial r}}} - \frac{1}{3}\frac{\partial \cdot V}{\partial r}\mathbf{I}$$

where **I** is the unit tensor.

$$\mathbf{I} = \begin{pmatrix} 1 & 0 & 0 \\ 0 & 1 & 0 \\ 0 & 0 & 1 \end{pmatrix}$$

The isotropic tensors belong to the type $k\mathbf{I}$.

Example

Pressure $p\mathbf{I}$.

Vector–tensor products

The tensorial product $a\mathbf{W}$ or abc is a third-order tensor $a_i W_{jk}$ or $a_i b_j c_k$.
The contracted tensor $\mathbf{a}.\mathbf{W}$ is obtained by setting $i = j$ and summing over i. The result is a vector with components $\sum\limits_{i} a_i W_{ik}$.

Examples

a) Vector $\frac{\partial \cdot \mathbf{P}}{\partial r}$ (components $\frac{\partial \cdot \mathbf{P}_x}{\partial r}, \frac{\partial \cdot \mathbf{P}_y}{\partial r}, \frac{\partial \cdot \mathbf{P}_z}{\partial r}$), where $\mathbf{P}_x$ (with components $\tau_{xx}, \tau_{xy}, \tau_{xz}$) is the vector associated with the first line of the tensor **P** equal to

$$\begin{pmatrix} \tau_{xx} & \tau_{xy} & \tau_{xz} \\ \tau_{yx} & \tau_{yy} & \tau_{yz} \\ \tau_{zx} & \tau_{zy} & \tau_{zz} \end{pmatrix}$$

b) Vector $\mathbf{V} \cdot \frac{\partial \mathbf{V}}{\partial r}$ with components $\mathbf{V} \cdot \frac{\partial u}{\partial r}, \mathbf{V} \cdot \frac{\partial v}{\partial r}, \mathbf{V} \cdot \frac{\partial w}{\partial r}$.

c) Scalar $\frac{\partial \cdot (\mathbf{P} \cdot \mathbf{V})}{\partial r} = \frac{\partial \cdot (\mathbf{P}_x \cdot \mathbf{V})}{\partial x} + \frac{\partial \cdot (\mathbf{P}_y \cdot \mathbf{V})}{\partial y} + \frac{\partial \cdot (\mathbf{P}_z \cdot \mathbf{V})}{\partial z}$.

Tensor–tensor product

The tensorial product $\mathbf{WW'}$ is a fourth-order tensor, the contracted product $\mathbf{W} \cdot \mathbf{W'}$ is a second-order tensor with components $\sum_i W_{ij} W'_{ij}$, and the doubly contracted tensor $\mathbf{W} : \mathbf{W'}$ is a scalar quantity (sum of the homologous components).

Example

$$\mathbf{P}: \frac{\partial \mathbf{V}}{\partial \mathbf{r}} = \tau_{xx} \frac{\partial u}{\partial x} + \cdots$$

Miscellaneous results

$$\mathbf{W} \cdot \mathbf{a} = \mathbf{a} \cdot \mathbf{W}$$

$$\mathbf{a} \cdot \mathbf{I} = \mathbf{I} \cdot \mathbf{a} = \mathbf{a}$$

$$\mathbf{I} : \mathbf{I} = 3$$

$$\mathbf{I} : \mathbf{W} = \mathbf{W} : \mathbf{I} = W_{xx} + W_{yy} + W_{zz}$$

$$\overset{0}{\mathbf{W}} = \mathbf{W} - \frac{1}{3}\mathbf{I}(\mathbf{I} : \mathbf{W})$$

$$\overset{0}{\mathbf{W}} : \overset{0}{\mathbf{W'}} = \overset{0}{\mathbf{W}} : \mathbf{W'} = \mathbf{W} : \overset{0}{\mathbf{W'}}$$

$$(\mathbf{ab}) \cdot \mathbf{d} = \mathbf{a} \cdot (\mathbf{bd})$$

$$\mathbf{d} \cdot (\mathbf{ab}) = (\mathbf{d} \cdot \mathbf{a})\mathbf{b}$$

$$\mathbf{b} \cdot (\mathbf{a} \cdot \mathbf{W}) = (\mathbf{ba}) : \mathbf{W}$$

$$(\mathbf{ab}) : (\mathbf{cd}) = (\mathbf{ac}) : (\mathbf{bd}) = (\mathbf{a} \cdot \mathbf{d})(\mathbf{c} \cdot \mathbf{b})$$

Appendix 1.2 Elements of molecular physics

Generalities

Only a few fundamental elements necessary for the understanding of the physics of non-equilibrium gaseous media are presented here. They essentially concern the properties of the molecules which constitute the most numerous particles of the media considered here, the other being atoms or, occasionally, ions and electrons.

The specific energy of a particle p on a level i, ε_{ip}, may be roughly considered as the sum of a translation energy $\varepsilon_{tp} = \frac{1}{2} m_p u^2$ and of an internal energy composed of a rotational energy $\varepsilon_{i_r p}$ and of a vibrational energy $\varepsilon_{i_v p}$; it may also include a reference energy ε_0. This does not exclude the interactions between the energy modes, generally included in the expressions of energies. Similarly, the particles are assumed to be in the fundamental electronic level; exceptions are given in Chapter 12, where molecules in different electronic levels are considered.

Of course, the molecules possess only rotational and vibrational energies. Thus

$$\varepsilon_{ip} = \varepsilon_{tp} + \varepsilon_{i_r p} + \varepsilon_{i_v p} + \varepsilon_0 \tag{1.62}$$

By multiplying Eqn. (1.62) by f_{ip}, integrating over the velocity space, and summing over levels i, we may obtain the mean energy per molecule p, E_p, that is

$$E_p = E_{Tp} + E_{Rp} + E_{Vp} + E_p^0 \tag{1.63}$$

The mean energy per particle of the mixture E is such that

$$nE = nE_T + \sum_P n_p (E_{Rp} + E_{Vp} + E_p^0) \tag{1.64}$$

The mean mass energy of the species p is

$$e_p = \frac{E_p}{m_p} = e_{Tp} + e_{Rp} + e_{Vp} + e_p^0 \tag{1.65}$$

If $c_p = \frac{\rho_p}{\rho}$ is the mass concentration of the species p, the mean mass energy of the mixture is

$$e = \sum_p c_p e_p \tag{1.66}$$

and ρe is the mean volume energy.

The possibility of separating the electronic and atomic wave functions (Born–Oppenheimer approximation) gives us a way to analyse the energy modes and states of the molecules separately.

Translational energy

We consider the translational mode of the molecules or atoms moving 'freely' without any outer force field. The analysis of the corresponding Schrödinger equation gives a spectrum of eigenvalues corresponding to translation energy levels $\varepsilon_t = \frac{1}{2}mu^2$ that are very close as compared to the quantity kT; thus, this energy may be considered non-quantified and may be classically written in the following form:

$$E_T = \frac{3}{2}kT, \quad e_T = \frac{3}{2}\frac{k}{m}T = \frac{3}{2}RT \tag{1.67}$$

with

$$R = \sum_p c_p R_p \quad \text{and} \quad R_p = \frac{\mathcal{R}}{M_p} \tag{1.68}$$

Rotational energy

A physical model must be defined for this energy mode. The simplest one is to consider the molecule as a rigid rotator independent of vibration; this hypothesis is essentially valid for diatomic molecules. Thus, from the Schrödinger equation, it may be deduced that the energy of the rotational levels is

$$\varepsilon_J = \frac{J(J+1)h^2}{8\pi^2 I} \tag{1.69}$$

where, traditionally, J is the rotation quantum number (denoted here also as i_r), and I is the moment of inertia of the molecule. It is usual to denote

$$\frac{h^2}{8\pi^2 I} = h\nu_r \tag{1.70}$$

where ν_r represents the rotation frequency of the molecule. Then, a 'rotational characteristic temperature' θ_R may be defined, so that

$$h\nu_r = k\theta_R$$

and

$$\varepsilon_J = k\theta_R J(J+1) \tag{1.71}$$

The rotational levels are close enough to have $\varepsilon_{J+1} - \varepsilon_J \ll kT$, which corresponds to very low characteristic temperatures θ_R of a few degrees K and to frequencies ν_r in the far infrared.

It is also found that every level J corresponds to $(2J+1)$ possible states (degeneracy) and therefore has a statistical weight $g_J = 2J+1$.

A more elaborate model, necessary for the interpretation of the emission spectra (Chapter 12), takes into account the molecular vibration which modifies the moment of inertia. As the corresponding period is generally much smaller than the rotation period ($v_r \ll v_v$), we have, for the diatomic molecules:

$$\varepsilon_J = J(J+1)B_V - J^2(J+1)^2 D_V \tag{1.72}$$

with

$$B_V = \frac{h^2}{8\pi^2 I} - \alpha\left(v + \frac{1}{2}\right) + \cdots$$

$$D_V = D_{V=0} - \beta\left(v + \frac{1}{2}\right) + \cdots$$

where I is the average moment of inertia, and α, β are constants.

Vibrational energy

The eigenvalues obtained from the Schrödinger equation, when considering molecules as vibrating springs (harmonic oscillator model) are such that

$$\varepsilon_v = \left(v + \frac{1}{2}\right) h v_v \tag{1.73}$$

where v is the vibration quantum number (denoted here also as i_v).

The energy levels are equidistant, and we have

$$\varepsilon_{v+1} - \varepsilon_v = h v_v = k \theta_V \tag{1.74}$$

where v_v represents the characteristic vibrational frequency and θ_V the 'vibrational characteristic temperature' of the considered species. This temperature is relatively high; thus, approximately, we have

$$\theta_{VN_2} \simeq 3395 \text{ K} \quad \theta_{VO_2} \simeq 2275 \text{ K}$$
$$\theta_{VCO} \simeq 3120 \text{ K} \quad \theta_{VCN} \simeq 2980 \text{ K}$$
$$\theta_{VNO} \simeq 2740 \text{ K} \quad \theta_{VH_2} \simeq 6335 \text{ K}$$

The frequencies v_v generally correspond to the near infrared spectrum (examples in Chapter 12). The reference level is $\varepsilon_0 = \frac{h v_v}{2}$.

For 'moderate' temperatures, only the lowest levels are significantly populated (Chapter 2), and the harmonic oscillator model is a good approximation, particularly for diatomic molecules. For more complex molecules, several vibrational modes coexist, the number of modes increasing with the atomicity of the molecule. For linear triatomic molecules, as a first approximation, the energetic contribution of each mode may be added. Thus, for CO_2, there are three vibrational modes (Appendix 10.4), and one of them (the bending mode) is degenerate ($g_V = 2$), with

$$\theta_{V1} \simeq 1980 \text{ K} \quad \text{(symmetrical mode } v_1\text{)}$$
$$\theta_{V2} \simeq 960 \text{ K} \quad \text{(bending mode } v_2\text{)}$$
$$\theta_{V3} \simeq 3380 \text{ K} \quad \text{(asymmetrical mode } v_3\text{)}$$

At higher temperatures, when $T \sim \theta_V$, or when the degree of dissociation is significant, the population of the high levels may be important, and an anharmonic oscillator model is necessary (Fig. 4). Then, taking into account a molecular Morse potential $V = d_e\{1 - \exp[-a(r - r_e)]\}$, we have for the vibrational energy

$$\varepsilon_V = h\nu_v \left[\left(v + \frac{1}{2}\right) - x_e \left(v + \frac{1}{2}\right)^2 \right] \tag{1.75}$$

with

$$x_e = \frac{h\nu_v}{4d_e}$$

Other higher-order terms $\left(v + \frac{1}{2}\right)^n$ may be added in Eqn. (1.75), corresponding to more realistic potentials.

When $r \sim r_e$ (low levels), the harmonic model is recovered, with $a = \pi \nu_v \sqrt{m/d_e}$ for homonuclear diatomic molecules. For heteronuclear molecules, m is replaced by m_r (reduced mass of particles).

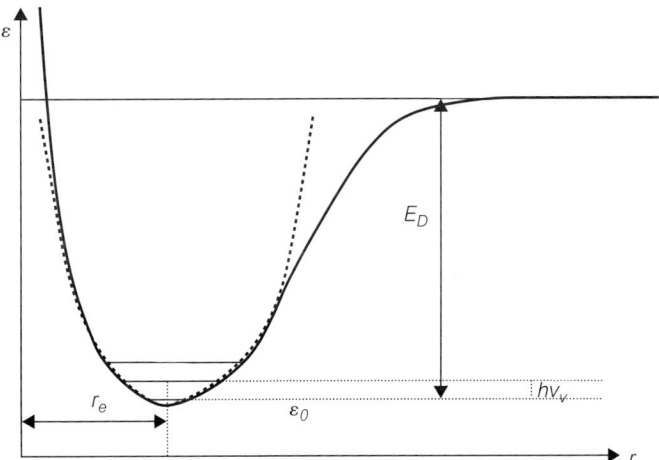

Figure 4. Energy diagram of the fundamental electronic level. · · · Harmonic oscillator, ⎯⎯ Anharmonic oscillator.

When $V \to d_e$, we have $r \to \infty$, which corresponds to the dissociation energy E_d. In the same way as we have $h\nu_v = k\theta_V$, we also have $d_e = k\theta_D$, where θ_D represents the 'dissociation characteristic temperature'. Thus, for example, we have $\theta_{DO_2} \simeq 59000$ K and $\theta_{DN_2} \simeq 113000$ K.

The vibrational levels, therefore, are closer and closer when approaching the dissociation threshold. Thus, E_D represents the formation energy of the corresponding atom (reference: $\varepsilon_0 = \frac{h\nu_v}{2}$).

Most molecules of interest include between 30 and 45 levels; thus, the molecule O_2 possesses 32 levels (but only 27 up to E_D with the harmonic oscillator model).

The vibrational and rotational levels of a diatomic molecule are schematically represented in Fig. 5. The difference between the vibrational levels is relatively important, and for each level the same rotational levels are present. One level i, therefore, represents one pair of levels (i_r, i_v). Then, the population of the level i_v of the species p is

$$n_{i_v p} = \sum_{i_r} n_{ip} = \sum_{i_r} \int_{v_p} f_{ip} d\boldsymbol{v}_p \tag{1.76}$$

Analogously, the total population of the level i_r is

$$n_{i_r p} = \sum_{i_v} n_{ip} = \sum_{i_v} \int_{v_p} f_{ip} d\boldsymbol{v}_p \tag{1.77}$$

Therefore, the population of the species p is

$$n_p = \sum_{i_v} n_{i_v p} = \sum_{i_r} n_{i_r p} \tag{1.78}$$

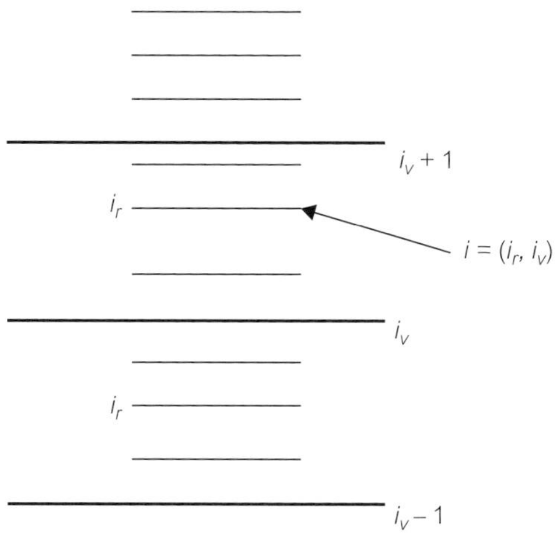

Figure 5. Scheme of molecular energy.

and the corresponding mean energies E_{Rp} and E_{Vp} are

$$n_p E_{Rp} = \sum_{i_r} n_{i_r p} \varepsilon_{i_r p} \quad \text{and} \quad n_p E_{Vp} = \sum_{i_v} n_{i_v p} \varepsilon_{i_v p} \quad (1.79)$$

Electronic energy

As indicated above, the results concerning the rotational and vibrational energies are valid only for the fundamental electronic level (state Σ). For the other electronic levels (Π, Δ ...), which may be occupied at high temperature, the potential curves are more complex than those represented in Fig. 4. Then, it is necessary to take into account the motion of electrons with particular quantum numbers, such as Λ, characteristic of the angular momentum of the electrons, and S, characteristic of the spin angular momentum. Similarly, the interaction with the rotation is represented by a special quantum number K: an example for CN is presented in Chapter 12, corresponding to transitions between two electronic states and to frequencies located in the 'visible' part of the spectrum. Then, an electronic spectrum is composed of rotation–vibration bands including numerous lines corresponding to 'possible' combinations of the quantum numbers.

However, the large majority of the flows or situations considered in this book (except in Chapter 12) do concern the fundamental electronic state, for which $\Lambda - S = 0$, so that only one single ensemble of vibrational levels is to be considered. Similarly, atomic particles are generally assumed to lie on the fundamental electronic level and to possess only translation energy.

Appendix 1.3 Mechanics of collisions

Spherical models of molecular interaction

As previously defined, we have

$$\mathbf{F}_{pq}(r) = -\mathbf{F}_{qp}(r) = -\frac{d\varphi}{d\mathbf{r}}$$

Model of rigid elastic spheres

Each molecule p is considered as a sphere of diameter d_p. The collision takes place when

$$r = \frac{d_p + d_q}{2} = d$$

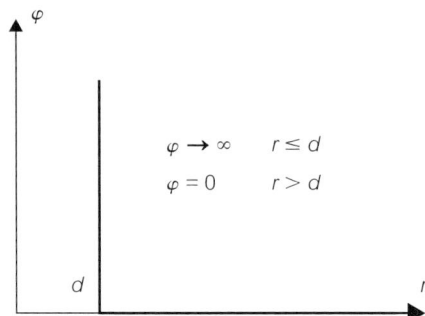

Figure 6. Rigid sphere model.

This potential is a crude representation of the repulsive forces, with the collision consisting of a specular elastic reflection of the molecules (Figs. 6 and 8). Here, d is of the order of 10^{-8} cm.

Model of repulsive centres

The potential has the following form:

$$\varphi = \frac{K}{r^{s-1}} \tag{1.80}$$

Though being approximate, this potential, however, is not as 'hard' as the previous one. The realistic values for s are between 9 and 12. The value $s = 2$ is characteristic of the Coulomb potential (plasmas). The potential corresponding to $s = 5$ is called the Maxwell potential and, though being non-realistic, is widely used because of its simplifying properties in many calculations (Appendix 3.4).

Lennard-Jones potential

This is the closest model to reality, taking the attractive and repulsive forces into account, i.e.

$$\varphi = \frac{d}{r^n} - \frac{e}{r^m} \tag{1.81}$$

For non-polar molecules, we have

$$\varphi = 4\varepsilon \left[\left(\frac{d}{r}\right)^{12} - \left(\frac{d}{r}\right)^{6} \right] \quad \text{(Fig. 7)} \tag{1.82}$$

For intermolecular collisions, it is clear that the spherical potential models are valid only for relatively large distances. However, the repulsive part may also be modelled by a sum of exponential terms roughly representing the interactions

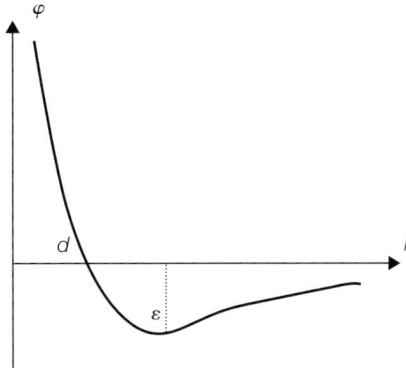

Figure 7. Lennard-Jones potential.

between all atoms of both interacting molecules. Thus, we have

$$\varphi = \varphi_0 \sum_{k,m} \exp\left(-\frac{r_{km}}{l}\right) \quad (1.83)$$

where φ_0 represents the spherical part, l a characteristic action distance, and r_{km} the respective distances between the atoms k of one molecule and the atoms m of the other.

Mechanism of the elastic collision

From the conservation equations (1.30), we have $\bm{G}' = \bm{G}, \bm{g}' = \bm{g}$, and $\bm{r} \wedge \bm{g} = \bm{K}$.

Then, the interaction may be represented in a plane perpendicular to a constant vector $\bm{K}$ including $\bm{g}$ and $\bm{r}$ (plane interaction with $\alpha = 0$).

In this plane, a coordinate system attached to the particle p is chosen, and the motion of the particle q relative to the particle p (fictitious particle of mass m_r), is analysed (Fig. 3).

In polar coordinates (r, θ), the relative velocity $\bm{g}$ has a normal component $\frac{dr}{dt}$ and a tangential component $r\frac{d\theta}{dt}$.

The conservation equations for the energy and kinetic momentum during the interaction are the following:

$$\frac{1}{2} m_r \left[\left(\frac{dr}{dt}\right)^2 + \left(r\frac{d\theta}{dt}\right)^2 \right] + \phi(r) = \frac{1}{2} m_r g^2 = \text{const.}$$

$$m_r (\bm{r} \wedge \bm{g}) = m_r r^2 \frac{d\theta}{dt} = m_r b g = m_r b' g' \quad (1.84)$$

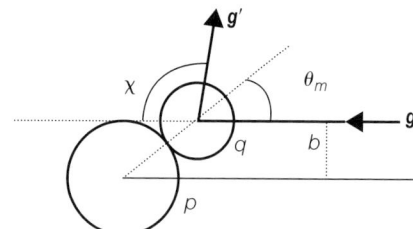

Figure 8. Collision of rigid elastic spheres.

Then $b' = b$, and eliminating the time t from these equations, we obtain the equation of the trajectory:

$$\theta = \int_\infty^r \left(\frac{r^4}{b^2} - r^2 - \frac{2\varphi r^4}{m_r b^2 g^2} \right) dr \qquad (1.85)$$

If the potential is purely repulsive, a minimum approach distance r_{min} is found corresponding to $\frac{dr}{d\theta} = 0$; then $\theta = \theta_{min}$, and the trajectory is symmetrical about this straight line.

The deviation χ is equal to $\pi - 2\theta_{min}$.

With the rigid elastic sphere model (Fig. 8), we have

$$\theta_{min} = \frac{\pi}{2} \quad \text{and} \quad \chi = 0 \quad \text{for } b > d$$

$$\sin \theta_{min} = \frac{b}{d} \quad \text{and} \quad \chi = 2\text{Arccos}\frac{b}{d} \quad \text{for } b \leq d$$

The differential cross section I is equal to $\frac{d^2}{4}$ and the total cross section C to πd^2. They are independent of the relative velocity g.

It is important to note that the equations governing the interaction are reversible, so that, for the 'inverse' collision $v'_p \to v_p$, $v'_q \to v_q$, we have

$$G' = G, \quad g' = g, \quad b' = b, \quad d\Omega' = d\Omega$$

and

$$d\boldsymbol{v}_p \, d\boldsymbol{v}_q = d\boldsymbol{v}'_p \, d\boldsymbol{v}'_q = d\boldsymbol{G} \, d\boldsymbol{g} \qquad (1.86)$$

because the Jacobian of the transformation is equal to 1.

Particular properties of the collisional term

If Ψ_{ip} is a quantity related to a molecule p in the level i, the total variation of the quantity Ψ due to a collision is equal to

$$\Psi'_{kp} + \Psi'_{lq} - \Psi_{ip} - \Psi_{jq} \quad \text{(denoted } [\Psi])$$

From the symmetry of the operators, we may deduce the following equality:

$$\sum_p \sum_i \int_v J\Psi \, dv = -\frac{1}{4} \sum_p \sum_i \int_v J[\Psi] \, dv \qquad (1.87)$$

In particular, if Ψ_{ip} is a collisional invariant, the total collisional balance is zero.

TWO

Equilibrium and Non-Equilibrium Collisional Regimes

2.1 Introduction

For particular physical systems, it is relatively easy to obtain expressions for the distribution function from which the macroscopic evolution of these systems may be determined. The conservation equations (1.26) can be closed and a solution sought. Thus, if we consider an 'isolated' medium exchanging no mass, momentum, or energy with the background, the evolution of this medium is due only to like collisions between the particles of the medium. This is approximately true for any system 'sufficiently' far from its boundaries. The collisional term of the Boltzmann equation is then dominant, and the evolution of the distribution function is determined by collisions. The boundaries of the system play a geometrical role only.

Thus, given some reference or observation time, if all types of collisions (elastic, inelastic, reactive, and so on) have characteristic times shorter than this reference time, they all participate in the determination of the distribution function. This corresponds to a 'general' collisional regime called the 'equilibrium state' of the system. If, however, only a few types of collision have this property, they provisionally determine the structure of the distribution function and the other types acting over a longer timescale, contributing to a further evolution of the system towards the equilibrium state. During this period of transition, the system is in a 'non-equilibrium state' but remains in the collisional regime.

In this chapter, we examine various equilibrium and non-equilibrium situations, each one depending on a particular dominant type of collision.

2.2 Collisional regimes: generalities

From a general point of view, if the Boltzmann equation (1.24) is written in a non-dimensional form, as done in Chapter 1, we have, without separating the various types of collision:

$$\frac{1}{\theta}\frac{df^*}{dt^*} = \frac{1}{\tau}J^* \quad \text{or} \quad \frac{df^*}{dt^*} = \frac{1}{\varepsilon}J^* \qquad (2.1)$$

where τ represents the characteristic time between collisions, θ is a reference time, and $\varepsilon = \tau/\theta$ is the Knudsen number.

As the non-dimensional starred quantities, including the operators, are generally of the order of 1, the relative order of magnitude of both terms of the Boltzmann equation is given by the ratio ε. Thus, if the time between collisions remains much smaller than the reference time, that is if $\varepsilon \ll 1$, the Boltzmann equation is reduced to:

$$J = 0 \quad \text{or} \quad J_P = J_G \quad \text{(balance null)} \qquad (2.2)$$

The collisional term therefore determines the distribution function. The regime is called the 'collisional regime', and the other influences, arising for example from the background, are negligible, except of course the geometrical conditions.

Conversely, the collisionless regime, called the 'free molecular regime', is such that $\frac{df}{dt} = 0$ and corresponds to rarefied media ($\varepsilon \gg 1$) (Appendix 6.6). The intermediate or 'transitional' regimes correspond to $\varepsilon \sim 1$.

Eqn. (2.2), which governs the collisional media, in spite of its simplicity hides a complex reality, arising from the different characteristic times of collisions, as discussed in Chapter 1. Thus, Figs 9 and 10 show the temperature variation of the characteristic times τ_T, τ_R, τ_V, and τ_D corresponding to collisions for the exchange of translational energy,[1] rotational energy,[10] vibrational energy,[11] and dissociation,[12] respectively, for pure gases N_2 and O_2. In these figures, we can see that the characteristic times generally decrease with temperature because of the increasing effectiveness of the collisions with temperature. It is also clear that there exists a separation of these times of at least one order of magnitude, which may serve to classify them in specific categories. As already indicated in Chapter 1, this property is used in the analysis of non-equilibrium flows. There is, however, an exception for O_2, whose characteristic times of vibrational relaxation and dissociation become relatively close at high temperature.

In the study of the collisional regimes, which is the subject of this chapter, we must proceed by stages; initially, the simplest systems, such as pure gases

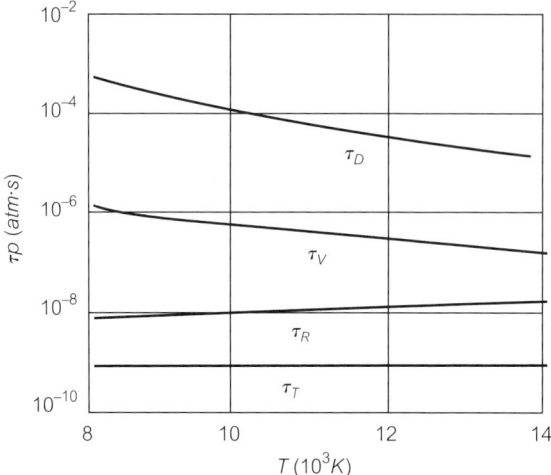

Figure 9. Translation, rotation, vibration, dissociation characteristic times (nitrogen).

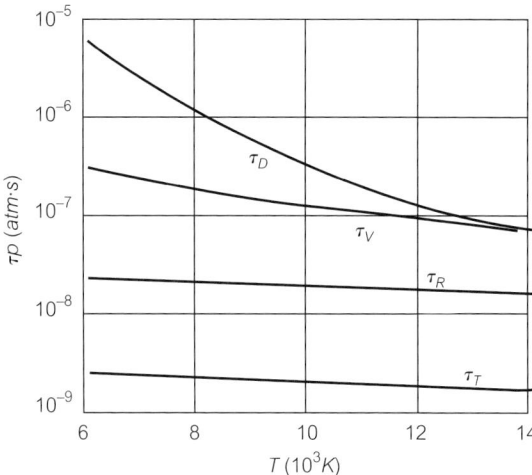

Figure 10. Translation, rotation, vibration, dissociation characteristic times (oxygen).

with only elastic collisions, are considered. This simple case is a classic example which gives concrete and precise results for the properties of the corresponding flows. Other more complex systems are then considered such as gas mixtures, polyatomic gases, relaxing and reactive mixtures, and so on.

2.3 Pure gases: equilibrium regimes

In this case, we have $p' = p = q' = q$, and the molecules may have different internal energies (Appendix 1.2). We consider here monatomic or diatomic gases, which constitute the great majority of physical situations. The generalization to polyatomic gases does not present fundamental differences (Chapter 10).

2.3.1 **Monatomic gases**

These particles exchange only translational energy, and the index designating the internal state is not required. The collisions are elastic if the relative energy in a binary collision is below the threshold for ionization. In the collisional regime, we have

$$J_{el} = 0 \qquad (2.3)$$

since

$$\varepsilon_{el} = \tau_{el}/\theta \ll 1$$

The properties of these collisions (Chapter 1) are such that Eqn. (2.3) can be written

$$\int_{\Omega, v_b} (f'_a f'_b - f_a f_b) I_{el} \, d\Omega \, dv_b = 0 \qquad (2.4)$$

The indices a and b are used to distinguish the two particles in a binary collision.

An obvious solution of Eqn. (2.4) is

$$f'_a f'_b = f_a f_b \qquad (2.5)$$

or

$$\text{Log} f'_a + \text{Log} f'_b = \text{Log} f_a + \text{Log} f_b$$

Therefore, $\text{Log} f$ is a 'collisional invariant', which is a linear combination of the standard invariants for this type of collision (Chapter 1), that is, m, mv, and $\frac{1}{2}mv^2$. Thus, we have

$$\text{Log} f = Am + \boldsymbol{B} \cdot m\boldsymbol{v} + C\frac{1}{2}mv^2$$

where A, $\boldsymbol{B}$, C depend only on r and t. These parameters are related to the macroscopic quantities n, V, and T, which are connected to f by their definitions (Chapter 1). Thus, taking into account the properties of the Eulerian integrals (Appendix 2.6), we find that

$$f = n \left(\frac{m}{2\pi kT}\right)^{3/2} \exp\left(-\frac{mu^2}{2kT}\right) \qquad (2.6)$$

where $\boldsymbol{u} = \boldsymbol{v} - \boldsymbol{V}$, and $n = n(r, t)$, $\boldsymbol{V} = \boldsymbol{V}(r, t)$, $T = T(r, t)$.

This Gaussian distribution is called a local Maxwellian distribution; it is spherically symmetric in $\boldsymbol{u}$. This form, known as 'normal', depends on the individual speed $\boldsymbol{v}$ via $\boldsymbol{u} = \boldsymbol{v} - \boldsymbol{V}$, and on $\boldsymbol{r}$ and t via n, $\boldsymbol{V}$, and T.

The conservation equations of the macroscopic quantities (1.26) do not contain additional unknown factors, since $\mathbf{P}$ and $\boldsymbol{q}$ may be expressed as functions of n, $\boldsymbol{u}$, and T, given the expression of f in Eqn. (2.6). Thus we have

$$\mathbf{P} = nkT\,\mathbf{I} \quad \text{and} \quad \boldsymbol{q} = 0$$

The quantity $p = nkT\,(= \rho RT)$ is the hydrostatic pressure, a scalar quantity representing only the 'internal' forces in the fluid. The heat flux is null: these results are in accordance with the assumption of an isolated system, characteristic of the collisional regime. Finally, the conservation equations are the Euler equations written in the following way:

$$\frac{\partial \rho}{\partial t} + \frac{\partial \cdot \rho \boldsymbol{V}}{\partial \boldsymbol{r}} = 0$$

$$\rho \frac{d\boldsymbol{V}}{dt} + \frac{\partial p}{\partial \boldsymbol{r}} = 0$$

$$\rho \frac{de}{dt} + p \frac{\partial \cdot \boldsymbol{V}}{\partial \boldsymbol{r}} = 0 \qquad (2.7)$$

with

$$p = \rho RT \quad \text{and} \quad e = \frac{3}{2} RT \qquad (2.8)$$

These equations are first-order equations in $\boldsymbol{r}$ and t, which follows from the assumption of isolated system. The evolution of the system is given by the above equations (2.7) together with specific initial and boundary conditions.

In summary, a pure gas with only elastic collisions and isolated (or little influenced by the external conditions) is governed by a Maxwellian distribution and the Euler equations. Such a system corresponding to a collisional regime is said to be in 'equilibrium'. It is shown (Appendix 2.1) that, if it is disturbed in a specified way, it evolves spontaneously to this state of equilibrium.

Remarks

- Energy includes energy of translation only.
- Here, $\frac{d}{dt} = \frac{\partial}{\partial t} + \boldsymbol{V} \cdot \frac{\partial}{\partial \boldsymbol{r}}$, and not, as in the Boltzmann equation, $\frac{\partial}{\partial t} + \boldsymbol{v} \cdot \frac{\partial}{\partial \boldsymbol{r}}$.
- Other properties of the Maxwellian distribution relating to the collision frequency, the mean velocities, and so on, are presented in Appendix 2.2.

- The ratio $\varepsilon_{el} = \tau_{el}/\theta$ is the Knudsen number, which, to some extent, determines the degree of rarefaction of the gas.

2.3.2 Diatomic gases

The collisions include elastic collisions (T), collisions with translation and rotation exchanges (R), and collisions with translation, rotation, and vibration exchanges (V), with different characteristic times. In the collisional regime, we have $\varepsilon_T, \varepsilon_R, \varepsilon_V \ll 1$, so that all characteristic times are much smaller than the reference time ($\tau_T, \tau_R, \tau_V \ll \theta$). Then

$$J_T + J_R + J_V = 0 \quad \text{or} \quad J_{TRV} = 0 \tag{2.9}$$

As in the monatomic case, taking into account the assumptions of Chapter 1, a solution of Eqn. (2.9), that is, $\text{Log} f_i$, consists of a linear combination of the corresponding collisional invariants, that is, m, $m\boldsymbol{v}$, and $\frac{1}{2}mv^2 + \varepsilon_{i_r} + \varepsilon_{i_v}$. This last invariant is a trivial invariant for the collisions T and R, for which ε_{i_v} remains constant. Using the definition of n, $\boldsymbol{V}$, and T as before, that is

$$n = \sum_{i_r, i_v} \int_{\boldsymbol{v}} f_i d\boldsymbol{v} \quad n\boldsymbol{V} = \sum_{i_r, i_v} \int_{\boldsymbol{v}} f_i \boldsymbol{v} \, d\boldsymbol{v} \quad \frac{3}{2} nkT = \sum_{i_r, i_v} \int_{\boldsymbol{v}} f_i \frac{1}{2} mu^2 d\boldsymbol{v}$$

we obtain for f_i

$$f_i = n \left(\frac{m}{2\pi kT}\right)^{3/2} \exp\left(-\frac{mu^2}{2kT}\right) \frac{g_{i_r} \exp\left(-\varepsilon_{i_r}/kT\right)}{Q_R} \frac{\exp\left(-\varepsilon_{i_v}/kT\right)}{Q_V} \tag{2.10}$$

where

$$Q_R = \sum_{i_r} g_{i_r} \exp\left(\frac{-\varepsilon_{i_r}}{kT}\right) \quad \text{and} \quad Q_V = \sum_{i_v} \exp\left(-\frac{\varepsilon_{i_v}}{kT}\right)$$

Here, Q_R and Q_V are, respectively, the partition functions for rotation and vibration (sum of the states), and g_{i_r} is the statistical weight of the i_r rotational level (Appendix 1.2).

The quantities n, $\boldsymbol{V}$, and T are given, as in the preceding case, by the Euler equations, which are formally identical to the above equations (2.7). However, the definition of the energy E appearing in these equations is different, because here it is the sum of all energies, that is:

$$\rho e = nE = \sum_{i_r, i_v} \int_{\boldsymbol{v}} f_i \left(\frac{1}{2} mu^2 + \varepsilon_{i_r} + \varepsilon_{i_v}\right) d\boldsymbol{v} = n(E_T + E_R + E_V)$$

with

$$nE_R = \sum_{i_r} \varepsilon_{i_r} \int_v f_i d\mathbf{v} = \sum_{i_r} n_{i_r} \varepsilon_{i_r}$$

$$nE_V = \sum_{i_v} \varepsilon_{i_v} \int_v f_i d\mathbf{v} = \sum_{i_v} n_{i_v} \varepsilon_{i_v}$$

However, from Eqn. (2.10), we can see that each type of energy E_T, E_R, E_V depends only on one temperature, that is the temperature of translation T, which thus becomes the temperature common to the three modes of translation, rotation, and vibration.

In these expressions, the populations n_{i_r}, n_{i_v} deduced from the expression of f_i (Eqn. (2.10)) are the following:

$$n_{i_r} = n \frac{g_{i_r} \exp\left(-\varepsilon_{i_r}/kT\right)}{Q_R} \quad \text{and} \quad n_{i_v} = n \frac{\exp\left(-\varepsilon_{i_v}/kT\right)}{Q_V}$$

Thus, the velocity distribution is Maxwellian, as for the monatomic case, and is called a Maxwell–Boltzmann distribution defined with one temperature. An H theorem may also be demonstrated in this case. We also have $p = nkT$ and $q = 0$.

Remarks

The expressions for rotational and vibrational energies can be developed by taking into account the models suggested in Appendix 1.2. Thus, the classical model for rotation leads to the well-known result (Appendix 2.3):

$$E_R = kT$$

If, for vibration, the harmonic oscillator model is chosen, we also find the following analytic result (Appendix 2.3):

$$E_V = \frac{k\theta_v}{\exp(\theta_v/T) - 1}$$

However, it is not always possible to use this model, in particular at high temperatures, when it is necessary to use the anharmonic model (Appendix 1.2).

2.4 Pure diatomic gases: general non-equilibrium regime

Diatomic gases dominate the physical systems encountered for non-equilibrium flows at high speeds and high temperatures. In many situations, indeed, the reference time θ can be relatively small; it can be, for example, the average transit time of the fluid particles along a model or in a duct, or the duration of an aerodynamic disturbance. Referring to typical characteristic times of Figs 9 and 10, θ can be of the order of τ_V or even smaller, while remaining generally higher than τ_T and τ_R, except in particular cases, such as with a very rarefied gas. If we thus consider this general case, and if we combine the collisions T and R, we can define a time τ_{TR} much lower than θ, so that $\varepsilon_{TR} \ll 1$, and

$$J_{TR} = 0 \qquad (2.11)$$

A solution of Eqn. (2.11) is determined as previously with the collisional invariants specific to TR collisions, that is:

$$m, \ mv, \ \text{and} \ \frac{1}{2}mv^2 + \varepsilon_{i_r}$$

so that we find,[1,13] as a solution of Eqn. (2.11):

$$f_i = n_{i_v} \left(\frac{m}{2\pi kT}\right)^{3/2} \exp\left(-\frac{mu^2}{2kT}\right) \frac{g_{i_r} \exp\left(-\epsilon_{i_r}/kT\right)}{Q_R} \qquad (2.12)$$

with

$$n = \sum_{i_v} n_{i_v}$$

The quantities n, $\mathbf{V}$ (or $\mathbf{u}$), and E are always given formally by the Euler equations (with $p = nkT$, $\mathbf{q} = 0$). However, in the expression for the total energy E, the vibrational part e_V (or E_V per molecule) cannot be expressed as a function of T and remains unknown at this stage, depending on the unspecified vibrational populations n_{i_v}, since

$$E_V = \frac{1}{n} \sum_{i_v} n_{i_v} \varepsilon_{i_v}$$

Thus, if the distribution (2.12) is an equilibrium distribution for translation and rotation (Maxwell–Boltzmann), with $E_T + E_R = \frac{5}{2}kT$, the same does not apply for the vibrational mode. In order to determine the vibrational

44 CHAPTER 2 EQUILIBRIUM AND NON-EQUILIBRIUM COLLISIONAL REGIMES

distribution, the two following cases should be distinguished:

- $\tau_V \gg \theta$, $(\varepsilon_V \gg 1)$. The vibrational population does not change during the reference time, and E_V remains constant ($\frac{dE_V}{dt} = 0$ in the Euler equations). The flow is said to be 'frozen'. This is the opposite case to the equilibrium regime.

- $\tau_V \sim \theta$, $(\varepsilon_V \sim 1)$. It is necessary to start again from the Boltzmann equation in which we have $J_{TR} = 0$. Since the collisional invariants of J_{TR} are not the same as those of J_V, we have

$$\frac{df_i}{dt} = J_V \qquad (2.13)$$

Both terms of Eqn. (2.13) have the same order of magnitude, and the flow is in vibrational non-equilibrium.

In this last case, if Eqn. (2.13) is integrated in the velocity space v and summed over the rotational levels i_r, we obtain the equations of evolution of the vibrational populations, that is:[14]

$$\frac{\partial n_{i_v}}{\partial t} + \frac{\partial \cdot n_{i_v} V}{\partial r} = \sum_{i_r,j,k,l} \int_{\Omega, v_i, v_j} (f'_k f'_l - f_i f_j) I^{k,l}_{i,j} g \, d\Omega \, dv_i \, dv_j \qquad (2.14)$$

We define

$$a^{k,l}_{i,j} = \frac{1}{n_i n_j} \int_{\Omega, v_i, v_j} f_i f_j I^{k,l}_{i,j} g \, d\Omega \, dv_i \, dv_j$$

and

$$a^{i,j}_{k,l} = \frac{1}{n_k n_l} \int_{\Omega, v_i, v_j} f'_k f'_l I^{k,l}_{i,j} g \, d\Omega \, dv_i \, dv_j$$

thus Eqn. (2.14) is written

$$\frac{\partial n_{i_v}}{\partial t} + \frac{\partial \cdot n_{i_v} V}{\partial r} = \sum_{i_r,j,k,l} n_k n_l a^{i,j}_{k,l} - n_i n_j a^{k,l}_{i,j} \qquad (2.15)$$

Here, $a^{k,l}_{i,j}$ and $a^{i,j}_{k,l}$ represent the equilibrium collision-rate coefficients corresponding to the transitions $i, j \rightleftarrows k, l$, and therefore are independent of the populations. These rates may also be defined independently of the assumption of the reversibility of the collisions, and other formulations are possible for Eqn. (2.15) (Appendix 2.4). Then, if the structure of the distribution function (2.12) is taken into account, a general relation between direct and inverse collision rates may be found, that is:

$$a^{i,j}_{k,l} = a^{k,l}_{i,j} \exp(\Delta \varepsilon) \qquad (2.16)$$

with

$$\Delta\varepsilon = \frac{\varepsilon_k + \varepsilon_l - \varepsilon_i - \varepsilon_j}{kT} \quad \text{(collisional balance of internal energy)}$$

Equation (2.16) is called the detailed balance condition.

If the rotational populations are assumed to be in equilibrium (2.12), Eqn. (2.15) may be written in a more symmetrical way[4] (Appendix 2.4):

$$\frac{\partial n_{i_v}}{\partial t} + \frac{\partial \cdot n_{i_v} V}{\partial r} = \sum_{j_v, k_v, l_v} n_{k_v} n_{l_v} a_{k_v, l_v}^{i_v, j_v} - n_{i_v} n_{j_v} a_{i_v, j_v}^{k_v, l_v} \quad (2.17)$$

where the summations over the rotational levels have been implicitly made.

These equations (one for each level), called relaxation equations, close the system of Euler equations in the non-equilibrium case ($\varepsilon_V \sim 1$). They formally give the evolution of the populations from which the vibrational energy E_V can be deduced. To complete these equations, it is necessary of course to use a physical model for the collision rates.

Harmonic oscillator model

As previously discussed, this model is valid only at 'moderate' temperatures, when the highest vibrational levels are not significantly populated. The transitions TV are 'active', as the transitions VV give a null vibrational balance. The rates of collision are thus of the type $a_{i_v}^{k_v}$ or $a_{k_v}^{i_v}$ and can be expressed as functions of a_1^0 (rate of de-excitation $1 \to 0$). Thus

$$a_{i_v+1}^{i_v} = (i_v + 1) a_1^0$$

and

$$a_{i_v}^{i_v+1} = a_{i_v+1}^{i_v} \exp\left(-\frac{\theta_v}{T}\right) \quad \text{(detailed balance)} \quad (2.18)$$

Moreover, the transitions are supposedly monoquantum, therefore $k_v = i_v \pm 1$.

With these assumptions, we find (Appendix 2.4) that the relaxation equations (2.17) may be replaced by only one equation concerning the energy of vibration E_V. This equation is obtained by multiplying Eqn. (2.17) by ε_{i_v} and summing over the levels. The equation is the Landau–Teller equation:[15]

$$\frac{dE_V}{dt} = \frac{\overline{E}_V - E_V}{\tau_V} \quad (2.19)$$

where $\overline{E}_V$ is the equilibrium vibrational energy at the local temperature T, and τ_V is a 'relaxation time' equal to

$$\left[na_1^0\left(1 - \exp\frac{\theta_v}{T}\right)\right]^{-1} \quad (2.20)$$

The variation of the vibrational energy is thus proportional to its local deviation from the equilibrium. The extreme cases may be easily deduced from Eqn. (2.19). Thus:

$$\text{If } \frac{\tau_V}{\theta} \to \infty \quad \text{then } \frac{dE_V}{dt} \to 0 \quad \text{(frozen case)}$$

$$\text{If } \frac{\tau_V}{\theta} \to 0 \quad \text{then } E_V \to \overline{E}_V \quad \text{(equilibrium case)}$$

Equation (2.19) thus makes it possible to close the Euler system, assuming that the vibrational relaxation time is known, either from experiment (Chapter 12) or from theory. According to Eqn. (2.20), τ_V is inversely proportional to n, and thus we can generally write

$$\tau_V = \frac{1}{P} f(T)$$

For a very large temperature range and many gases, we have[11,15,16]

$$\tau_V p \sim T^{-\frac{1}{3}} \tag{2.21}$$

The order of magnitude of τ_V is given in Chapters 9 and 12.

Remarks

The non-equilibrium solutions are not 'stable' solutions within the meaning of the H theorem but correspond to transitional solutions and take account of the selected timescale. If, after a non-equilibrium period (caused for example by a disturbance), the system is left to itself, it gradually finds again an equilibrium Maxwell–Boltzmann distribution (isolated system).

Another cause of the variation of the vibrational populations is the spontaneous emission of radiation, but this variation is much lower owing to a very different characteristic time.

2.5 Pure diatomic gases: specific non-equilibrium regimes

Collisions involving vibrational exchanges do not have the same probability (Chapter 9): thus, TV exchanges in which only one molecule changes its level are generally dominant at high temperature and for the highest levels, whereas VV exchanges are more probable at low temperature and for the low levels (we are then close to the conditions of validity of the harmonic oscillator model). As

2.5 PURE DIATOMIC GASES: SPECIFIC NON-EQUILIBRIUM REGIMES

for the resonant collisions Vr, which represent a particular case of VV collisions, they are the most probable in the majority of situations. These various cases and their consequences are successively examined below.

2.5.1 Dominant TV collisions

These collisions, of course, include the T and R collisions (trivial TV collisions). Then:

$$J_{TV} = 0 \quad (2.22)$$

The corresponding collisional invariants are the usual invariants only, m, mv and $\frac{1}{2}mv^2 + \varepsilon_{i_r} + \varepsilon_{i_v}$, so that for f_i we find the equilibrium Maxwell–Boltzmann distribution (Eqn. (2.10)), with the corresponding Euler equations (Eqn. (2.7)): TV collisions are thus sufficient to establish an equilibrium regime.[14]

2.5.2 Dominant VV collisions

In this case, we have

$$J_{VV} = 0 \quad (2.23)$$

As in the previous case, the VV collisions include the T and R collisions. Then, the collisional invariants are not only the usual invariants m, mv, and $\frac{1}{2}mv^2 + \varepsilon_{i_r} + \varepsilon_{i_v}$, but also the quantum number itself, i_v. If we assume, as is generally the case, that the transitions are isoquantum (and generally monoquantum, with $n = 1$), we have

$$i_v + j_v = (i_v \pm n) + (j_v \mp n)$$

The consequence is a Maxwell–Boltzmann distribution for translation and rotation, as in the previous cases, and a Treanor distribution for the vibrational population:[17]

$$n_{i_v} = n \frac{\exp\left(-\varepsilon_{i_v}/kT + Ki_v\right)}{\sum_{i_v} \exp\left(-\varepsilon_{i_v}/kT + Ki_v\right)} \quad (2.24)$$

The macroscopic parameter K is unknown, as are n, V, T, and requires a single relaxation equation to close the Euler system. Thus, starting again from the Boltzmann equation with $J_{VV} = 0$, we have

$$\frac{df_i}{dt} = J_{TV}$$

Multiplying this equation by the invariant i_v, integrating over the velocity space, and summing over all levels, we obtain the following relaxation equation, in which only the monoquantum transitions are taken into account:

$$\frac{dI_V}{dt} = [\exp(-K) - 1] \sum_{i_v} a_{i_v+1}^{i_v} n_{i_v+1} \qquad (2.25)$$

Here, I_V represents a 'mean quantum number', defined by the following relation:

$$nI_V = \sum_{i_v} n_{i_v} i_v$$

Therefore, I_V is connected to the parameter K by this definition. The Euler system and the relaxation equation (2.25), in which a physical model must be introduced, constitute a closed system giving the quantities n, V, T, as well as the populations n_{i_v} and the vibrational energy E_V.

The Treanor distribution has the property to present a minimum. It is therefore theoretically possible to obtain a 'population inversion' when $n_{i_v+1} > n_{i_v}$. This condition is widely known to produce an inversion resulting in laser action. The inversion conditions are thus fulfilled when

$$K > \frac{\varepsilon_{i_v+1} - \varepsilon_{i_v}}{kT}$$

This inversion, in fact, occurs for high levels with a small energy gap and generally occurs to a significant degree in the case of a mixture in which one of the components 'feeds' the high levels of another component by VV exchange (CO_2/N_2 mixture for example; Chapter 10).

Returning to the physical model, if we adopt the harmonic oscillator model, with $\varepsilon_{i_v} = i_v h\nu_v$, we may write

$$K - \frac{h\nu_v}{kT} = -\frac{h\nu_v}{kT_V}$$

because generally we have $K \ll \frac{h\nu_v}{kT}$.

Thus, a particular 'vibrational temperature' T_V is defined, so that it is also possible to define a Boltzmann distribution such as

$$n_{i_v} = n \frac{\exp\left(-\varepsilon_{i_v}/kT_V\right)}{\sum_{i_v} \exp\left(-\varepsilon_{i_v}/kT_V\right)} \qquad (2.26)$$

The Treanor distribution is thus reduced to a Boltzmann non-equilibrium distribution at a temperature T_V different from T. This distribution does not give any population inversion. The relaxation equation (2.25) is reduced to the Landau–Teller form of equation (2.19), with, moreover, $E_V = \bar{E}_V(T_V)$ and $\bar{E}_V = \bar{E}_V(T)$; the equations giving I_V and E_V are equivalent, ε_{i_v} becoming a collisional invariant (it is proportional to i_v).

The parameter K is thus related to anharmonicity and non-equilibrium, which thus highlights the importance of the physical model in the analysis of non-equilibrium flows. Simple models, such as the standard harmonic oscillator model, if they are useful in giving a qualitative description, can sometimes mask significant aspects of these flows.

2.5.3 Dominant resonant collisions

Here, the anharmonic oscillator model is considered.

These collisions, denoted Vr, are a particular type of VV collisions. Here, $i_v \rightarrow j_v$ and $j_v \rightarrow i_v$, which represents a strict vibrational exchange. This type of exchange is very probable in particular at low temperature. Thus we have

$$J_{Vr} = 0 \tag{2.27}$$

The corresponding collisional invariants are m, $m\mathbf{v}$, $\frac{1}{2}mv^2 + \varepsilon_{i_r} + \varepsilon_{i_v}$, and ε_{i_v}. Here, i_v is also invariant, but not independent, since if ε_{i_v} is invariant, i_v is also invariant, the reverse not being true, except with the assumption of the harmonic oscillator. The corresponding distribution is, as previously discussed, a Maxwell–Boltzmann distribution for the translation and rotation; for the vibration, we have

$$n_{i_v} = n \frac{\exp\left(-\varepsilon_{i_v}/kT_V\right)}{\sum_{i_v} \exp\left(-\varepsilon_{i_v}/kT_V\right)} \tag{2.28}$$

This distribution is identical to that of the preceding case, which included the assumption of the harmonic oscillator, but here it does not depend on the physical model. Thus it is possible to define a vibrational temperature independently of the model. The relaxation equation necessary to close the Euler system is obtained by multiplying the Boltzmann equation by the invariant ε_{i_v} and integrating and summing. This equation is written

$$n\frac{dE_V}{dt} = \sum_{i_v} \varepsilon_{i_v} \int_v (J_{TV} + J_{VV}) d\mathbf{v} \tag{2.29}$$

with

$$E_V = \frac{1}{n} \sum_{i_v} \frac{\varepsilon_{i_v} \exp\left(-\varepsilon_{i_v}/kT_V\right)}{Q_V(T_V)} \tag{2.30}$$

and

$$Q_V(T_V) = \sum_{i_v} \exp\left(-\varepsilon_{i_v}/kT_V\right)$$

In the J_{VV} term, the Vr collisions are absent. The result, developed for monoquantum transitions,[13] is presented in Appendix 2.5.

2.5.4 Physical applications of the results

The preceding results can be applied to flows for which the gas presents an important degree of non-equilibrium (downstream from shock waves, supersonic expansions, and so on; see Chapter 7); however, before adopting a model, it is necessary to consider its field of validity in practical cases. Thus, for example, at low temperature, as already discussed, VV collisions are most probable for the low levels. For the higher levels, TV collisions are more probable at high temperature (Chapter 9, Figs 43 and 44).

This example shows the limit of the preceding results deduced from a too rigid scheme. The separation into groups of levels inevitably presents some arbitrariness and may not be easily derived from a general method. In the same way, in a spatial or temporal evolution, the state of the system can be successively dominated by different types of collisions. Finally, the various types of collision can have similar probabilities, prohibiting any separation into different groups. In these cases, it is necessary to use the general relaxation equations (2.17) with collision rates deduced from theory or reliable experiments.

2.6 Gas mixtures: equilibrium regimes

As previously discussed, only those gases that have a simple structure, that is, essentially mono and diatomic gases in binary mixture, are considered in order to point out the main characteristics of a gas mixture flow. A generalization to more complex mixtures is examined within the framework of examples in Chapters 9 and 10.

2.6.1 Mixtures of monatomic gases

For a mixture of two species p and q with only elastic collisions, the Boltzmann equation specific to the species p is written

$$\frac{df_p}{dt} = J_{Tpp} + J_{Tpq} \tag{2.31}$$

where the collisions pp and pq are taken into account. There is a similar equation for the species q.

In the collisional regime ($\theta \gg \tau_{Tpp}, \tau_{Tpq}$), we therefore have

$$J_{Tpp} + J_{Tpq} = 0 \tag{2.32}$$

The solutions of Eqn. (2.32) are the classical collisional invariants m_p, $m_p \boldsymbol{v}_p$, and $\frac{1}{2} m_p v_p^2$. Taking into account the definitions of the usual macroscopic quantities (Chapter 1), we obtain for f_p the following Maxwellian distribution:

$$f_p = n_p \left(\frac{m_p}{2\pi kT}\right)^{3/2} \exp\left(-\frac{m_p u_p^2}{2kT}\right) \tag{2.33}$$

A similar expression for the species q is obtained. The Euler equations (2.7) remain valid for the mixture, with

$$p = p_p + p_q = (n_p + n_q)kT = nkT$$
$$q = 0$$
$$\boldsymbol{U}_p = 0$$

In particular, there is no diffusion of species, and the conservation equation of the species is written

$$\frac{\partial n_p}{\partial t} + \frac{\partial \cdot n_p \boldsymbol{V}}{\partial \boldsymbol{r}} = 0 \quad \text{or} \quad \frac{\partial \rho_p}{\partial t} + \frac{\partial \cdot \rho_p \boldsymbol{V}}{\partial \boldsymbol{r}} = 0 \tag{2.34}$$

There is of course a similar equation for the species q.

For mixtures of monatomic gases with components of very different mass (plasmas, for example), this solution may be not valid, and transitional non-equilibrium situations are possible.

2.6.2 Mixtures of diatomic gases

Without repeating in detail the preceding reasoning, if all characteristic times of translation, rotation, and vibration specific to each species or interspecies are much smaller than the reference time, the distribution of the species p is an equilibrium Maxwell–Boltzmann distribution:

$$f_{ip} = n_{ip} \left(\frac{m_p}{2\pi kT}\right)^{3/2} \exp\left(-\frac{m_p u_p^2}{2kT}\right) \tag{2.35}$$

with

$$n_{ip} = n_p \frac{\exp\left(-\varepsilon_{ip}/kT\right)}{Q_R(T) Q_V(T)} \tag{2.36}$$

The Euler equations and the species conservation equations (2.34) are still valid. In the definition of the total energy E, all types of energy are present.

The mixture of a monatomic gas and a diatomic gas in the collisional regime possesses respective distributions identical to the cases above. The classical example is that of a gas of molecules and atoms arising from the dissociation of these molecules in the equilibrium regime (Chapter 6).

2.7 Mixtures of diatomic gases in vibrational non-equilibrium

Typical non-equilibrium situations involve the vibrational mode, as in the case for pure gases. However, even for a binary mixture, many cases are possible, depending on the relative scale of characteristic times. Thus, for the component p, there are two TV characteristic times, corresponding respectively to collisions pp and pq, i.e. τ_{pp}^{TV} and τ_{pq}^{TV}, and two VV characteristic times, i.e. τ_{pp}^{VV} and τ_{pq}^{VV}.

Generally, the probabilities of the VVpp and VVpq collisions and particularly of the resonant collisions Vrpp and Vrqq are high, as for pure gases. Thus, a vibrational temperature T_{Vp} for the component p, and a temperature T_{Vq} for the component q, can be defined. This results from the solution of the following equations:

$$J_{Vrpp} = 0 \quad \text{and} \quad J_{Vrqq} = 0 \tag{2.37}$$

Taking into account the specific invariants of these collisions (2.27), we have

$$f_{ip} = n_p \left(\frac{m_p}{2\pi kT}\right)^{3/2} \exp\left(-\frac{m_p u_p^2}{2kT}\right) \frac{g_{i_rp} \exp\left(-\varepsilon_{i_rp}/kT\right) \exp\left(-\varepsilon_{i_vp}/kT_{vp}\right)}{Q_R(T) Q_V(T_{Vp})} \tag{2.38}$$

and a similar function for f_{iq}.

The relaxation to equilibrium is due to TVpp and TVpq collisions, which may have different characteristic times, and also to the VVpq collisions, ensuring a coupling between the components. This coupling may also accelerate the relaxation to equilibrium if τ_{pq}^{VV} is very short. Then, there is a quasi-single relaxation (with $T_{Vp} \simeq T_{Vq}$). If not, both gases evolve quasi-independently. Two examples illustrating these two cases are discussed in Chapters 9 and 12 (N_2/O_2 and CO/N_2). With the definition of vibrational temperatures, it is not necessary to solve the equations (2.17), but only the global relaxation equation giving E_{Vp}, that is:

$$n_p \frac{dE_{Vp}}{dt} = \sum_{i_{vp}} \varepsilon_{i_{vp}} \int_{v_p} J_{Vp} d\boldsymbol{v}_p = \sum_{i_{vp}} \varepsilon_{i_{vp}} \int_{v_p} \left(J_{TVpp} + J_{TVpq} + J_{VVpq} \right) d\boldsymbol{v}_p$$

(2.39)

This equation should be coupled with the Euler equations, also valid in this case. With the assumption of the harmonic oscillator model,[15,18] we obtain an analytic expression for this equation given in Appendix 2.5, which depends on all involved relaxation times.

The generalization to a mixture of several mono and diatomic gases (for example, the case of air at high temperature) generally requires simplifications (Chapter 9).

2.8 Mixtures of reactive gases

Here, we successively examine the cases of mixtures of reacting gases without and with internal energy modes.

2.8.1 Reactive gases without internal modes

Only reactive collisions and elastic collisions take place. As, generally, $\tau_T \ll \tau_C$ (Figs 9 and 10) in the collisional regime, we have for the species p

$$J_{Tp} = 0 \tag{2.40}$$

and the corresponding Maxwellian distribution function, i.e.

$$f_p = n_p \left(\frac{m_p}{2\pi kT} \right)^{3/2} \exp\left(-\frac{m_p u_p^2}{2kT} \right) \tag{2.41}$$

The Euler equations are therefore valid, and the evolution equations for the species p are formally given by Eqn. (1.28), that is:

$$\frac{\partial n_p}{\partial t} + \frac{\partial \cdot n_p V}{\partial \boldsymbol{r}} = \int_{v_p} J_{Cp} d\boldsymbol{v}_p \tag{2.42}$$

Or, after multiplying by m_p:

$$\frac{\partial \rho_p}{\partial t} + \frac{\partial \cdot \rho_p V}{\partial \boldsymbol{r}} = m_p \int_{v_p} J_{Cp} d\boldsymbol{v}_p = \dot{w}_p$$

with

$$\sum_p m_p \int_{\boldsymbol{v}_p} J_{Cp} d\boldsymbol{v}_p = \sum_p \dot{w}_p = 0$$

If the characteristic reaction time is short compared to the reference time, Eqn. (2.42) is reduced to

$$\frac{\dot{w}_p}{m_p} = \int_{\boldsymbol{v}_p} J_{Cp} d\boldsymbol{v}_p = 0 \qquad (2.43)$$

which corresponds to zero balance for the creation of each species: we are then also in the case of equilibrium for the chemical reactions.

The development of the preceding relations can be made only for particular cases, given the diversity of the possible reactions (see Chapters 9 and 10, for example).

We consider the example of only one chemical reaction corresponding to collisions between two components p and q that gives two other components p' and q', i.e.

$$p + q \rightleftarrows p' + q'$$

The chemical balance of the species p can be written

$$\frac{\dot{w}_p}{m_p} = \int_{\boldsymbol{v}_p} J_{Cp} d\boldsymbol{v}_p = k_r n_{p'} n_{q'} - k_f n_p n_q$$

with the following definitions for the forward k_f and backward k_r rate coefficients:

$$k_f = \int_{\Omega,\boldsymbol{v}_p,\boldsymbol{v}_q} \frac{f_p f_q}{n_p n_q} I_{p,q}^{p',q'} g_{pq} d\Omega \, d\boldsymbol{v}_p d\boldsymbol{v}_q = \int_{\boldsymbol{v}_p,\boldsymbol{v}_q} \frac{f_p f_q}{n_p n_q} C_{p,q}^{p',q'} g_{pq} d\boldsymbol{v}_p d\boldsymbol{v}_q$$

$$k_r = \int_{\Omega',\boldsymbol{v}_{p'},\boldsymbol{v}_{q'}} \frac{f_{p'} f_{q'}}{n_{p'} n_{q'}} I_{p',q'}^{p,q} g_{p'q'} d\Omega' d\boldsymbol{v}_{p'} d\boldsymbol{v}_{q'} = \int_{\boldsymbol{v}_{p'},\boldsymbol{v}_{q'}} \frac{f_{p'} f_{q'}}{n_{p'} n_{q'}} C_{p',q'}^{p,q} d\boldsymbol{v}_{p'} d\boldsymbol{v}_{q'}$$

where $C_{p,q}^{p',q'}$ and $C_{p',q'}^{p,q}$ represent the total cross sections of the direct and reverse reactions respectively.

The rate coefficients k_f and k_r depend on the reactive collision cross sections and on the translational distribution functions but not on the species concentrations. They can also be called Arrhenius rate constants, which depend only on the temperature T and the reaction model. Thus we have

$$\frac{\partial n_p}{\partial t} + \frac{\partial \cdot n_p \boldsymbol{V}}{\partial \boldsymbol{r}} = k_r n_{p'} n_{q'} - k_f n_p n_q \qquad (2.44)$$

In the case of chemical equilibrium ($\dot{w}_p = 0$), we have

$$\frac{k_f}{k_r} = \frac{n_{p'} n_{q'}}{n_p n_q} = K_c(T) \tag{2.45}$$

Here, $K_c(T)$ is the equilibrium constant of the reaction, dependent only on the temperature.

Equations (2.44) or (2.45), coupled with the Euler equations, enable us to determine the concentrations of the components.

2.8.2 Reactive gases with internal modes

In this case, the species conservation equation is written

$$\frac{\partial n_p}{\partial t} + \frac{\partial \cdot n_p V}{\partial r} = \sum_{i_p} \int_{v_p} J_{Cp} dv_p$$

With the same example as in the preceding case, we have

$$\sum_{i_p} \int_{v_p} J_{Cp} dv_p = k_r n_{p'} n_{q'} - k_f n_p n_q$$

with

$$k_f = \sum_{i_p j_q k_{p'} l_{q'}} \xi_{i_p} \xi_{j_q} k_{i_p j_q}^{k_{p'} l_{q'}} \quad \text{and} \quad k_r = \sum_{i_p j_q k_{p'} l_{q'}} \xi_{k_{p'}} \xi_{l_{q'}} k_{k_{p'} l_{q'}}^{i_p j_q} \tag{2.46}$$

Here, $k_{i_p j_q}^{k_{p'} l_{q'}}$ and $k_{k_{p'} l_{q'}}^{i_p j_q}$ are the reaction-rate constants per level. They are independent of the populations.

If the distribution is a Boltzmann distribution at a temperature T, then k_f and k_r depend only on that temperature. The same of course applies for the equilibrium constant $K_c = \frac{k_f}{k_r}$, and the relation (2.45) remains valid. If not, k_f and k_r depend on the populations and thus possibly on non-equilibrium. If, as usual, the rotational mode is in equilibrium but not the vibrational mode, we have (omitting the indices of species):

$$\frac{n_i}{n_{i_v}} = \frac{n_{i_r}}{n} = \frac{\exp\left(-\frac{\varepsilon_{i_r}}{kT}\right)}{Q_R}$$

and

$$k_f = \sum_{i_v j_v k_v l_v} \xi_{i_v} \xi_{j_v} k_{i_v j_v}^{k_v l_v} \tag{2.47}$$

where
$$k_{i_v j_v}^{k_v l_v} = \sum_{i_r j_r k_r l_r} \frac{\exp\left(-\frac{\varepsilon_{i_r}+\varepsilon_{j_r}}{kT}\right)}{(\overline{Q}_R)^2} k_{ij}^{kl}$$

is the vibrational reaction constant, independent of the populations. We have similar relations for k_r.

Finally, therefore, the constants k_r and k_f may depend on non-equilibrium and in particular on the vibrational temperatures T_{Vp} and T_{Vq}. A related example is given by the dissociation–recombination reactions (Chapter 5).

Appendix 2.1 The H theorem

A geometrically closed domain D is considered. It contains N identical particles of a pure gas undergoing elastic collisions and specular reflections with the wall. Thus, the distribution function of the reflected particles differs from that of the incident particles only by its sign; there is thus no 'exchange' with the external medium. Among N particles, N_a have velocity v_a at the coordinate r_a in the elementary volume $\delta v = dv_a dr_a$. The generalized element of volume in a $6N$ dimension space is $(\delta v)^N$, and the probability of finding this distribution is equal to

$$W = \frac{N!}{\prod_a N_a!}(\delta v)^N \qquad (2.48)$$

where $\frac{N!}{\prod_a N_a!}$ represents all possible combinations for identical particles.

With Stirling's approximation, we have

$$\text{Log} W = -\sum_a N_a \text{Log} N_a$$

As there is a large number of particles, the summations may be replaced by integrations, so that we have

$$\text{Log} W = -\int_D \int_{v_a} f_a \text{Log} f_a \, dr_a \, dv_a = -H \qquad (2.49)$$

From this relation, the entropy S_D of the domain may be defined as

$$S_D = k \, \text{Log} W$$

then

$$S_D = -kH$$

Equation (2.49) depends only on the time, so that $H = H(t)$. In order to know its temporal evolution, the Boltzmann equation is multiplied by $(1 + \text{Log} f)$ and integrated in the domain D and the velocity space. Taking into account the symmetry properties (1.87), we find

$$\frac{dH}{dt} = -\frac{1}{4} \int_D \int_{v_a, v_b} \left(\text{Log}\frac{f'_a f'_b}{f_a f_b}\right) (f'_a f'_b - f_a f_b)\, gI\, d\Omega\, d\boldsymbol{r}\, d\boldsymbol{v}_a\, d\boldsymbol{v}_b \quad (2.50)$$

It can easily be seen that the integral is always positive or null, so that if there is an initial disturbance in the system, H always decreases until cancelling out: the corresponding 'stable' state is obtained for

$$f'_a f'_b = f_a f_b$$

We then have a Maxwellian distribution (2.6), and this state may be considered the 'equilibrium' state of the system. It is of course possible to generalize this result to the case of polyatomic gases, to mixtures, reactive or not, and so on.

Appendix 2.2 Properties of the Maxwellian distribution

In the case of a pure gas dominated by elastic collisions (Maxwellian regime), we have

$$f = n\left(\frac{m}{2\pi kT}\right)^{3/2} \exp\left(-\frac{mu^2}{2kT}\right)$$

In addition to the Euler equations, characteristic of this regime, particular quantities used in the following chapters can be calculated.

Average peculiar velocity U

The mean peculiar velocity is zero, but the average of U is

$$U = \frac{1}{n}\int_v fu\, d\boldsymbol{v}$$

From the properties of the Eulerian integrals (Appendix 2.6), we have

$$U = \left(\frac{8kT}{\pi m}\right)^{1/2} \quad (2.51)$$

This value is independent of n.

Average quadratic velocity W

We have

$$nW^2 = \int_v fu^2 d\boldsymbol{v}$$

so that

$$W = \left(\frac{3kT}{m}\right)^{1/2} = 1.086\, U \qquad (2.52)$$

These velocities U and W may be compared to the sound speed a in an ideal gas (Chapter 8), that is:

$$a = \left(\frac{\gamma kT}{m}\right)^{1/2} \qquad (2.53)$$

with $\gamma = \frac{C_T + k}{C_T}$, the ratio of specific heats in a gas with elastic collisions, for which

$$C_T = \frac{dE_T}{dT} = \frac{3}{2}k, \quad \gamma = \frac{5}{3}, \quad \text{and} \quad a = \left(\frac{5kT}{3m}\right)^{1/2}$$

Comparing Eqns (2.51) and (2.53), we see that U and a have comparable values. This is not surprising, since macroscopic disturbances (arising for example from external media) can be propagated only by collisions.

Collision frequency Z

For a pure gas, we have (1.56),

$$nZ = \int_{\Omega, \boldsymbol{v}_a, \boldsymbol{v}_b} f_a f_b g I \, d\Omega \, d\boldsymbol{v}_a \, d\boldsymbol{v}_b$$

With the variable change $\boldsymbol{v}_a, \boldsymbol{v}_b \to \boldsymbol{g}, \boldsymbol{G}$, taking into account the properties of the Eulerian integrals (Appendix 2.6) and using the rigid elastic sphere model $\left(I = \frac{d^2}{4}\right)$, we find that

$$Z = 4nd^2 \left(\frac{\pi kT}{m}\right)^{1/2} = \tau_{el}^{-1} \qquad (2.54)$$

The order of magnitude of the mean free path λ may thus be found, i.e.

$$\lambda \sim \frac{U}{Z} = \left(\sqrt{2}\pi nd^2\right)^{-1} \qquad (2.55)$$

With this model, the total number of collisions per second Z_0 is

$$Z_0 = 2n^2 d^2 \left(\frac{\pi kT}{m}\right)^{1/2} = \frac{n}{2} Z \qquad (2.56)$$

In the same way, the total number of collisions per second in a binary mixture p, q is equal to

$$Z_0 = n_p n_q \left(\frac{d_p + d_q}{2}\right)^2 \left(\frac{8\pi kT}{m_r}\right)^{1/2} \qquad (2.57)$$

Appendix 2.3 Models for internal modes

Rotational mode: rigid rotator model

The equilibrium Boltzmann distribution for a diatomic gas is

$$\overline{E_R} = \sum_{i_r} \frac{n_{i_r} \varepsilon_{i_r}}{n} = \frac{g_{i_r} \varepsilon_{i_r} \exp\left(-\varepsilon_{i_r}/kT\right)}{Q_R(T)} \qquad (2.58)$$

with (Appendix 1.2):

$$g_{i_r} = 2i_r + 1$$

$$\varepsilon_{i_r} = h\gamma_r i_r (i_r + 1) = k\theta_r i_r (i_r + 1)$$

$$Q_R = \sum_{i_r} g_{i_r} \exp\left(-\varepsilon_{i_r}/kT\right) = \sum_{i_r} (2i_r + 1) \exp\left[-\frac{\theta_r}{T} i_r (i_r + 1)\right]$$

If we assume a continuous rotational spectrum because the rotational levels are close, we have, setting $x = i_r(i_r + 1)$:

$$\overline{E_R} = k\theta_r \left[\int_0^\infty x \exp\left(-\frac{\theta_r}{T} x\right) dx\right] \left[\int_0^\infty \exp\left(-\frac{\theta_r}{T} x\right) dx\right]^{-1}$$

so that we find

$$\overline{E_R} = kT \qquad (2.59)$$

The rotational specific heat C_R is then

$$C_R = \frac{d\overline{E_R}}{dT} = k$$

Vibrational mode: harmonic oscillator model

With the ground state $\varepsilon_0 = h\nu_v/2$, and $\varepsilon_{i_v} = i_v h\nu_v = i_v k\theta_v$, $E_V = \sum_{i_v} \frac{n_{i_v}}{n} \varepsilon_{i_v}$, we have

$$\overline{E}_V = k\theta_v \left[\sum_{i_v} i_v \exp\left(-i_v \frac{\theta_v}{T}\right) \right] [Q_V(T)]^{-1}$$

where

$$Q_V(T) = \sum_{i_v} \exp\left(-i_v \frac{\theta_v}{T}\right)$$

Setting $i_v = n$ and $\exp\left(-\theta_v/T\right) = x$, we have $\overline{E}_V = k\theta_v \frac{\sum_n n x^n}{\sum_n x^n}$.

When $n \to \infty$, then $\sum_n x^n \to (1-x)^{-1}$, $\sum_n n x^n \to x/(1-x)^2$, and then

$$\overline{E}_V = \frac{k\theta_v}{\exp\left(\theta_v/T\right) - 1} \tag{2.60}$$

For $T \gg \theta_v$, we obtain the classical limit

$$\overline{E}_V \to kT \tag{2.61}$$

The corrections brought to relations (2.59) and (2.61) by the effects of rotation–vibration interaction (vibrating rotator, anharmonic oscillator; see Appendix 1.2) are sufficiently weak so that we can neglect them here.

Appendix 2.4 General vibrational relaxation equation

Effects of rotational transitions

The general relaxation equation giving the vibrational populations (Eqn. (2.15)) is

$$\frac{\partial n_{i_v}}{\partial t} + \frac{\partial \cdot n_{i_v} V}{\partial r} = \sum_{i_r,j,k,l} \left(n_k n_l a_{k,l}^{i,j} - n_i n_j a_{i,j}^{k,l} \right)$$

As the rotational distribution is in equilibrium, we have

$$\frac{n_i}{n_{i_v}} = \frac{n_{i_r}}{n} = \frac{\exp\left(-\varepsilon_{i_r}/kT\right)}{Q_R(T)} \tag{2.62}$$

Substituting (2.62) in (2.15), and separating the summations over both modes, we obtain for the left-hand side of this equation:

$$\sum_{j_v,k_v,l_v} \sum_{i_r,j_r,k_r,l_r} \left[n_{k_v} n_{l_v} \exp\left(-\frac{\varepsilon_{k_r} + \varepsilon_{l_r}}{kT}\right) a_{k_r,k_v,l_r,l_v}^{i_r,i_v,j_r,j_v} \right.$$

$$\left. - n_{i_v} n_{j_v} \exp\left(-\frac{\varepsilon_{i_r} + \varepsilon_{j_r}}{kT}\right) a_{i_r,i_v,j_r,j_v}^{k_r,k_v,l_r,l_v} \right] [Q_R(T)]^{-2}$$

Finally, the relaxation equation (2.15) may be written in the following symmetrical form (Eqn. (2.17)):

$$\frac{\partial n_{i_v}}{\partial t} + \frac{\partial \cdot n_{i_v} V}{\partial r} = \sum_{j_v,k_v,l_v} \left(n_{k_v} n_{l_v} a_{k_v,l_v}^{i_v,j_v} - n_{i_v} n_{j_v} a_{i_v,j_v}^{k_v,l_v} \right)$$

with

$$a_{k_v,l_v}^{i_v,j_v} = \left[\sum_{i_r,j_r,k_r,l_r} \exp\left(-\frac{\varepsilon_{k_r} + \varepsilon_{l_r}}{kT}\right) a_{k_r,k_v,l_r,l_v}^{i_r,i_v,j_r,j_v} \right] [Q_R(T)]^{-2}$$

and a similar expression for $a_{i_v,j_v}^{k_v,l_v}$.

In Eqn. (2.17), the vibrational collision rates take into account all possible rotational transitions.

Various forms of the vibrational relaxation general equation

The previous form (Eqn. (2.17)) is simple and symmetrical. It also has the advantage of allowing us to separate the terms related to the populations (n_i, n_j, n_k, n_l) from those depending on the collision itself ($a_{i,j}^{k,l}$, $a_{k,l}^{i,j}$). Other forms that are sometimes physically clearer are also possible.

Later on, in order to simplify the form of the expressions, the vibrational indices are omitted.

Thus, if $Q_{i,j}^{k,l}$ is the mean probability of an inelastic collision, and Z is the collision frequency, we have

$$a_{i,j}^{k,l} = \frac{Z}{n} Q_{i,j}^{k,l} \qquad (2.63)$$

For moderate temperatures, Z may be considered the frequency of elastic collisions, so that $Z_{i,j}^{k,l} = ZQ_{i,j}^{k,l}$ represents the frequency of inelastic collisions.

The production term $\sum_{j,k,l} n_i n_j a_{i,j}^{k,l}$ may then be written $\sum_{j,k,l} \frac{Z}{n} n_i n_j Q_{i,j}^{k,l}$, where

$\frac{Z}{n}$ is the mean number of collisions undergone by one particle per second

$\frac{Z}{n} n_j$ is the mean collision frequency between one particle and particles n_j

$\frac{Z}{n} n_j n_i$ is the mean collision frequency between particles n_i and particles n_j

$\sum_{j,k,l} \frac{Z}{n} n_i n_j Q_{i,j}^{k,l}$ is the mean number of particles n_i having undergone a transition $i \to k$ per collision and per second.

Appendix 2.5 Specific vibrational relaxation equations

Landau–Teller equation

If we only take into account the monoquantum transitions, the general relaxation equation (2.17) after multiplication by ε_{i_v} and summation over i_v may be written

$$n\frac{dE_V}{dt} = \sum_{i_v, j_v} \varepsilon_{i_v} \left(n_{i_v+1} n_j a_{i_v+1}^{i_v} - n_{i_v} n_{j_v} a_{i_v}^{i_v+1} + n_{i_v-1} n_{j_v} a_{i_v-1}^{i_v} - n_{i_v} n_{j_v} a_{i_v}^{i_v-1} \right)$$

(2.64)

Taking into account Eqn. (2.18), we have

$$\frac{1}{a_1^0} \frac{dE_V}{dt} = \left[\exp\left(-\frac{h\nu_v}{kT}\right) - 1 \right] \sum_{i_v} i_v h\nu_v n_{i_v} + nh\nu_v \exp\left(-\frac{h\nu_v}{kT}\right)$$

As $E_V = \frac{1}{n} \sum_{i_v} i_v h\nu_v$, and $\overline{E}_V = \frac{h\nu_v \exp\left(-\frac{h\nu_v}{KT}\right)}{1-\exp\left(-\frac{h\nu_v}{KT}\right)}$, we find

$$\frac{dE_V}{dt} = na_1^0 \left[1 - \exp\left(-\frac{h\nu_v}{kT}\right) \right] (\overline{E}_V - E_V) \qquad (2.65)$$

This is the Landau–Teller equation.

Relaxation equation with dominant VV collisions

The relaxation equation is the following:

$$n\frac{dI_V}{dt} = \sum_{i_v} i_v \int_v J_{TV}\, d\mathbf{v}$$

Keeping only the monoquantum transitions, we find

$$\int_v J_{TV}\, d\mathbf{v} = n^2[1 - \exp(-K)]\left(n_{i_v+1} a_{i_v+1}^{i_v} - n_{i_v} a_{i_v}^{i_v-1}\right) \qquad (2.66)$$

Equation (2.25) is easily deduced from Eqn. (2.66), i.e.

$$\frac{dI_V}{dt} = [\exp(-K) - 1]\sum_{i_v} a_{i_v+1}^{i_v} n_{i_v+1}$$

Dominant resonant collisions

Starting from the following relaxation equation:

$$n\frac{dE_V}{dt} = \sum_{i_v} \varepsilon_{i_v} \int_v (J_{TV} + J_{VV})\, d\mathbf{v}$$

and keeping only the monoquantum transitions, we find

$$n\frac{dE_V}{dt} = n\sum_{i_v} n_{i_v+1} a_{i_v+1}^{i_v} \left(\varepsilon_{i_v+1} - \varepsilon_{i_v}\right)$$

$$\times \left\{ \exp\left[-(\varepsilon_{i_v+1} - \varepsilon_{i_v})\left(\frac{1}{kT} - \frac{1}{kT_V}\right)\right] - 1 \right\}$$

$$- \sum_{i_v, j_v} n_{i_v+1} n_{j_v} a_{i_v+1, j_v}^{i_v, j_v+1} \left(\varepsilon_{i_v+1} - \varepsilon_{i_v}\right)$$

$$\times \left\{ \exp\left[-(\varepsilon_{i_v+1} + \varepsilon_{j_v} - \varepsilon_{i_v} - \varepsilon_{j_v+1})\left(\frac{1}{kT} - \frac{1}{kT_V}\right)\right] - 1 \right\} \quad (2.67)$$

The first term of the right-hand side of Eqn. (2.67) represents the balance of the TV collisions and the second that of the other VV collisions.

If we adopt the harmonic oscillator model, we again find the Landau–Teller equation.

Relaxation equations for a binary mixture of diatomic gases

With the harmonic oscillator model for each component p and q, we have the relations (Eqn. (2.18)) for the collision rates TVpp and TVpq. For the collision rates VVpq, we have the following relation:

$$a_{i_p,j_q}^{i_p+1,j_q-1} = j_q(i_p+1)a_{0_p,1_q}^{1_p,0_q} \qquad (2.68)$$

The vibrational energy balance is zero for the collisions VVpp and VVqq, which enables us to define vibrational temperatures specific to each component T_{Vp} and T_{Vq} and thus distribution functions of type (2.28).

The general relaxation equation for the component p is

$$n_p \frac{dE_{Vp}}{dt} = \sum_{i_{vp}} \varepsilon_{i_{vp}} \int_v \left(J_{TVpp} + J_{TVpq} + J_{VVpq}\right) d\mathbf{v} \qquad (2.69)$$

and there is a similar equation for the component q. Finally, we have

$$\frac{dE_{Vp}}{dt} = \left(\frac{\xi_p}{\tau_{pp}^{TV}} + \frac{\xi_q}{\tau_{pq}^{TV}}\right)\left(\overline{E}_{Vp} - E_{Vp}\right)$$

$$+ \frac{\xi_q}{k\theta_{vq}\tau_{pq}^{VV}} \left[\begin{array}{l} E_{Vq}\left(E_{Vp} + k\theta_{vp}\right) \exp\left(\frac{\theta_{vq}-\theta_{vp}}{T}\right) \\ -E_{Vp}\left(E_{Vq} + k\theta_{vq}\right) \end{array}\right]$$

$$\frac{dE_{Vq}}{dt} = \left(\frac{\xi_p}{\tau_{qp}^{TV}} + \frac{\xi_q}{\tau_{qq}^{TV}}\right)\left(\overline{E}_{Vq} - E_{Vq}\right)$$

$$+ \frac{\xi_p}{k\theta_{vp}\tau_{pq}^{VV}} \left[\begin{array}{l} E_{Vq}\left(E_{Vp} + k\theta_{vp}\right) \exp\left(\frac{\theta_{vq}-\theta_{vp}}{T}\right) \\ -E_{Vp}\left(E_{Vq} + k\theta_{vq}\right) \end{array}\right] \qquad (2.70)$$

For the relaxation times, we have the following relations:

$$\tau_{pp}^{TV} = \left\{na_{1p,p}^{0p,p}\left[1 - \exp\left(-\frac{\theta_{vp}}{T}\right)\right]\right\}^{-1}$$

$$\tau_{pq}^{TV} = \left\{na_{1p,q}^{0p,q}\left[1 - \exp\left(-\frac{\theta_{vp}}{T}\right)\right]\right\}^{-1} \qquad (2.71)$$

and similar expressions for τ_{qq}^{TV} and $\tau_{qp}^{TV} \neq \tau_{pq}^{TV}$:

$$\tau_{pq}^{VV} = \tau_{qp}^{VV} = \frac{1}{na_{0p,1p}^{1p,0q}} \exp\left(\frac{\theta_{vq} - \theta_{vp}}{T}\right) \qquad (2.72)$$

Knowledge of these five relaxation times is thus necessary to solve the system of Euler and relaxation equations. Examples are presented in Chapter 9.

For mixtures of several gases, there is an important number of possible collisions, so that approximations become necessary (Chapters 9 and 10).

Appendix 2.6 Properties of the Eulerian integrals

Integrals including only scalar quantities

$$\int_{-\infty}^{+\infty} x^{2n} \exp\left(-kx^2\right) dx = \frac{(2n-1)(2n-3)\cdots 5.3}{2^n} \left(\frac{\pi}{k^{2n+1}}\right)^{1/2}$$

$$= 2 \int_0^{\infty} x^{2n} \exp\left(-kx^2\right) dx$$

$$\int_0^{\infty} x^{2n+1} \exp\left(-kx^2\right) dx = \frac{n!}{2k^{n+1}}$$

Integrals including vectors and second-order tensors

(Vector u: components u_x, u_y, u_z. Tensor B independent of u.)

$$\int_{-\infty}^{+\infty} f(u)\, du = 0 \quad \text{if } f(u) \text{ is an odd function}$$

$$\int_{-\infty}^{+\infty} f(u)\, u_x^2\, du = \int_{-\infty}^{+\infty} f(u)\, u_y^2\, du = \int_{-\infty}^{+\infty} f(u)\, u_z^2\, du = \frac{1}{3} \int_{-\infty}^{+\infty} f(u)\, u^2\, du$$

$$\int_{-\infty}^{+\infty} f(u)\, \boldsymbol{uu}\, du = \left(\frac{1}{3} \int_{-\infty}^{+\infty} u^2 f(u)\, du\right) \mathbf{I} \quad \text{and} \quad \int_{-\infty}^{+\infty} f(u)\, \overset{0}{\boldsymbol{uu}}\, du = 0$$

$$\int_{-\infty}^{+\infty} f(u)\, u_x^2 u\, du = \frac{1}{3} \int_{-\infty}^{+\infty} f(u)\, u^4\, du \quad \text{and} \quad \int_{-\infty}^{+\infty} f(u)\, u_x^4\, du = \frac{1}{5} \int_{-\infty}^{+\infty} f(u)\, u^4\, du$$

$$\int_{-\infty}^{+\infty} f(u)\, u_x^2 u_y^2\, du = \frac{1}{15} \int_{-\infty}^{+\infty} f(u)\, u^4\, du \quad \text{and} \quad \int_{-\infty}^{+\infty} f(u)\, \boldsymbol{uu} \left(\overset{0}{\boldsymbol{uu}} : \mathbf{B}\right) du$$

$$= \frac{2}{15} \overset{0}{\mathbf{B}} \int_{-\infty}^{+\infty} f(u)\, u^4\, du$$

THREE

Transport and Relaxation in Quasi-Equilibrium Regimes: Pure Gases

3.1 Introduction

The solutions presented in the preceding chapter concern only purely collisional regimes, but may include non-equilibrium flows. These regimes do not take into account the influence of the exchanges with the background: this means that these flows are relatively far from the boundaries or interfaces limiting the domain. As already discussed, these boundaries play a geometrical role only.

This is of course not entirely realistic for non-isolated media limited by boundaries or interfaces. However, in continuum flows, the influence of the background is generally important in the immediate neighbourhood of these boundaries and is relatively weak elsewhere. Therefore, we may consider that in these zones the collisions between particles remain dominant for the determination of the distribution function, but that this function is somewhat influenced by the outer medium. From these considerations, it seems logical to expand the distribution function in a series of a 'small' parameter, chosen as the ratio of a characteristic time between collisions and a reference time, that is, the parameter $\varepsilon = \tau/\theta$, defined in Chapter 2, and to search for solutions of higher order in ε.

The difficulty lies, however, in the fact that there are generally several characteristic times with different orders of magnitude, so that non-equilibrium situations may occur. Thus, we will examine a number of these situations representative of realistic flows and will try to obtain sufficiently general solutions.

3.2 Expansion of the distribution function

3.2.1 Definition of flow regimes

If we consider only two characteristic times[14] (we will see below that this is generally sufficient) τ_I and τ_{II}, which differ by a different order of magnitude, and if we name J_I and J_{II} the respective corresponding collisional balances, the Boltzmann equation may be written in the following way:

$$\frac{df_{ip}^*}{dt^*} = \frac{1}{\varepsilon_I}J_I^* + \frac{1}{\varepsilon_{II}}J_{II}^* \tag{3.1}$$

where

$$\varepsilon_I = \frac{\tau_I}{\theta} \quad \text{and} \quad \varepsilon_{II} = \frac{\tau_{II}}{\theta}$$

As we consider regimes close to the collisional regime, we set $\varepsilon_I \ll 1, (\tau_I \ll \theta)$. Collisions of type I are then dominant, that is they take place during the shortest timescale. The 'small parameter' ε is therefore equal to ε_I, $\varepsilon = \varepsilon_I$.

Expanding the distribution function in a series of this parameter and stopping at the first order, we have

$$f_{ip}^* = f_{ip}^{*0} + \varepsilon f_{ip}^{*1}$$

or in a dimensional form

$$f_{ip} = f_{ip}^0(1 + \varphi_{ip}) \tag{3.2}$$

with

$$\varphi_{ip} \ll 1$$

At the zeroth order of the distribution function, we therefore have

$$J_I^0 = 0 \tag{3.3}$$

The Maxwell–Boltzmann–Euler (MBE) solutions are deduced from this equation.

The order of magnitude of τ_{II} as compared to θ remains to be defined, without necessarily comparing it to τ_I. Three cases are possible:

$$\tau_{II} \gg \theta, \quad \tau_{II} \sim \theta, \quad \tau_{II} \ll \theta$$

Case 1: $\tau_{II} \gg \theta$, $\left(\varepsilon_{II} \sim \frac{1}{\varepsilon} \gg 1\right)$

In Eqn. (3.1), the term ε_{II} does not appear at zero order or first order in the expansion of the distribution function; we then have the following system giving the zero-order and the first-order solutions successively:

$$\begin{aligned} J_I^0 &= 0 \\ \frac{df_{ip}^0}{dt} &= J_I^1 \end{aligned} \tag{3.4}$$

The regime is frozen for collisions of type II, and only collisions of type I are efficient in the determination of f_{ip}. They are dominant for the determination of the zero-order solution (solution of type MBE) and also in the first-order linearized equation (3.4), which must be solved to obtain φ_{ip}.

Case 2: $\tau_{II} \ll \theta$, $(\varepsilon_{II} \sim \varepsilon \ll 1)$

In Eqn. (3.1), the term ε_{II} appears at the same level as ε_{I}, and therefore at the zero and first-order levels. Thus, we have to solve the following system:

$$(J_I + J_{II})^0 = 0$$
$$\frac{df_{ip}^0}{dt} = (J_I + J_{II})^1 \tag{3.5}$$

Both types of collision appear at the same level in the zero and first-order solutions. Therefore, they play the same role in the determination of the zero and first-order solutions. At zero order, we find again the equilibrium MBE solutions of the preceding chapter. Here, the first-order solution of the linearized equation (φ_{ip}) remains to be determined.

Case 3: $\tau_{II} \sim \theta$, $(\varepsilon_{II} \sim 1)$

The system of equations that corresponds to zero and first-order solutions may be written

$$J_I^0 = 0$$
$$\frac{df_{ip}^0}{dt} = J_I^1 + J_{II}^0 \tag{3.6}$$

Typically, the zero-order solution is out of equilibrium for type II collisions, as described in Chapter 2 However, these collisions do influence the first-order solution, which we have yet to find here.

Whatever the zero-order solution is, the first-order solution is out of equilibrium. In this third case, in which the zero-order solution is already out of equilibrium, it brings a supplementary non-equilibrium. This non-equilibrium remains weak because of the linearization. From a mathematical point of view, the linearization presents advantages because of the use of efficiently tested computational methods.

3.2.2 Classification of flow regimes

The three cases presented above may hide complex situations. First, the indices I and II represent various types of collisions, either simple (T, R, V, C), complex (TR, TRV, and so on), or particularized (VV, resonant, and so on); thus the specific situations are numerous.

It is therefore necessary to distinguish zero-order solutions describing purely collisional regimes from those requiring a first-order expansion because of the influence of the background. Thus, at zero order, as seen in Chapter 2, there are equilibrium regimes for which all types of collisions are efficient during the considered timescale, and there are non-equilibrium regimes for which only some of them are efficient during the same timescale. Then, as the first-order solutions themselves bring a weak non-equilibrium, we distinguish the 'strong non-equilibrium' regimes (SNE), corresponding to regimes already out of equilibrium at zero order, and the 'weak non-equilibrium' regimes (WNE), corresponding to regimes in equilibrium at zero order.[19]

Furthermore, each type of regime may apply to one or several types of collisions. For example, we call $(WNE)_I+(SNE)_{II}$ a weak non-equilibrium regime for type I collisions and a strong non-equilibrium for type II collisions. If, as a standard example, we consider the case of a pure gas in a strong vibrational non-equilibrium, we call this case $(WNE)_{TR}+(SNE)_V$. In the same way, if the chemical non-equilibrium is significant at zero order and if the vibrational non-equilibrium is less important, we call this regime $(WNE)_{TRV}+(SNE)_C$, or simply $(WNE)_V+(SNE)_C$, and so on.

3.3 First-order solutions

For the above three cases, we therefore must try to determine the perturbation solution φ_{ip} of the first-order linearized equation. The first step is to replace the 'anonymous' type I and II collisions with realistic collisions T, R, V, C, and so on, and to place them into the above schemes (WNE, SNE). The second step is to examine all possible cases, which are numerous and sometimes complex.

We proceed with increasing complexity by first analysing cases 1 and 2 (frozen and WNE cases for pure gases) in this chapter. Then the same cases for mixtures are analysed in the following chapter (Chapter 4), and the SNE cases (case 3) are studied in Chapter 5.

We begin with pure gases with elastic collisions T, constituting the collisions of type I, in absence of any other collision. This classical case, from a historical point of view, is the methodology generally used for the determination of the first-order distribution function (Chapman–Enskog method). Strictly, this case concerns only monatomic gases. Then the case of diatomic gases including only one internal active mode (rotation) is examined, which corresponds to a moderate temperature range for these gases, assuming an equilibrium distribution at the zeroth order of the distribution function (WNE case). The same case (WNE) for diatomic gases with two internal modes (rotation and vibration) is finally treated; the general transport terms and their usual approximations are also presented.

3.3.1 Pure gases with elastic collisions: monatomic gases

Chapman–Enskog Method

Only one type of collision is considered here, and the corresponding index is omitted. The above first system (Eqn. (3.4)) is then written

$$\begin{aligned} J^0 &= 0 \\ \frac{df^0}{dt} &= J^1 \end{aligned} \tag{3.7}$$

with $f^0 = n\left(\frac{m}{2\pi kT}\right)^{3/2} \exp\left(-\frac{mu^2}{2kT}\right)$ (Eqn. (2.6)), which is the solution of the equation $J^0 = 0$, and with

$$J^1 = J^1[\varphi] = \int_{\Omega, v_b} f_a^0 f_b^0 (\varphi_a' + \varphi_b' - \varphi_a - \varphi_b) I_{el} g \, d\Omega \, dv_b \tag{3.8}$$

However, in the expression for f^0, the macroscopic quantities n, V, T are defined with f^0 and not with f. Thus, we have

$$n = \int_v f^0 dv, \quad nV = \int_v f^0 v \, dv, \quad \text{and} \quad \frac{3}{2} nkT = \int_v f^0 \frac{1}{2} mu^2 dv \tag{3.9}$$

From a physical point of view, this implies that, from the local values of n, V, T (when measured, for example), a fictitious equilibrium distribution is rebuilt for the computation of the perturbation φ. From a mathematical point of view, this introduces constraints in the first-order solution, that is:

$$\int_v f^0 \varphi \, dv = 0, \quad \int_v f^0 \varphi v \, dv = 0, \quad \text{and} \quad \int_v f^0 \varphi \frac{1}{2} mu^2 dv = 0 \tag{3.10}$$

The first-order linearized equation of the system (Eqn. (3.7)) is a Fredholm equation with the following integration conditions:

$$\int_v \Psi \frac{df^0}{dt} dv = 0 \tag{3.11}$$

This means that the solutions Ψ of the homogeneous equation $J^1 = 0$ must be orthogonal to the inhomogeneous part of the complete equation of the system (Eqn. (3.7)). It is easy to see that these solutions are the classical invariants of elastic collisions and that therefore the conditions (3.11) are nothing else than the Euler equations, so that these integration conditions are fulfilled.

The calculation of $\frac{df^0}{dt}$ is obtained by using the Euler equations (Eqn. (2.7)) to eliminate the time derivatives, so that we obtain the following expression:

$$\frac{1}{f^0}\frac{df^0}{dt} = \left(\frac{mu^2}{2kT} - \frac{5}{2}\right)\frac{1}{T}\frac{\partial T}{\partial \boldsymbol{r}}\cdot \boldsymbol{u} + \frac{m}{kT}\frac{\partial \boldsymbol{V}}{\partial \boldsymbol{r}} : \overset{0}{\boldsymbol{uu}} \qquad (3.12)$$

With the conditions (3.10) and the expression (3.12) for $\frac{df^0}{dt}$, the unknown φ in J^1 may be written in the form

$$\varphi = A\frac{1}{T}\frac{\partial T}{\partial \boldsymbol{r}}\cdot \boldsymbol{u} + B\frac{\partial \boldsymbol{V}}{\partial \boldsymbol{r}} : \overset{0}{\boldsymbol{uu}} \qquad (3.13)$$

where $A(u, \boldsymbol{r}, t)$ and $B(u, \boldsymbol{r}, t)$ are unknown scalar quantities which are respective solutions of the following equations:

$$\begin{aligned} J^1[A\boldsymbol{u}] &= f^0\left(\frac{mu^2}{2kT} - \frac{5}{2}\right)\boldsymbol{u} \\ J^1\left[B\overset{0}{\boldsymbol{uu}}\right] &= f^0\frac{m}{kT}\overset{0}{\boldsymbol{uu}} \end{aligned} \qquad (3.14)$$

The general method of solving the above equations (3.14) consists[2] of expanding A and B in eigenfunctions of the operator J^1. It may be shown that these eigenfunctions constitute an orthogonal basis and that the eigenvalues are negative[4] (Appendix 3.1). However, they are not known for an arbitrary interaction potential, except for a Maxwellian potential $V(r) \sim r^{-4}$ (not really realistic): in this case, the (discrete) eigenfunctions are the Sonine–Laguerre polynomials[5] (Appendix 3.1). Then, the best we can do is to expand A and B in this basis, taking into account the expressions of the right hand side of the above equations (3.14). Thus

$$\begin{aligned} A &= \sum_{m=1}^{\infty} a_m(\boldsymbol{r}, t) S_{3/2}^m\left(\frac{mu^2}{2kT}\right) \\ B &= \sum_{m=0}^{\infty} b_m(\boldsymbol{r}, t) S_{5/2}^m\left(\frac{mu^2}{2kT}\right) \end{aligned} \qquad (3.15)$$

From the conditions (3.10), we deduce: $a_0 = 0$.

On the other hand, we have: $S_{3/2}^1\left(\frac{mu^2}{2kT}\right) = \frac{5}{2} - \frac{mu^2}{2kT}$ and $S_{5/2}^0\left(\frac{mu^2}{2kT}\right) = 1$. It is then clear that the first term of each expansion is dominant, the others giving only corrective terms after important calculations.

The terms a_m and b_m can be computed in a classical way for the orthogonal expansions (Appendix 3.1). Thus, they may be written as functions of 'collisional integrals' $\alpha_m^{m'}$ and $\beta_m^{m'}$, i.e.

$$\alpha_m^{m'} = \int_v J^1\left[S_{3/2}^m \boldsymbol{u}\right] \cdot S_{3/2}^{m'}\boldsymbol{u}\, du$$

$$\beta_m^{m'} = \int_v J^1\left[S_{5/2}^m \overset{0}{\boldsymbol{uu}}\right] : S_{5/2}^{m'} \overset{0}{\boldsymbol{uu}}\, du \tag{3.16}$$

with $\alpha_m^{m'} = \beta_m^{m'} = 0$ for $m \neq m'$.

Stopping the expansions at the first term, we have

$$\alpha_1^1 = -8n^2 \left(\frac{kT}{m}\right) \langle \gamma^4 \sin^2 \chi \rangle$$

$$\beta_0^0 = -16n^2 \left(\frac{kT}{m}\right)^2 \langle \gamma^4 \sin^2 \chi \rangle \tag{3.17}$$

with the following notation:

$$\langle \cdots \rangle = \left(\frac{kT}{\pi m}\right)^{1/2} \int_{\Omega,\gamma} \exp(-\gamma^2)\gamma^3 (\cdots) I\, d\Omega\, d\gamma \tag{3.18}$$

where $\gamma = \sqrt{\frac{m}{2kT}} g$ (non-dimensional relative velocity).

We find[13]

$$a_1 = -\frac{15}{2}\frac{nkT}{m}\frac{1}{\alpha_1^1} = -\frac{15}{16n}\langle \gamma^4 \sin^2 \chi \rangle^{-1}$$

$$b_0 = -10\frac{nkT}{m}\frac{1}{\beta_0^0} = -\frac{5}{8n}\left(\frac{m}{kT}\right)\langle \gamma^4 \sin^2 \chi \rangle^{-1} \tag{3.19}$$

The first-order distribution function may then be developed and, at the first order of the expansion of the Sonine–Laguerre polynomials, we have for φ:

$$\varphi = -\frac{5}{8n}\frac{1}{\langle \gamma^4 \sin^2 \chi \rangle}\left[\frac{3}{2}\left(\frac{mu^2}{2kT}-\frac{5}{2}\right)\boldsymbol{u}\cdot\frac{1}{T}\frac{\partial T}{\partial r} + \frac{m}{kT}\overset{0}{\boldsymbol{uu}} : \frac{\partial V}{\partial r}\right] \tag{3.20}$$

Remarks

- We indeed find a negative sign for the eigenvalues (inverse of coefficients a_m and b_m).

- In the collisional integrals $\langle \ldots \rangle$, the parameters of the collision do appear, so that these integrals depend on the collisional model and, of course, on the 'state' of the gas (species and temperature).

- The use of two indices for the collisional integrals seems unnecessary here, because of the orthogonal properties of the expansion terms. It is, however, necessary in the case with internal modes (Section 3.3.2).

- The coefficients a_1 and b_0 depend on the same collisional integral $\langle \gamma^4 \sin^2 \chi \rangle$.

Transport terms: Navier–Stokes equations

The transport terms in the conservation equations (1.26) may now be calculated in order to obtain a closed system in the same way as the Euler system at zero order.

The mass flux is clearly null (pure gas).
For the momentum flux, we have

$$\mathbf{P} = \int_v f^0 (1+\varphi) m\mathbf{u}\mathbf{u} \, dv = p\mathbf{I} + m \int_v f^0 \varphi \mathbf{u}\mathbf{u} \, dv = p\mathbf{I} + \mathbf{P}' \qquad (3.21)$$

With the properties of this type of integral (Appendix 2.6), we obtain for $\mathbf{P}'$:

$$\mathbf{P}' = -\frac{5}{4} \frac{kT}{\langle \gamma^4 \sin^2 \chi \rangle} \overline{\frac{\partial \mathbf{V}}{\partial r}}^0 = -2\mu \overline{\frac{\partial \mathbf{V}}{\partial r}}^0 \qquad (3.22)$$

This relation shows that there is a linear relationship between a part of the stress tensor $\mathbf{P}'$ and the strain rate tensor $\overline{\frac{\partial \mathbf{V}}{\partial r}}^0$ (Newton's law). Here, μ is the viscosity coefficient, equal to

$$\mu = \frac{5}{8} \frac{kT}{\langle \gamma^4 \sin^2 \chi \rangle} \qquad (3.23)$$

or

$$\mu = -R p T b_0 = 10 \left(\frac{nkT}{m}\right)^2 kT (\beta_0^0)^{-1} \qquad (3.24)$$

It depends on the nature of the gas, its temperature, and the collision model.
For the heat flux, we have

$$\mathbf{q} = \int_v f^0 \varphi \frac{1}{2} m u^2 \mathbf{u} \, dv \qquad (3.25)$$

With the properties of this integral (Appendix 2.6), we obtain

$$q = -\frac{75}{32} \frac{k}{\langle \gamma^4 \sin^2 \chi \rangle} \frac{kT}{m} \frac{\partial T}{\partial r} = -\lambda \frac{\partial T}{\partial r} \tag{3.26}$$

This relation also shows the linear relationship between the heat flux (translation energy flux) and the temperature gradient (Fourier's law). Here, λ is the conductivity coefficient, equal to

$$\lambda = \frac{75}{32} \frac{k}{\langle \gamma^4 \sin^2 \chi \rangle} \frac{kT}{m} \tag{3.27}$$

or

$$\lambda = -\frac{5}{2} R p a_1 = \frac{75}{4} \left(\frac{nkT}{m}\right)^2 k \left(\alpha_1^1\right)^{-1} \tag{3.28}$$

After putting these expressions into the conservation equations (1.26), we obtain a closed system (Navier–Stokes system), which may be written in the following form:

$$\frac{\partial \rho}{\partial t} + \frac{\partial \cdot \rho V}{\partial r} = 0$$

$$\rho \frac{dV}{dt} = -\frac{\partial p}{\partial r} - \frac{\partial}{\partial r} \left(2\mu \overline{\frac{\partial V}{\partial r}}^0\right)$$

$$\rho \frac{de}{dt} = \frac{\partial}{\partial r} \cdot \left(\lambda \frac{\partial T}{\partial r}\right) - p \frac{\partial \cdot V}{\partial r} + 2\mu \overline{\frac{\partial V}{\partial r}}^0 : \frac{\partial V}{\partial r} \tag{3.29}$$

with $p = nkT$ and $e = \frac{3}{2}\frac{kT}{m}$. Here, μ and λ are given by Eqns (3.23) and (3.27) respectively.

This is a second-order system, which takes into account the influence of the background by means of the 'boundary conditions'.

For a complete determination of the system, an interaction potential model must be defined (Appendix 1.3) and introduced in the coefficients μ and λ. In fact, their ratio does not depend on this model, or on the temperature, at least at the chosen level of approximation. Thus, we have

$$\frac{\lambda}{\mu} = \frac{15}{4} \frac{k}{m} \tag{3.30}$$

As the translational specific heat is equal to $\frac{3}{2}\frac{k}{m}$ $\left(= \frac{de}{dT}\right)$, the Prandtl number P (Chapter 7) is such that

$$P = \frac{\mu}{\lambda} \left(\frac{3}{2} \frac{k}{m} + \frac{k}{m}\right) = \frac{2}{3} \tag{3.31}$$

Experiments have validated this remarkably constant value.

3.3 FIRST-ORDER SOLUTIONS

As for the influence of the collision model on the coefficients μ and λ, two models only are considered here: the rigid sphere model below and the Maxwellian interaction potential in Appendix 3.4. Thus, for the rigid sphere model, we have

$$\langle \gamma^4 \sin^2 \chi \rangle = \left(\frac{kT}{\pi m}\right)^{1/2} 2\pi d^2$$

therefore

$$\mu = \frac{5}{16 d^2} \left(\frac{mkT}{\pi}\right)^{1/2}$$

and

$$\lambda = \frac{75 k}{64 d^2} \left(\frac{kT}{\pi m}\right)^{1/2} \tag{3.32}$$

It is interesting to note that from the measurement of a macroscopic parameter, generally μ, we can obtain the other (λ) and, at least, the order of magnitude of a microscopic parameter d ($d \sim 10^{-8}$ cm).

Thus, these coefficients vary with temperature as $T^{1/2}$ for an interaction potential considered to be 'too hard', and as T for a Maxwellian potential which is considered 'too soft' (Appendix 3.4). Of course, the reality is between these two extreme cases.

Finally, these coefficients, depending on the temperature of the medium but not on its density, are connected to the response of this medium to external dynamic and thermal perturbations. Therefore, they are related to the relaxation times of the system. As in the present case there are only elastic collisions, these coefficients depend on τ_{el}. Thus, as an example, with the rigid sphere model, assuming that the value of τ_{el} remains close to its value in the Maxwellian regime, that is, $\frac{1}{4nd^2}\left(\frac{mkT}{\pi}\right)^{1/2}$, we have

$$\mu = \frac{5}{4} \tau_{el} p$$

and

$$\lambda = \frac{75 k}{16 m} \tau_{el} p \tag{3.33}$$

3.3.2 Pure diatomic gases with one internal mode

As seen above, the rotational and vibrational characteristic times may be quite different. First, we consider the case of a single excited internal mode, that is, the rotational mode, assuming that the vibrational mode is frozen: this corresponds to realistic situations at 'moderate' temperatures. The approach of the problem is also more gradual. Thus, assuming $\tau_{TR} \ll \theta$, we have

$$J_{TR}^0 = 0$$
$$\frac{df_i^0}{dt} = J_{TR}^1 \qquad (3.34)$$

with $i = i_r$ only.

As in Chapter 2, (Eqn. (2.12)), a Maxwell–Boltzmann distribution f_i^0 is the zero-order solution, i.e.

$$f_i^0 = n \left(\frac{m}{2\pi kT}\right)^{3/2} \exp\left(-\frac{mu^2}{2kT}\right) \frac{g_{i_r} \exp\left(-\varepsilon_{i_r}/kT\right)}{\overline{Q}_R}$$

The Euler equations are still valid, with

$$E = E_T + E_R = \sum_i \int_v f_i^0 \left(\frac{1}{2}mu^2 + \varepsilon_{i_r}\right) dv = \overline{E}(T) = \frac{5}{2}kT$$

The temperature T is a 'measure' of the total energy, defined with the zero-order distribution function, like n and $\mathbf{V}$.

Extension of the Chapman–Enskog method

Using the expression of f_i^0 (Eqn. (2.12)) and, as above, eliminating the time derivatives by means of the Euler equations, we have

$$\frac{1}{f_i^0}\frac{df_i^0}{dt} = \left(\frac{mu^2}{2kT} - \frac{5}{2} + \frac{\varepsilon_{i_r} - \overline{E}_R}{kT}\right)\frac{1}{T}\frac{\partial T}{\partial r} \cdot \mathbf{u} + \frac{m}{kT}\frac{\partial \mathbf{V}}{\partial r} : \overset{0}{\mathbf{uu}}$$
$$+ \left[\frac{2}{3}\frac{C_R}{C_{TR}}\left(\frac{mu^2}{2kT} - \frac{3}{2}\right) - \frac{k}{C_{TR}}\frac{\varepsilon_{i_r} - \overline{E}_R}{kT}\right]\frac{\partial \cdot \mathbf{V}}{\partial r}$$

Comparing with the case of elastic collisions, a supplementary term with $\frac{\partial \cdot \mathbf{V}}{\partial r}$ appears. Then, the perturbation φ_i may be written in the general form

$$\varphi_i = A_i \frac{1}{T}\frac{\partial T}{\partial r} \cdot \mathbf{u} + B_i \frac{\partial \mathbf{V}}{\partial r} : \overset{0}{\mathbf{uu}} + D_i \frac{\partial \cdot \mathbf{V}}{\partial r} \qquad (3.35)$$

3.3 FIRST-ORDER SOLUTIONS

so that we have to solve the following equations:

$$J^1[A_i u] = f_i^0 \left(\frac{mu^2}{2kT} - \frac{5}{2} + \frac{\varepsilon_{i_r} - \bar{E}_R}{kT} \right) u$$

$$J^1\left[B_i \overset{0}{uu}\right] = f_i^0 \frac{m}{kT} \overset{0}{uu}$$

$$J^1[D_i] = f_i^0 \left[\frac{2}{3} \frac{C_R}{C_{TR}} \left(\frac{mu^2}{2kT} - \frac{3}{2} \right) - \frac{k}{C_{TR}} \frac{\varepsilon_{i_r} - \bar{E}_R}{kT} \right] \quad (3.36)$$

As before, the unknowns $X_i = A_i, B_i, D_i$ are expanded in orthogonal functions close to the eigenfunctions of the operator J^1, taking into account the term on the right-hand side of each of the above equations (3.36). The inclusion of an internal mode modifies the treatment used for monatomic gases with only elastic collisions. Thus, we add a set of polynomials $P_i^n\left(\frac{\varepsilon_{i_r}}{kT}\right)$, introduced by Wang-Chang and Uhlenbeck[20] for the internal degrees, to the Sonine–Laguerre polynomials S. They are built on the same model (Appendix 3.1), so that the complete basis is

$$X_i = \sum_{m,n} x_{mn} \Psi_{mn}^r \quad (3.37)$$

with

$$\Psi_{mn}^r = S_r^m \left(\frac{mu^2}{2kT} \right) P_i^n \left(\frac{\varepsilon_{i_r}}{kT} \right) \quad (3.38)$$

The indices m and n respectively correspond to the translation and rotation modes.

The computation of the coefficients x_{mn} is the same as above (Appendix 3.1). Thus, a_{mn}, b_{mn}, d_{mn}, respectively, are functions of the collisional integrals $\alpha_{mn}^{m'n'}$, $\beta_{mn}^{m'n'}$, $\delta_{mn}^{m'n'}$, defined in the following way:

$$\alpha_{mn}^{m'n'} = \sum_i \int_v J^1\left[\Psi_{mn}^{3/2} u\right] \cdot \Psi_{m'n'}^{3/2} u \, dv = \alpha_{m'n'}^{mn}$$

$$\beta_{mn}^{m'n'} = \sum_i \int_v J^1\left[\Psi_{mn}^{5/2} \overset{0}{uu}\right] \cdot \Psi_{m'n'}^{5/2} \overset{0}{uu} \, dv = \beta_{m'n'}^{mn}$$

$$\delta_{mn}^{m'n'} = \sum_i \int_v J^1\left[\Psi_{mn}^{1/2}\right] \Psi_{m'n'}^{1/2} \, dv = \delta_{m'n'}^{mn} \quad (3.39)$$

Keeping only the first term of the expansions Ψ_{00}, Ψ_{01}, and Ψ_{10}, the integration conditions resulting from the definition of n, V, and T give

$$a_{00} = d_{00} = 0$$

78 CHAPTER 3 QUASI-EQUILIBRIUM REGIMES: PURE GASES

and the first non-zero terms $a_{10}, a_{01}, b_{00}, d_{10}$, and d_{01} are given by the following expressions:

$$a_{10}\alpha_{10}^{10} + a_{01}\alpha_{01}^{10} = -\frac{15}{2}\frac{nkT}{m}$$

$$a_{10}\alpha_{10}^{01} + a_{01}\alpha_{01}^{01} = 3\frac{C_R}{k}\frac{nkT}{m}$$

$$b_{00} = -10\frac{nkT}{m}\left(\beta_{00}^{00}\right)^{-1}$$

$$d_{10} = \frac{C_R}{C_T}d_{01} = -n\left(\frac{C_R}{C_{TR}}\right)^2 \left(\delta_{10}^{10}\right)^{-1} \quad (3.40)$$

Then, the expression for φ_i is

$$\varphi_i = \left[a_{10}\left(\frac{5}{2} - \frac{mu^2}{2kT}\right) + a_{01}\left(\frac{\varepsilon_{i_r} - \overline{E_R}}{kT}\right)\right] \boldsymbol{u} \cdot \frac{1}{T}\frac{\partial T}{\partial \boldsymbol{r}} + b_{00}\overset{0}{\overline{\boldsymbol{u}\boldsymbol{u}}} : \frac{\partial \boldsymbol{V}}{\partial \boldsymbol{r}}$$

$$+ \left[d_{10}\left(\frac{3}{2} - \frac{mu^2}{2kT}\right) + d_{01}\left(\frac{\varepsilon_{i_r} - \overline{E_R}}{kT}\right)\right]\frac{\partial \cdot \boldsymbol{V}}{\partial \boldsymbol{r}}$$

The expressions for the collisional integrals $\alpha_{10}^{10}, \alpha_{01}^{10} \,(= \alpha_{10}^{01}), \alpha_{01}^{01}, \beta_{00}^{00}, \delta_{10}^{10}$ are given in Appendix 3.3 as functions of integrals $\langle \cdots \rangle$, with

$$\langle \cdots \rangle = \left(\frac{kT}{\pi m}\right)^{1/2} \sum_{i,j,k,l} \frac{\bar{n}_i \bar{n}_j}{n^2} \int_{\Omega,\gamma} \exp\left(-\gamma^2\right) \gamma^3 (\ldots) I_{i,j}^{k,l} \, d\Omega \, d\gamma \quad (3.41)$$

Transport terms: Navier–Stokes equations

As before, the mass flux is null.
For the momentum flux, we have

$$\mathbf{P} = p\mathbf{I} + \mathbf{P}'$$

with

$$\mathbf{P}' = -2\mu \overline{\frac{\partial \boldsymbol{V}}{\partial \boldsymbol{r}}}^0 - \eta \frac{\partial \cdot \boldsymbol{V}}{\partial \boldsymbol{r}} \mathbf{I} \quad (3.42)$$

Here, μ is the dynamic viscosity coefficient:

$$\mu = \frac{5}{8} \frac{kT}{\langle \gamma^4 \sin^2 \chi - \Delta\varepsilon_r \gamma^2 \sin^2 \chi + \frac{1}{3}(\Delta\varepsilon_r)^2 \rangle} \quad (3.43)$$

or

$$\mu = -Rp T b_{00} = 10 \left(\frac{nkT}{m}\right)^2 kT \left(\beta_{00}^{00}\right)^{-1} \quad (3.44)$$

with $\Delta\varepsilon_r = \frac{\varepsilon_{k_r}+\varepsilon_{l_r}-\varepsilon_{j_r}-\varepsilon_{i_r}}{kT}$ (non-dimensional collisional balance of internal energy).

The bulk viscosity coefficient η is:

$$\eta = \frac{kT}{2n}\left(\frac{C_R}{C_{TR}}\right)^2 \frac{1}{\langle(\Delta\varepsilon_r)^2\rangle} \quad (3.45)$$

or

$$\eta = p d_{10} = -nkT \left(\frac{C_R}{C_{TR}}\right)^2 \left(\delta_{10}^{10}\right)^{-1} \quad (3.46)$$

Comparing the dynamic viscosity due to TR collisions (3.43) with that due only to collisions T (3.23), we have

$$\frac{\mu_{TR}}{\mu_T} = \frac{b_{00}}{b_0} = \frac{\beta_0^0}{\beta_{00}^{00}} = \frac{\langle\gamma^4 \sin^2\chi\rangle_T}{\langle\gamma^4 \sin^2\chi - (\Delta\varepsilon_r)\gamma^2 \sin^2\chi + \frac{11}{8}(\Delta\varepsilon_r)^2\rangle_{TR}} \quad (3.47)$$

In the stress tensor (3.42) there is a term of dynamic viscosity that is similar (but not equal) to that due to elastic collisions (3.23), and also a term specific to the internal energy (bulk viscosity η). Therefore η is directly connected to inelastic collisions.

The heat flux q is the sum of a translational flux q_T and a rotational flux q_R, thus

$$q = \sum_{i_r}\int_v f_i^0 \varphi_i \frac{1}{2}mu^2 u \, dv + \sum_i \varepsilon_{i_r}\int_v f_i^0 \varphi_i u \, dv = q_T + q_R$$

Finally, we obtain

$$q_T = -\lambda_T \frac{\partial T}{\partial r} \quad \text{and} \quad q_R = -\lambda_R \frac{\partial T}{\partial r} \quad (3.48)$$

where λ_T and λ_R are the translational and rotational conductivity coefficients respectively, such that

$$\lambda_T = -\frac{5}{2} Rp a_{10} \quad \text{and} \quad \lambda_R = \frac{pC_R}{m} a_{01} \quad (3.49)$$

where a_{10} and a_{01} are given by the system (3.40) and are functions of the integrals $\alpha_{10}^{10}, \alpha_{01}^{01}$, and $\alpha_{10}^{01} (= \alpha_{01}^{10})$, given in Appendix 3.3.

The transport coefficients μ, η, λ_T, and λ_R are in fact functions of only three collisional integrals and not five, as specified above (Appendix 3.3).

The Navier–Stokes equations, deduced from the conservation equations (1.26), are different from the monatomic case, and they are

$$\frac{\partial \rho}{\partial t} + \frac{\partial \cdot \rho V}{\partial r} = 0$$

$$\rho \frac{dV}{dt} = -\frac{\partial p}{\partial r} - \frac{\partial}{\partial r}\left(2\mu \overset{0}{\overline{\frac{\partial V}{\partial r}}} + \eta \frac{\partial \cdot V}{\partial r}\right)$$

$$\rho \frac{de}{dt} = \frac{\partial}{\partial r} \cdot \left(\lambda \frac{\partial T}{\partial r}\right) - p \frac{\partial \cdot V}{\partial r} + 2\mu \overset{0}{\overline{\frac{\partial V}{\partial r}}} : \frac{\partial V}{\partial r} + \eta \left(\frac{\partial \cdot V}{\partial r}\right)^2 \quad (3.50)$$

with $p = nkT$, $e = \frac{5}{2}\frac{kT}{m}$, $\lambda = \lambda_T + \lambda_R$, and μ, η, λ_T, and λ_R given respectively by Eqns (3.43), (3.45), and (3.49).

Rotational non-equilibrium: characteristic times

In the Navier–Stokes system, as in the Euler system, one single temperature is used common to the translational and rotational modes, and the system is indeed closed. However, if we compute separately the translational and rotational energies E_T and E_R, we obtain the following expressions:

$$E_T = \frac{1}{n}\sum_i \int_v f_i^0 (1 + \varphi_i) \frac{1}{2} m u^2 \, d\mathbf{v} = \bar{E}_T \left(1 - d_{10}\frac{\partial \cdot V}{\partial r}\right)$$

$$E_R = \frac{1}{n}\sum_i \int_v f_i^0 (1 + \varphi_i) \varepsilon_{i_r} \, d\mathbf{v} = \bar{E}_R \left(1 + d_{01}\frac{\partial \cdot V}{\partial r}\right) \quad (3.51)$$

Of course, we again have $E_T + E_R = \bar{E}_T + \bar{E}_R$, since

$$d_{10} = \frac{C_R}{C_T} d_{01} = \frac{2}{3} d_{01}$$

However, at the first order of the expansion, a non-equilibrium appears between the translational and the rotational modes (3.51); then a distinct temperature may be defined for each mode, i.e. T_T and T_R, so that

$$E_T = \frac{3}{2}kT \quad \text{and} \quad E_R = kT_R \quad (3.52)$$

In principle, because of the linearization, this non-equilibrium is weak and is given by the above relations (3.51) and not by means of a relaxation equation (Chapter 2). These relations are independent of the Navier–Stokes system and may be calculated after having solved this system.

3.3 FIRST-ORDER SOLUTIONS

As easily seen, the non-equilibrium is related to the bulk viscosity η, since $\eta = pd_{10}$, so that this coefficient is representative of this non-equilibrium, owing to the definition of one single temperature T for both modes at the zeroth order of the distribution function. Thus, at first order, a (weak) non-equilibrium appears by means of this transport coefficient. This non-equilibrium is sensitive to the acceleration of the fluid (proportional to $\frac{\partial \cdot \mathbf{V}}{\partial r}$). Finally, as d_{10} is homogeneous to a time (like d_{01}), it seems logical to connect it to a phenomenological relaxation time τ_R, used for example in the interpretation of data with a Landau–Teller type equation[21] (Chapter 2):

$$\frac{dE_R}{dt} = \frac{\overline{E}_R(T_T) - E_R}{\tau_R} \tag{3.53}$$

with $\overline{E}_R(T_T) = kT_T$ and $E_R = \overline{E}_R(T_R) = kT_R$

Thus, we have

$$\frac{dT_R}{dt} = \frac{T_T - T_R}{\tau_R}$$

An equation of this type may also be deduced from the Boltzmann equation in an approximate way by multiplying this equation by ε_{i_r} then integrating and summing. Thus, we obtain

$$n\frac{dE_R}{dt} + \frac{\partial \cdot \mathbf{q}_R}{\partial r} = \sum_{i_r} \int_v (J^0 + J^1)\, dv \tag{3.54}$$

If we neglect the flux term and assume that the system (3.34) may be applied to Eqn. (3.54), i.e. $J^0 = 0$ and $J^1 = \frac{df_i^0}{dt}$, we obtain from the expression of $\frac{df_i^0}{dt}$:

$$\frac{dE_R}{dt} = -kT\frac{C_R}{C_{TR}}\frac{\partial \cdot \mathbf{V}}{\partial r} \tag{3.55}$$

Eliminating $\frac{\partial \cdot \mathbf{V}}{\partial r}$ with one of the above relations (3.51), we obtain

$$\frac{dE_R}{dt} = \frac{k}{C_{TR}}\frac{\overline{E}_R(T) - E_R}{d_{01}} \tag{3.56}$$

Thus, $\frac{C_{TR}}{k} d_{01} = \frac{5}{2} d_{01}$ may be considered a phenomenological relaxation time τ'_R. However, the reference temperature for $\overline{E}_R$ is not the same in Eqns. (3.56) and (3.53). This may be easily corrected, because we may write

$$\overline{E}_R(T) - E_R = \frac{C_T}{C_{TR}}\left(\overline{E}_R(T_T) - E_R\right) = \frac{3}{5}\left(\overline{E}_R(T_T) - E_R\right) \tag{3.57}$$

so that Eqn. (3.56) may be rewritten in the following way:

$$\frac{dE_R}{dt} = \frac{\overline{E}_R(T_T) - E_R}{\tau_R}$$

This equation is identical to Eqn. (3.53), with

$$\tau_R = \frac{C_{TR}^2}{kC_T}d_{01} = \frac{C_{TR}^2}{kC_R}d_{10} = \frac{C_{TR}^2}{kC_R}\frac{\eta}{p} \qquad (3.58)$$

or

$$\tau_R = \frac{25}{6}d_{01} = \frac{25}{4}d_{10} = \frac{25}{4}\frac{\eta}{p}$$

Here, d_{10}, d_{01}, and η have been connected to a relaxation time τ_R. This time is also connected to the collisional integral $\langle(\Delta\varepsilon_R)^2\rangle$, since

$$\tau_R = \frac{C_R}{2nk}\langle(\Delta\varepsilon_r)^2\rangle^{-1}$$

and finally, the rotational non-equilibrium may be written

$$\frac{E_R - \bar{E}_R}{\bar{E}_R} = d_{01}\frac{\partial \cdot V}{\partial r} = \frac{kC_T}{C_{TR}}\frac{\partial \cdot V}{\partial r}\tau_R \qquad (3.59)$$

A relaxation equation strictly identical to Eqn. (3.53) may be obtained by linearizing the right-hand side of Eqn. (3.54) and retaining only the zero-order terms (Appendix 3.5).

3.3.3 Pure diatomic gases with two internal modes

When the temperature is higher than in the above cases, the vibrational mode is excited, and assuming that the collisions T, R, and V have characteristic times much smaller than θ, we have the following system, without any assumptions regarding the relative orders of magnitude of τ_R and τ_V:

$$J_{TRV}^0 = 0$$
$$\frac{df_i^0}{dt} = J_{TRV}^1 \qquad (3.60)$$

The zero-order solution is the MBE distribution function (Eqn. (2.10)), with a unique temperature for the three modes.

The expression for φ_i is written in Appendix 3.2, as well as the equations giving the coefficients A_i, B_i, D_i.

Generalities and transport

The first-order solution is not entirely developed here; structurally it does not differ from the solution corresponding to one internal mode. The addition of the vibrational mode adds only one more term, similar to the rotational term where D_R and D_V are respectively given by the following relations:

$$D_R = \frac{3}{8}\frac{C_R T}{nm}\left[\left\langle\left(\frac{\varepsilon_{i_r}-\overline{E}_R}{kT}\right)\left(\frac{\varepsilon_{i_r}-\varepsilon_{j_r}}{kT}\right)\gamma^2 - \left(\frac{\varepsilon_{k_r}-\varepsilon_{l_r}}{kT}\right)\boldsymbol{\gamma}\cdot\boldsymbol{\gamma}'\right\rangle\right]^{-1} \tag{3.70}$$

$$D_V = \frac{3}{8}\frac{C_V T}{nm}\left[\left\langle\left(\frac{\varepsilon_{i_v}-\overline{E}_V}{KT}\right)\left(\frac{\varepsilon_{i_v}-\varepsilon_{j_v}}{kT}\right)\gamma^2 - \left(\frac{\varepsilon_{k_v}-\varepsilon_{l_v}}{kT}\right)\boldsymbol{\gamma}\cdot\boldsymbol{\gamma}'\right\rangle\right]^{-1} \tag{3.71}$$

These quantities are called 'energy diffusion coefficients', for rotation and vibration respectively, by analogy with the mass 'self-diffusion' coefficient defined in Chapter 4, i.e.

$$D = \frac{3kT}{8nm}\left(\langle\gamma^2 - \boldsymbol{\gamma}\cdot\boldsymbol{\gamma}'\rangle\right)^{-1} \tag{3.72}$$

A still more drastic hypothesis consists of also neglecting the terms that include $\frac{1}{\tau_R}$ and $\frac{1}{\tau_V}$ and assimilating the energy diffusion coefficients to the self-diffusion coefficient. Then

$$\lambda_T = \frac{5}{2}\mu\frac{C_T}{m}, \quad \lambda_R = \rho D\frac{C_R}{m}, \quad \text{and} \quad \lambda_V = \rho D\frac{C_V}{m} \tag{3.73}$$

These expressions constitute the 'modified Eucken corrections'.[22,23] Thus, schematically, the heat flux may be considered to be the sum of a translational heat flux, closely related to the momentum flux (viscosity), as for monatomic gases and rotational and vibrational heat fluxes, which have a diffusive character.

As for the other transport coefficients, represented by μ and η, in the present framework, μ is rather close to μ_T (Eqn. (3.47)), which is not surprising, since the process of momentum transfer is related to these coefficients. As for η, it has already been related to the relaxation times τ_R and τ_V (Eqn. 3.68)).

Finally, from Eqns (3.43), (3.45), and (3.62), we have $\frac{\mu}{\eta} \sim \frac{\tau_T}{\tau_R}$ or $\frac{\mu}{\eta} \sim \frac{\tau_T}{\tau_V}$, depending on the degree of excitation of the internal modes.

These approximations, without being entirely justified, give more than qualitative expressions for the transport terms, particularly in the case of a weak internal non-equilibrium (WNE case), as considered here.

Appendix 3.1 Orthogonal bases

Eigenvalues and eigenfunctions of the operator J^1

If we consider only the case of elastic collisions, the eigenvalues x_m^{-1} of the operator $\frac{1}{f^0}J^1[\Psi]$ and the corresponding eigenfunctions Ψ^m are defined by the equation

$$\frac{1}{f^0} J^1 [\Psi^n] = x_n^{-1} \Psi^n$$

Multiplying by $f^0 \Psi^m$ and integrating with respect to $\boldsymbol{u}$, we have

$$\int_u \Psi^m J^1 [\Psi^n] \, d\boldsymbol{u} = \int_{\Omega, \boldsymbol{u}_a, \boldsymbol{u}_b} f_a^0 f_b^0 \Psi_a^m \left(\Psi_a'^n + \Psi_b'^n - \Psi_a^n - \Psi_b^n \right) gI \, d\Omega \, d\boldsymbol{u}_a d\boldsymbol{u}_b$$

With integral symmetry properties, this term may be written

$$\int_u \Psi^n J^1 [\Psi^m] \, d\boldsymbol{u}$$

and therefore J^1 is self-adjoint, and the eigenvalues are real. We have

$$x_m^{-1} \int_u f^0 \Psi^m \Psi^n \, d\boldsymbol{u} = x_n^{-1} \int_u f^0 \Psi^n \Psi^m \, d\boldsymbol{u}$$

which shows the orthogonality of the eigenfunctions Ψ with the weight function f^0.

We also have

$$\int_u \Psi^m J^1 [\Psi^m] \, d\boldsymbol{u} = x_m^{-1} \int_u f^0 \left(\Psi^m \right)^2 d\boldsymbol{u}$$

$$= -\frac{1}{4} \int_{\Omega, \boldsymbol{u}_a, \boldsymbol{u}_b} f_a^0 f_b^0 \left(\Psi_a'^m + \Psi_b'^m - \Psi_a^m - \Psi_b^m \right)^2 gI \, d\Omega \, d\boldsymbol{u}_a \, d\boldsymbol{u}_b$$

This shows that the eigenvalues x_m^{-1} are negative, like the coefficients $a_m, b_m \ldots$.

Bases of orthogonal polynomials

Sonine–Laguerre polynomials

The definition of the Sonine–Laguerre polynomials is

$$S_r^m (x) = \sum_{p=0}^m \frac{(-x)^p (r+m)^{m-p}}{p! (m-p)!}$$

They satisfy the following orthogonality relation:

$$\int_0^\infty x^n \exp(-x) S_n^m S_{n'}^{m'} \, dx = \frac{\Gamma(n+m+1)}{\Gamma(m+1)} \delta_{mm'}$$

Thus, for example, we have

$$S^0_r(x) = 1, \quad S^1_{1/2}(x) = \frac{3}{2} - x^2, \quad \text{and} \quad S^1_{3/2}(x) = \frac{5}{2} - x^2$$

Wang-Chang and Uhlenbeck polynomials

These polynomials are derived with the general formula

$$P^n_i(\varepsilon_i) = \varepsilon_i P^{n-1}_i - \sum_{t=0}^{n-1} \frac{\|\varepsilon_i P^{n-1}_i P^t_i\|}{a_t} P^t_i$$

with $a_t = \|(P^t_i)^2\|$

Thus, the zero and first-order polynomials are

$$P^0_i\left(\frac{\varepsilon_i}{kT}\right) = 1 \quad \text{and} \quad P^1_i\left(\frac{\varepsilon_i}{kT}\right) = \frac{\varepsilon_i - \overline{E_{int}}}{kT}$$

$$\text{with } a_0 = 1 \quad \text{and} \quad a_1 = \|(P^1_i)^2\| = \frac{C_{int}}{k}$$

They satisfy the following orthogonality relation:

$$\sum_i P^n_i P^{n'}_i \exp(-\varepsilon_i) = \frac{a_n}{Q(T)} \partial_{nn'}$$

They are derived from the model of Sonine–Laguerre polynomials ($P^0_i = S^0_r = 1$). At first order, they are proportional to the non-equilibrium for the concerned quantity.

Coefficients of the expansions and collisional integrals: calculated example

As an example,[13] we take the expansion of the unknown A (Eqn. (3.15)):

$$A = \sum_m a_m \Psi^m$$

The first equation of the system (3.14) is written

$$J^1[A\boldsymbol{u}] = J^1\left[\sum_m a_m \Psi^m \boldsymbol{u}\right] = f_0\left(\frac{mu^2}{2kT} - \frac{5}{2}\right)\boldsymbol{u}$$

Multiplying both sides of this equation by $\Psi_m u$ and taking into account the orthogonality of the basis function, we obtain for the coefficient a_m:

$$a_m = \frac{\int_u f^0 \left(\frac{mu^2}{2kT} - \frac{5}{2}\right) u \cdot \Psi^m u \, du}{\int_u J^1 [\Psi^m u] \cdot \Psi^m u \, du}$$

Thus, for the coefficient a_1 (or a_1^1), with $\Psi^1 = S_{3/2}^1 \left(\frac{mu^2}{2kT}\right)$, we have

$$a_1^1 \alpha_1^1 = \int_u f^0 \left(\frac{mu^2}{2kT} - \frac{5}{2}\right)^2 du$$

with

$$\alpha_1^1 = \int_u J^1 \left[\left(\frac{mu^2}{2kT} - \frac{5}{2}\right) u\right] \cdot \left(\frac{mu^2}{2kT} - \frac{5}{2}\right) u \, du$$

or

$$\alpha_1^1 = \int_{\Omega, u_a, u_b} f_a^0 f_b^0 \frac{m}{2kT} \left(-u_b'^2 u_b' - u_a'^2 u_a' + u_b^2 u_b + u_a^2 u_a\right)$$

$$\times \left(\frac{5}{2} - \frac{mu_a^2}{2kT}\right) u_a I g \, d\Omega \, du_a du_b$$

After the change of coordinates, $u_a, u_b \to G, g$, we develop $f_a^0 f_b^0$, and integrate over G (Eulerian integrals). We thus obtain

$$\alpha_1^1 = -\frac{8n^2}{\sqrt{\pi}} \left(\frac{kT}{m}\right)^{3/2} \int_0^\infty \exp\left(-\gamma^2\right) \gamma^7 I \sin^2 \chi \, d\Omega \, d\gamma$$

with

$$\gamma = \sqrt{\frac{m}{kT} \frac{g}{2}} \quad \text{(non-dimensional relative velocity)}$$

The numerator of a_1^1 is equal to $\frac{15}{2} \frac{nkT}{m}$, so that

$$a_1^1 = -\frac{15}{16n} \langle \gamma^4 \sin^2 \chi \rangle^{-1}$$

with the notation

$$\langle \cdots \rangle = \left(\frac{kT}{\pi m}\right)^{1/2} \int_{\Omega, \gamma} \exp\left(-\gamma^2\right) \gamma^3 (\cdots) I \, d\Omega \, d\gamma$$

The other coefficients $a, b, d \ldots$ and the collisional integrals $\alpha, \beta, \delta \cdots$ are computed in the same way.

Appendix 3.2 Systems of equations for a, b, d coefficients

Diatomic gases

One internal mode (rotation)

Cramer system for the coefficients a_{mn}

$$a_{10}\alpha_{10}^{10} + a_{01}\alpha_{10}^{01} = -\frac{15}{2}\left(\frac{nkT}{m}\right)$$

$$a_{10}\alpha_{10}^{01} + a_{01}\alpha_{01}^{01} = 3\frac{C_R}{k}\left(\frac{nkT}{m}\right)$$

Coefficients b_{mn} and d_{mn}

$$b_{00}\beta_{00}^{00} = -10\left(\frac{nkT}{m}\right)$$

$$d_{10}\delta_{10}^{10} = -n\left(\frac{C_R}{C_{TR}}\right)^2 = d_{01}\delta_{10}^{10}\left(\frac{C_R}{C_T}\right)$$

Two internal modes (rotation, vibration)

Expressions of $\frac{df_i^0}{dt}$ and φ_i

With the relation (2.10), we have

$$\frac{1}{f_i^0}\frac{df_i^0}{dt} = \left(\frac{mu^2}{2kT} - \frac{5}{2} + \frac{\varepsilon_{i_r} - \bar{E}_R}{kT} + \frac{\varepsilon_{i_v} - \bar{E}_V}{kT}\right)\frac{1}{T}\frac{\partial T}{\partial r}\cdot u + \frac{m}{kT}\frac{\partial V}{\partial r} : \overset{0}{uu}$$

$$+ \left[\frac{2}{3}\frac{C_{RV}}{C_{TRV}}\left(\frac{mu^2}{2kT} - \frac{3}{2}\right) - \frac{k}{C_{TRV}}\left(\frac{\varepsilon_{i_r} - \bar{E}_R}{kT} + \frac{\varepsilon_{i_v} - \bar{E}_V}{kT}\right)\right]\frac{\partial \cdot V}{\partial r}$$

Then

$$\varphi_i = \left[a_{100}\left(\frac{5}{2} - \frac{mu^2}{2kT}\right) + a_{010}\left(\frac{\varepsilon_{i_r} - \bar{E}_R}{kT}\right) + a_{001}\left(\frac{\varepsilon_{i_v} - \bar{E}_V}{kT}\right)\right]\frac{1}{T}\frac{\partial T}{\partial r}\cdot u$$

$$+ b_{000}\frac{\partial V}{\partial r} : \overset{0}{uu} + \left[d_{100}\left(\frac{3}{2} - \frac{mu^2}{2kT}\right) + d_{010}\left(\frac{\varepsilon_{i_r} - \bar{E}_R}{kT}\right)\right.$$

$$\left. + d_{001}\left(\frac{\varepsilon_{i_v} - \bar{E}_V}{kT}\right)\right]\frac{\partial \cdot V}{\partial r}$$

Cramer system for the coefficients a_{mnq}

$$a_{100}\alpha_{100}^{100} + a_{010}\alpha_{100}^{010} + a_{001}\alpha_{100}^{001} = -\frac{15}{2}\left(\frac{nkT}{m}\right)$$

$$a_{100}\alpha_{010}^{100} + a_{010}\alpha_{010}^{010} + a_{001}\alpha_{010}^{001} = 3\frac{C_R}{k}\left(\frac{nkT}{m}\right)$$

$$a_{100}\alpha_{001}^{100} + a_{010}\alpha_{001}^{010} + a_{001}\alpha_{001}^{001} = 3\frac{C_V}{k}\left(\frac{nkT}{m}\right)$$

For b_{000}, we have

$$b_{000}\beta_{000}^{000} = -10\left(\frac{nkT}{m}\right)$$

Cramer system for the coefficients d_{mnq}

$$d_{100}\delta_{100}^{100} + d_{010}\delta_{100}^{010} + d_{001}\delta_{100}^{001} = -n\frac{C_R}{C_{TRV}}$$

$$d_{100}\delta_{010}^{100} + d_{010}\delta_{010}^{010} + d_{001}\delta_{010}^{001} = -n\frac{C_V}{C_{TRV}}$$

The third equation necessary for solving this system is not independent of the two above, but it is replaced by the following equation arising from the integration conditions.

$$C_T d_{100} = C_R d_{010} + C_V d_{001}$$

Appendix 3.3 Expressions of the collisional integrals

Monatomic case

$$\alpha_m^{m'} = \int_u J^1\left[S_{3/2}^m u\right] \cdot S_{3/2}^{m'} u \, du \quad \text{and} \quad \beta_m^{m'} = \int_u J^1\left[S_{5/2}^m \overset{0}{uu}\right] : S_{5/2}^{m'} \overset{0}{uu} \, du$$

In particular, we have

$$\alpha_1^1 = -8n^2\left(\frac{kT}{m}\right)\langle \gamma^4 \sin^2 \chi \rangle$$

$$\beta_0^0 = -16n^2\left(\frac{kT}{m}\right)^2 \langle \gamma^4 \sin^2 \chi \rangle$$

with $\quad \langle \cdots \rangle = \left(\frac{kT}{\pi m}\right)^{1/2} \int_{\Omega,\gamma} \exp(-\gamma^2) \gamma^3 (\cdots) I_{el} \, d\Omega \, d\gamma$

Diatomic case
One internal mode (rotation)

$$\alpha_{mn}^{m'n'} = \sum_i \int_u J^1 \left[\Psi_{mn}^{3/2}\boldsymbol{u}\right] \cdot \Psi_{m'n'}^{3/2} \boldsymbol{u}\, d\boldsymbol{u} = \alpha_{m'n'}^{mn}$$

$$\beta_{mn}^{m'n'} = \sum_i \int_u J^1 \left[\Psi_{mn}^{5/2} \overset{0}{\boldsymbol{uu}}\right] : \Psi_{m'n'}^{5/2} \overset{0}{\boldsymbol{uu}}\, d\boldsymbol{u} = \beta_{m'n'}^{mn}$$

$$\delta_{mn}^{m'n'} = \sum_i \int_u J^1 \left[\Psi_{mn}^{1/2}\right] \Psi_{m'n'}^{1/2}\, d\boldsymbol{u} = \delta_{m'n'}^{mn}$$

Thus, we have

$$\alpha_{10}^{10} = -8n^2 \left(\frac{kT}{m}\right) \left\langle \gamma^4 \sin^2\chi - (\Delta\varepsilon_r)\gamma^2 \sin^2\chi + \frac{11}{8}(\Delta\varepsilon_r)^2 \right\rangle$$

$$\alpha_{10}^{01} = \alpha_{01}^{10} = -5n^2 \left(\frac{kT}{m}\right) \langle(\Delta\varepsilon_r)^2\rangle$$

$$\alpha_{01}^{01} = -8n^2 \left(\frac{kT}{m}\right) \left\langle \frac{\varepsilon_{i_r} - \overline{E_R}}{kT}\left(\frac{\varepsilon_{i_r} - \varepsilon_{j_r}}{kT}\gamma^2 - \frac{\varepsilon_{k_r} - \varepsilon_{l_r}}{kT}\boldsymbol{\gamma}\cdot\boldsymbol{\gamma}' - \frac{3}{2}(\Delta\varepsilon_r)\right) \right\rangle$$

$$\beta_{00}^{00} = -16n^2 \left(\frac{kT}{m}\right)^2 \left\langle \gamma^4 \sin^2\chi - (\Delta\varepsilon_r)\sin^2\chi + \frac{1}{3}(\Delta\varepsilon_r)^2 \right\rangle$$

$$\delta_{10}^{10} = -2n^2\langle(\Delta\varepsilon_r)^2\rangle$$

with

$$\langle\cdots\rangle = \left(\frac{kT}{\pi m}\right)^{1/2} \sum_{i,j,k,l} \frac{\bar{n}_i \bar{n}_j}{n^2} \int_{\Omega,\gamma} \exp\left(-\gamma^2\right)\gamma^3 (\ldots) I_{i,j}^{k,l}\, d\Omega\, d\gamma$$

There are relations between these five integrals. Thus,

$$\alpha_{10}^{10} = \frac{m}{2kT}\beta_{00}^{00} + \frac{5}{3}\alpha_{10}^{01}$$

and

$$\delta_{10}^{10} = \frac{2}{5}\frac{m}{kT}\alpha_{10}^{01}$$

Two internal modes (rotation, vibration)

The α integrals

$$\alpha_{mnq}^{m'n'q'} = \sum_i \int_u J^1 \left[\Psi_{mnq}^{3/2}\boldsymbol{u}\right] \cdot \Psi_{m'n'q'}^{3/2}\boldsymbol{u}\, d\boldsymbol{u} = \alpha_{m'n'q'}^{mnq}$$

Thus, we have

$$\alpha^{100}_{100} = -8n^2 \left(\frac{kT}{m}\right) \left\langle \gamma^4 \sin^2 \chi - (\Delta\varepsilon_r + \Delta\varepsilon_v) \gamma^2 \sin^2 \chi + \frac{11}{8} (\Delta\varepsilon_r + \Delta\varepsilon_v)^2 \right\rangle$$

$$\alpha^{010}_{100} = -5n^2 \left(\frac{kT}{m}\right) \langle (\Delta\varepsilon_r)^2 \rangle$$

$$\alpha^{001}_{100} = -5n^2 \left(\frac{kT}{m}\right) \langle (\Delta\varepsilon_v)^2 \rangle$$

$$\alpha^{010}_{010} = 8n^2 \left(\frac{kT}{m}\right) \left\langle \frac{\varepsilon_{i_r} - \overline{E_R}}{kT} \frac{3}{2}(\Delta\varepsilon_r) - \gamma^2 \frac{\varepsilon_{i_r} - \varepsilon_{j_r}}{kT} + \boldsymbol{\gamma} \cdot \boldsymbol{\gamma}' \frac{\varepsilon_{k_r} - \varepsilon_{l_r}}{kT} \right\rangle$$

$$\alpha^{001}_{001} = 8n^2 \left(\frac{kT}{m}\right) \left\langle \frac{\varepsilon_{i_v} - \overline{E_V}}{kT} \left(\frac{3}{2}\Delta\varepsilon_v - \gamma^2 \frac{\varepsilon_{i_v} - \varepsilon_{j_v}}{kT} + \boldsymbol{\gamma} \cdot \boldsymbol{\gamma}' \frac{\varepsilon_{k_v} - \varepsilon_{l_v}}{kT}\right) \right\rangle$$

$$\alpha^{001}_{010} = -8n^2 \left(\frac{kT}{m}\right)$$

$$\times \left[\left\langle \frac{3}{2}\langle \Delta\varepsilon_r \Delta\varepsilon_v \rangle + \left\langle \frac{\varepsilon_{i_r} - \overline{E_R}}{kT} \frac{\varepsilon_{i_v} - \overline{E_V}}{kT} \gamma^2 + \frac{\varepsilon_{k_r} - \overline{E_R}}{kT} \frac{\varepsilon_{k_v} - \overline{E_V}}{kT} \gamma'^2 \right\rangle \right. \right.$$
$$\left. \left. - \left\langle \left(\frac{\varepsilon_{i_r} - \overline{E_R}}{kT} \frac{\varepsilon_{k_v} - \overline{E_V}}{kT} + \frac{\varepsilon_{k_r} - \overline{E_R}}{kT} \frac{\varepsilon_{i_v} - \overline{E_V}}{kT}\right) \boldsymbol{\gamma} \cdot \boldsymbol{\gamma}' \right\rangle \right],$$

with

$$\langle \cdots \rangle = \left(\frac{kT}{\pi m}\right)^{1/2} \sum_{i,j,k,l} \frac{\bar{n}_{i_r} \bar{n}_{j_r} \bar{n}_{i_v} \bar{n}_{j_v}}{n^4} \int_{\Omega,\gamma} \exp(-\gamma^2) \gamma^3 (\ldots) I^{k,l}_{i,j} \, d\Omega \, d\gamma$$

The β integral

$$\beta^{m'n'q'}_{mnq} = \sum_i \int_{\boldsymbol{u}} J^1 \left[\Psi^{5/2}_{mnq} \overset{0}{\overline{\boldsymbol{uu}}}\right] : \overset{0}{\overline{\boldsymbol{uu}}} \, d\boldsymbol{u} = \beta^{mnq}_{m'n'q'}$$

Thus we have

$$\beta^{000}_{000} = -16n^2 \left(\frac{kT}{m}\right)^2 \left\langle \gamma^4 \sin^2 \chi - (\Delta\varepsilon_r + \Delta\varepsilon_v) \gamma^2 \sin^2 \chi + \frac{1}{3}(\Delta\varepsilon_r + \Delta\varepsilon_v)^2 \right\rangle$$

The δ integrals

$$\delta_{mnq}^{m'n'q'} = \sum_i \int_u J^1 \left[\Psi_{mnq}^{1/2}\right] \Psi_{m'n'q1}^{1/2} d\boldsymbol{u}$$

$$\delta_{100}^{100} = -2n^2 \langle (\Delta\varepsilon_r + \Delta\varepsilon_v)^2 \rangle$$
$$\delta_{100}^{010} = -2n^2 \left[\langle (\Delta\varepsilon_r)^2 \rangle + \langle (\Delta\varepsilon_r \Delta\varepsilon_v) \rangle\right]$$
$$\delta_{100}^{001} = -2n^2 \left[\langle (\Delta\varepsilon_v)^2 \rangle + \langle (\Delta\varepsilon_r \Delta\varepsilon_v) \rangle\right]$$
$$\delta_{010}^{010} = -2n^2 \langle (\Delta\varepsilon_r)^2 \rangle$$
$$\delta_{010}^{001} = -2n^2 \langle (\Delta\varepsilon_r \Delta\varepsilon_v) \rangle$$
$$\delta_{001}^{001} = -2n^2 \langle (\Delta\varepsilon_v)^2 \rangle$$

There are clear relations between the integrals δ, so that only three of them are independent.

Appendix 3.4 Influence of the collisional model on the transport terms

In the case of elastic collisions, we compare the expressions obtained for the viscosity coefficient in two extreme cases: the rigid sphere model and the Maxwellian potential model (Appendix 1.3). This comparison is also valid for the heat conductivity coefficient.

The case of the rigid sphere model has already been examined in Chapter 3, where we found (Eqn. (3.32)):

$$\mu \sim T^{1/2}$$

More generally, for a purely repulsive potential $\varphi = \frac{K}{r^{s-1}}$, the deviation χ may be written, from Eqn. (1.36):

$$\chi = \pi - 2 \int_0^{\beta_{min}} \frac{d\beta}{\left[1 - \beta^2 - \frac{1}{s-1}\left(\frac{\beta}{\delta}\right)^{s-1}\right]^{1/2}}$$

where

$$\beta = \frac{b}{r} \quad \text{and} \quad \delta = \frac{bmg^2}{4(s-1)K}$$

Here, β_{min} corresponds to the minimum distance r_{min} given by the equation

$$1 - \beta_{min}^2 - \frac{1}{s-1}\left(\frac{\beta_{min}}{\delta}\right)^{s-1} = 0$$

Then, the collisional integral $J = \int_{b,\varepsilon,\boldsymbol{v}_b} (f'_a f'_b - f_a f_b)\, gb\, db\, d\varepsilon\, d\boldsymbol{v}_b$ becomes

$$J = \left(\frac{4(s-1)K}{m}\right)^{\frac{2}{s-1}} \int_{\delta,\varepsilon,\boldsymbol{v}_b} (f'_a f'_b - f_a f_b)\, g^{\frac{s-5}{s-1}} \delta\, d\delta\, d\varepsilon\, d\boldsymbol{v}_b$$

It is clear that, for a Maxwellian potential ($s = 5$), the relative velocity g disappears from the integral and that, therefore, the expression $gb\, db\, d\varepsilon = Ig\, d\Omega$ does not depend on g.

We have also seen (Eqn. (3.23)) that

$$\mu \sim \frac{T}{\langle \gamma^4 \sin^2 \chi \rangle} \sim \frac{T^{1/2}}{\int_{\Omega,\gamma} \exp(-\gamma^2)\, \gamma^7 \sin^2 \chi I\, d\Omega\, d\gamma}$$

which may also be written

$$\mu \sim \frac{T^{1/2}}{\int_{\Omega,\gamma} \exp(-\gamma^2)\, \frac{\gamma^7}{g} Ig\, d\Omega\, d\gamma}$$

We know that $Ig\, d\Omega$ does not depend on g, and we have $g \sim \frac{T^{-1/2}}{\gamma}$; therefore

$$\mu \sim T$$

In the same way, we have

$$\lambda \sim T$$

As seen above, with the rigid sphere model, we find that μ and λ are proportional to $T^{1/2}$ (Eqn. (3.32)). As already emphasized, the reality is of course between these extreme models.

Appendix 3.5 Linearization of the relaxation equation

The general relaxation equation giving the populations (2.14) is

$$\frac{\partial n_i}{\partial t} + \frac{\partial \cdot n_i \boldsymbol{V}}{\partial \boldsymbol{r}} = \sum_{j,k,l} \int_v J d\boldsymbol{v}$$

APPENDIX 3.5 LINEARIZATION OF THE RELAXATION EQUATION

Considering first only one internal mode, we have at zero order (setting K for the term $\sum_{j,k,l} \int_v J^0 d\mathbf{v}$):

$$K = \sum_{j,k,l} n_k n_l \int_{\Omega, \mathbf{u}_i, \mathbf{u}_j} \exp\left[-\frac{m}{2kT}\left(u_k^2 + u_l^2\right)\right] I_{i,j}^{k,l} g \, d\Omega \, d\mathbf{u}_i d\mathbf{u}_j$$

$$- \sum_{j,k,l} n_i n_j \int_{\Omega, \mathbf{u}_i, \mathbf{u}_j} \exp\left[-\frac{m}{2kT}\left(u_i^2 + u_j^2\right)\right] I_{i,j}^{k,l} g \, d\Omega \, d\mathbf{u}_i d\mathbf{u}_j$$

With $\Delta\varepsilon = \frac{\varepsilon_k + \varepsilon_l - \varepsilon_i - \varepsilon_j}{kT}$, we have

$$K = \sum_{j,k,l} \left[n_k n_l \exp(\Delta\varepsilon) - n_i n_j\right] \int_{\Omega, \mathbf{u}_i, \mathbf{u}_j} \exp\left[-\frac{m}{2kT}\left(u_i^2 + u_j^2\right)\right] I_{i,j}^{k,l} g \, d\Omega \, d\mathbf{u}_i d\mathbf{u}_j$$

After the change of variables $(\mathbf{u}_i, \mathbf{u}_j) \to (\mathbf{G}, \mathbf{g})$, we find

$$u_i^2 + u_j^2 = 2G^2 + \frac{g^2}{2}$$

and

$$\int_0^\infty \exp\left(-\frac{m}{kT}G^2\right) G^2 \, dG = \frac{\sqrt{\pi}}{4}\left(\frac{kT}{m}\right)^{3/2} \quad \text{(Appendix 2.6)}$$

Finally:

$$K = 4\pi \left(\frac{nkT}{m}\right)^{3/2} \sum_{j,k,l} \left[n_k n_l \exp(\Delta\varepsilon) - n_i n_j\right] \int_{\Omega, g} \exp\left(-\frac{mg^2}{4kT}\right) I_{i,j}^{k,l} g^3 \, d\Omega dg$$

Now, we assume a Boltzmann distribution at a temperature T_{int} close to T, and we linearize the terms that include $T_{int} - T$; independently, we obtain a relaxation equation (giving $E_{int} = \frac{1}{n}\sum_i n_i \varepsilon_i$) by multiplying Eqn. (2.14) by ε_i and summing over the levels. Setting $C_{int} = \frac{dE_{int}}{dT_{int}}$, for this equation we have

$$\frac{dE_{int}}{dt} = \frac{E_{int} - \bar{E}_{int}}{C_{int}} \frac{4\pi n}{TQ}\left(\frac{m}{4\pi kT}\right)^{3/2} \sum_{i,j,k,l} \varepsilon_i (\Delta\varepsilon) \exp\left(-\frac{\varepsilon_i + \varepsilon_j}{kT}\right)$$

$$\times \int_{\Omega, g} \exp\left(-\frac{mg^2}{4kT}\right) g^3 I_{i,j}^{k,l} \, d\Omega \, dg$$

Using the property (1.87),

$$\sum_{i,j,k,l} \varepsilon_i \ldots \int_{\Omega, g} \ldots = -\frac{1}{4} \sum_{i,j,k,l} kT (\Delta\varepsilon) \ldots \int_{\Omega, g} \ldots$$

and setting, as before, $\gamma = \left(\frac{m}{2kT}\right)^{1/2} g$, we obtain a relaxation equation identical to Eqn. (3.53), that is:

$$\frac{dE_{int}}{dt} = \frac{\bar{E}_{int} - E_{int}}{\tau_{int}}$$

with

$$\frac{1}{\tau_{int}} = \left(\frac{kT}{\pi m}\right)^{1/2} \frac{2nk}{C_{int}} \sum_{i,j,k,l} \frac{\bar{n}_i \bar{n}_j}{n^2} (\Delta \varepsilon)^2 \int_0^\infty \exp\left(-\gamma^2\right) \gamma^3 I_{i,j}^{k,l} \, d\Omega \, d\gamma$$

This expression is identical to the relation (3.59) obtained for the rotational mode. It is of course possible to apply this equation to the vibrational mode, with the rotational mode in equilibrium.

Appendix 3.6 Vibrational non-equilibrium distribution

The non-equilibrium vibrational populations may be determined from the general formula

$$n_{i_v} = \sum_{i_r} \int_v f_i^0 (1 + \varphi_i) \, d\boldsymbol{v}$$

or, after some calculation:

$$n_{i_v} = \bar{n}_{i_v} \left[1 + d_{001} \left(\frac{\varepsilon_{i_v} - \bar{E}_V}{kT}\right) \frac{\partial \cdot \boldsymbol{V}}{\partial \boldsymbol{r}}\right]$$

It is important to note that an integration condition is related to the definition of n but not to the definition of n_{i_v}, since this definition is

$$n = \sum_i \int_v f_i^0 \, d\boldsymbol{v}$$

so that

$$\sum_{i_v} \left(n_{i_v} - \bar{n}_{i_v}\right) = 0$$

Using Eqn. (3.63) to eliminate d_{001}, we may write n_{i_v} as a function of $E_V - \bar{E}_V$, that is:

$$n_{i_v} = \bar{n}_{i_v} \left[1 + \left(\frac{E_V - \bar{E}_V}{C_V T}\right) \left(\frac{\varepsilon_{i_v} - \bar{E}_V}{kT}\right)\right] \quad (3.74)$$

Thus, with an equilibrium value $\bar{E}_V(T)$ assumed between ε_n and ε_{n+1}, if the non-equilibrium corresponds to a vibrational temperature $T_V > T$ (that is, $E_V > \bar{E}_V$), the levels lower than (or equal to) ε_n are underpopulated compared with an equilibrium distribution, and of course, vice versa for $T_V < T$. However, we must be careful in the application of the formula (3.74), as it represents a linearized non-equilibrium and is therefore valid only close to equilibrium.

From Eqn. (3.66), we can also write

$$n_{i_v} = \bar{n}_{i_v}\left[1 + \frac{kC_{TR}}{C_{TRV}^2}\tau_V \frac{\partial \cdot \mathbf{V}}{\partial \mathbf{r}}\left(\frac{\varepsilon_{i_v} - \bar{E}_V}{kT}\right)\right]$$

The population of the rotational levels could be described in the same way.

FOUR

Transport and Relaxation in Quasi-Equilibrium Regimes: Gas Mixtures

4.1 Introduction

Quasi-equilibrium regimes (WNE regimes), discussed in the previous chapter, correspond to first-order solutions for the distribution function of pure monatomic and diatomic gases. This distribution function has a zero-order solution which represents complete equilibrium. Similarly, in this chapter we start from zero-order equilibrium distributions and develop the first-order solutions for binary gas mixtures. The complexity of computations is of course increased and becomes tedious. The principles and methods remain based on an extension of the Chapman–Enskog method.[1-3,24,25] However, alternative methods are briefly developed in Appendices 4.4 and 4.5.

We first consider monatomic gas mixtures and then extend the analysis to mixtures of diatomic gases with internal energy modes which can be significantly excited. Approximate expressions for transport and relaxation terms are also given. These expressions are generally sufficiently accurate and provide a clear physical interpretation. Finally, the case of reactive mixtures, in equilibrium at zero order, is also examined in the framework of well-established approximations.

4.2 Gas mixtures with elastic collisions

4.2.1 Chapman–Enskog method

We consider the mixture of two monatomic gases p and q. As discussed in Chapter 2, the Boltzmann equation for the distribution function of component

4.2 GAS MIXTURES WITH ELASTIC COLLISIONS

p is written

$$\frac{df_p}{dt} = J_p \tag{4.1}$$

where

$$J_p = J_{pp} + J_{pq}$$

If we assume that the characteristic times for $p-p$ collisions and $p-q$ collisions are much smaller than the reference time, the expansion of the function $f_p = f_p^0(1+\varphi_p)$ gives the following system:

$$J_p^0 = 0$$
$$\frac{df_p^0}{dt} = J_p^1 \tag{4.2}$$

and we have a similar system for the component q.

The zero-order solution of the system (4.2), f_p^0 (as well as f_q^0) is the Maxwellian distribution (2.33), completely defined by the Euler equations and the conservation equation for the species (2.34). The macroscopic quantities n_p, n_q, V, and T are defined using f_p^0 and f_q^0.

The first-order solution is given by

$$J_p^1[\varphi_p] = J_{pp}^1 + J_{pq}^1$$

where

$$J_{pp}^1 = \int_{\Omega, v_{pb}} f_{pa}^0 f_{pb}^0 \left(\varphi_{pa}' + \varphi_{pb}' - \varphi_{pa} - \varphi_{pb}\right) I_{el}^{pp} g_{pp} d\Omega\, dv_{pb}$$

$$J_{pq}^1 = \int_{\Omega, v_{qb}} f_{pa}^0 f_{qb}^0 \left(\varphi_{pa}' + \varphi_{qb}' - \varphi_{pa} - \varphi_{qb}\right) I_{el}^{pq} g_{pq} d\Omega\, dv_{qb} \tag{4.3}$$

As before, the subscripts a and b correspond to interacting molecules. Of course, we have similar terms for $J_q^1[\varphi_q]$.

After elimination of time derivatives, we have

$$\frac{1}{f_p^0}\frac{df_p^0}{dt} = \left(\frac{m_p u_p^2}{2kT} - \frac{5}{2}\right)\frac{1}{T}\frac{\partial T}{\partial r}\cdot u_p + \frac{m_p}{kT}\frac{\partial V}{\partial r}:\overset{0}{u_p u_p} + \frac{1}{\xi_p}u_p \cdot t_p \tag{4.4}$$

with the terms in $\frac{\partial T}{\partial r}$ and $\frac{\partial V}{\partial r}$ similar to those for pure gases. These terms are at the origin of conduction and viscosity phenomena, respectively. There is also a supplementary term, i.e. t_p, with

$$t_p = \frac{\partial \xi_p}{\partial r} - (\xi_p - c_p)\frac{1}{p}\frac{\partial p}{\partial r} \tag{4.5}$$

If φ_p is written as

$$\varphi_p = A_p \frac{1}{T}\frac{\partial T}{\partial \boldsymbol{r}}\cdot \boldsymbol{u}_p + B_p \frac{\partial \boldsymbol{V}}{\partial \boldsymbol{r}} : \overset{0}{\boldsymbol{u}_p \boldsymbol{u}_p} + L_p \boldsymbol{u}_p \cdot \boldsymbol{t}_p \qquad (4.6)$$

the coefficients A_p, B_p, L_p are given by the following integral equations:

$$J_p^1\left[A_p \boldsymbol{u}_p\right] = f_p^0 \left(\frac{m_p u_p^2}{2kT} - \frac{5}{2}\right)\boldsymbol{u}_p$$

$$J_p^1\left[B_p \overset{0}{\boldsymbol{u}_p \boldsymbol{u}_p}\right] = f_p^0 \frac{m_p}{kT}\overset{0}{\boldsymbol{u}_p \boldsymbol{u}_p}$$

$$J_p^1\left[L_p \boldsymbol{u}_p\right] = f_p^0 \frac{n}{n_p}\boldsymbol{u}_p \qquad (4.7)$$

As in the case for pure gases, A_p and B_p are expanded in the basis set of Sonine polynomials (Eqn. (3.15)) and L_p in the same way as A_p, that is:

$$L_p = \sum_{m=0}^{\infty} l_{pm}(\boldsymbol{r},t) S_{p3/2}^m\left(\frac{m_p u_p^2}{2kT}\right) \qquad (4.8)$$

The coefficients a_{pm}, b_{pm}, l_{pm} are expressed as functions of collisional integrals corresponding to collisions $p-p$ and $p-q$, that is (Appendix 4.2):

$$\left(\alpha_{pm}^{pm}\right)_{pp}, \left(\alpha_{pm}^{qm}\right)_{pq}, \left(\alpha_{qm}^{pm}\right)_{pq} \quad \text{for the coefficients } a_{pm}$$

$$\left(\beta_{pm}^{pm}\right)_{pp}, \left(\beta_{pm}^{qm}\right)_{pq}, \left(\beta_{qm}^{pm}\right)_{pq} \quad \text{for the coefficients } b_{pm}$$

$$\left(\lambda_{pm}^{pm}\right)_{pp}, \left(\lambda_{pm}^{qm}\right)_{pq}, \left(\lambda_{qm}^{pm}\right)_{pq} \quad \text{for the coefficients } l_{pm}$$

We have of course similar expressions for the coefficients a_{qm}, b_{qm}, l_{qm}.

If the expansion is truncated at the first term, we obtain the following system for the coefficients a_{p1} and a_{q1}, since we have $a_{p0} = a_{q0} = 0$:

$$a_{p1}\left[\left(\alpha_{p1}^{p1}\right)_{pp} + \left(\alpha_{p1}^{p1}\right)_{pq}\right] + a_{q1}\left(\alpha_{q1}^{p1}\right)_{pq} = -\frac{15}{2}\frac{n_p kT}{m_p}$$

$$a_{q1}\left[\left(\alpha_{q1}^{q1}\right)_{qq} + \left(\alpha_{q1}^{q1}\right)_{pq}\right] + a_{p1}\left(\alpha_{p1}^{q1}\right)_{pq} = -\frac{15}{2}\frac{n_q kT}{m_q} \qquad (4.9)$$

For b_{p0} and b_{q0}, we have

$$b_{p0}\left[\left(\beta_{p0}^{p0}\right)_{pp} + \left(\beta_{p0}^{p0}\right)_{pq}\right] + b_{q0}\left(\beta_{q0}^{p0}\right)_{pq} = 10\frac{n_p kT}{m_p}$$

$$b_{q0}\left[\left(\beta_{q0}^{q0}\right)_{qq} + \left(\beta_{q0}^{q0}\right)_{pq}\right] + b_{p0}\left(\beta_{p0}^{q0}\right)_{pq} = 10\frac{n_q kT}{m_q} \qquad (4.10)$$

and for l_{p0} and l_{q0}

$$l_{p0}\left[\left(\lambda_{p0}^{p0}\right)_{pp} + \left(\lambda_{p0}^{q0}\right)_{pq}\right] + l_{q0}\left(\lambda_{q0}^{p0}\right)_{pq} = 3\frac{nkT}{m_p}$$

$$l_{q0}\left[\left(\lambda_{q0}^{q0}\right)_{qq} + \left(\lambda_{q0}^{q0}\right)_{pq}\right] + l_{p0}\left(\lambda_{p0}^{q0}\right)_{pq} = 3\frac{nkT}{m_q} \quad (4.11)$$

The definition of the collisional integrals α, β, λ is given in Appendix 4.1.

4.2.2 Transport terms: Navier–Stokes equations

As in the case for pure gases, momentum flux **P** and heat flux **q** are developed in order to close the conservation equations (1.26), but the term U_p (diffusion velocity of the species p), appearing in the conservation equation of this species, is also used, i.e.

$$\frac{\partial \rho_p}{\partial t} + \frac{\partial \cdot \rho_p V_p}{\partial r} = 0$$

with

$$V_p = V + U_p$$

Thus, for the momentum flux, we have

$$\mathbf{P} = \sum_p m_p \int_{v_p} f_p^0 (1 + \varphi_p) \mathbf{u}_p \mathbf{u}_p d\mathbf{v}_p$$

or

$$\mathbf{P} = nkT\mathbf{I} + 2(kT)^2 \left(\frac{n_p}{m_p} b_{p0} + \frac{n_q}{m_q} b_{q0}\right) \overline{\frac{\partial V}{\partial r}}^0 \quad (4.12)$$

which may also be written as

$$\mathbf{P} = \sum_p \left[p_p \mathbf{I} + 2(kT)^2 \frac{n_p}{m_p} b_{p0} \overline{\frac{\partial V}{\partial r}}^0 \right] \quad (4.13)$$

where $p_p = n_p kT$ represents the partial pressure of component p.

Thus, as in the case for pure gases, we can write

$$\mathbf{P} = p\mathbf{I} - 2\mu \overline{\frac{\partial V}{\partial r}}^0 \quad (4.14)$$

with

$$p = \sum_p p_p = nkT \quad \text{(static pressure)}$$

and
$$\mu = (kT)^2 \sum_p \frac{n_p}{m_p} b_{p0} \quad \text{(dynamic viscosity coefficient)}$$

where the coefficients b_{p0} are deduced from the solution of system (4.10).
For the mass flux $\boldsymbol{j}_p = \rho_p \boldsymbol{U}_p$, we have

$$\boldsymbol{U}_p = \frac{1}{n_p} \int_{\boldsymbol{v}_p} f_p^0 \varphi_p \boldsymbol{u}_p d\boldsymbol{v}_p \tag{4.15}$$

and with the expression for φ_p (Eqn. (4.6)), we find

$$\boldsymbol{U}_p = \frac{kT}{m_p} l_{p0} \boldsymbol{t}_p = \frac{kT}{m_p} l_{p0} \left[\frac{\partial \xi_p}{\partial r} - (\xi_p - c_p) \frac{1}{p} \frac{\partial p}{\partial r} \right] \tag{4.16}$$

In high-temperature gas dynamics, we may generally assume that, for mixtures of gases that have comparable molecular masses,[63] the pressure gradient term in Eqn. (4.16) is negligible in comparison with the term of the concentration gradient. Thus, we can write

$$\boldsymbol{U}_p = \frac{kT}{m_p} l_{p0} \frac{\partial \xi_p}{\partial r} \tag{4.17}$$

This expression is generally written in terms of the binary diffusion coefficient D_{pq}, so that

$$\rho_p \boldsymbol{U}_p = -n \frac{m_p m_q}{m} D_{pq} \frac{\partial \xi_p}{\partial r} \quad \text{(Fick's law)} \tag{4.18}$$

where

$$D_{pq} = -\frac{\rho}{n^2} \frac{n_p kT}{m_p m_q} l_{p0} \tag{4.19}$$

As we have $\rho_p \boldsymbol{U}_p + \rho_q \boldsymbol{U}_q = 0$, then

$$D_{pq} = D_{qp} \tag{4.20}$$

and
$$n_p l_{p0} = n_q l_{q0}$$

with
$$\xi_p + \xi_q = 1$$

The coefficients l_{p0} are deduced from the solution of the system (4.11) and from the relation (4.20), arising from the definition of V (Appendix 4.1).
For the heat flux, we have

$$q = \sum_p \int_{v_p} f_p^0 \varphi_p \frac{1}{2} m_p u_p^2 \boldsymbol{u}_p d\boldsymbol{v}_p$$

4.2 GAS MIXTURES WITH ELASTIC COLLISIONS

and we find that

$$q = -\frac{5}{2}(kT)^2 \left[\left(\frac{n_p}{m_p} a_{p1} + \frac{n_q}{m_q} a_{q1} \right) \frac{1}{T} \frac{\partial T}{\partial r} - \left(\frac{n_p}{m_p} l_{p0} - \frac{n_q}{m_q} l_{q0} \right) t_p \right] \quad (4.21)$$

with

$$t_p = -t_q = \frac{\partial \xi_p}{\partial r}$$

Thus, the heat flux is the sum of a conduction term that is proportional to the temperature gradient and of a diffusion term (since $t_p \sim U_p$), essentially because of the concentration gradient. We write

$$q = -\lambda \frac{\partial T}{\partial r} - \rho D \sum_p h_p \frac{\partial c_p}{\partial r} \quad (4.22)$$

with

$$h_p = e_p + \frac{p_p}{\rho_p} = \frac{3}{2} \frac{kT}{m_p} + \frac{kT}{m_p} \quad \text{(enthalpy per unit mass of the component } p\text{)}$$

$$(4.23)$$

$$c_p = \frac{m_p}{m} \xi_p \quad \text{(mass concentration of the component } p\text{)} \quad (4.24)$$

$$\lambda = \sum_p \frac{5}{2} \frac{n_p}{m_p} (kT)^2 \frac{a_{p1}}{T} \quad \text{(thermal conductivity coefficient)} \quad (4.25)$$

and

$$D = D_{pq} = D_{qp} = -\frac{\rho}{n^2} \frac{n_p kT}{m_p m_q} l_{p0} \quad \text{(binary diffusion coefficient)} \quad (4.26)$$

The coefficients a_{p1} are deduced from the solution of the system (4.9) (Appendix 4.1).

Remarks

- The diffusion flux $q_d = -\rho D \sum_p h_p \frac{\partial \xi_p}{\partial r}$ is also written

$$q_d = \sum_p h_p j_p = \sum_p \rho_p h_p U_p \quad (4.27)$$

This flux corresponds to the total available energy (enthalpy) 'transported' by diffusion.

- The first-order expansion of L_p (4.7) gives null terms, but at the second order there is a term that includes $\frac{\partial T}{\partial r}$, corresponding to the 'thermal diffusion'. Here this term is therefore neglected, as is the term that includes $\frac{\partial p}{\partial r}$.

- For D, we find (Appendix 4.1)

$$D = D_{pq} = \frac{3kT}{16nm_r} \langle \gamma^2 - \boldsymbol{\gamma} \cdot \boldsymbol{\gamma}' \rangle^{-1} \qquad (4.28)$$

where $m_r = \frac{m_p m_q}{m_p + m_q}$ (reduced mass). Thus, if $p = q$ (pure gas), we can define a 'self-diffusion' coefficient D_{pp}, such as

$$D_{pp} = \frac{3kT}{8nm_p} \langle \gamma^2 - \boldsymbol{\gamma} \cdot \boldsymbol{\gamma}' \rangle^{-1} \qquad (4.29)$$

even in the absence of diffusion velocity. This coefficient, however, presents some similarity with the diffusion coefficients of internal energy (Eqns (3.70) and (3.71)), explaining the approximations made in Chapter 3 (Eqn. (3.73)).

With the definition of various quantities for a mixture (Chapter 1) and that of the above transport terms (μ, λ, D), the first two Navier–Stokes equations (conservation of mass and momentum) are identical to those for a pure gas (3.29), but the conservation equations of energy and of species p are respectively written

$$\rho \frac{de}{dt} = \frac{\partial \cdot}{\partial r} \left(\lambda \frac{\partial T}{\partial r} + \rho D \sum_p h_p \frac{\partial c_p}{\partial r} \right) - p \frac{\partial \cdot V}{\partial r} + 2\mu \, \overline{\frac{\partial V}{\partial r}}^0 : \frac{\partial V}{\partial r} \qquad (4.30)$$

$$\rho \frac{dc_p}{dt} = \frac{\partial \cdot}{\partial r} \left(\rho D \frac{\partial c_p}{\partial r} \right) \qquad (4.31)$$

Remark

The above results may be generalized to the case of a multinary mixture by considering that, for diffusion, only two types of particles are taken into account (for example, heavy and light particles). Then, one single diffusion coefficient is used, as long as the condition $\sum_p \rho_p U_p = 0$ is satisfied. Then, the Navier–Stokes equations remain identical to the case of a binary mixture. This approximation is widely used in the case of complex mixtures (Chapter 7).

4.3 Binary mixtures of diatomic gases

4.3.1 One internal mode

Transport terms

As the situation with an internal mode can be treated in the same way as in the preceding sections, the analysis and results are not discussed in detail here. Thus, we analyse only briefly the case in which the rotational energy mode alone is excited.

In the same way, the temperature non-equilibrium appearing at the first order is treated only for the vibrational mode (Section 4.3.2), since the translational and rotational temperatures remain close to each other (Chapter 3).

The zero-order distribution is given by Eqn. (2.35), so that we can write the perturbation φ_{ip} in the form

$$\varphi_{ip} = \left[a_{p10} \left(\frac{5}{2} - \frac{m_p u_p^2}{2kT} \right) + a_{p01} \left(\frac{\varepsilon_{i_rp} - \overline{E}_{Rp}}{kT} \right) \right] \mathbf{u}_p \cdot \frac{1}{T} \frac{\partial T}{\partial \mathbf{r}} + b_{p00} \overset{0}{\mathbf{u}_p \mathbf{u}_p} : \frac{\partial \mathbf{V}}{\partial \mathbf{r}}$$

$$+ \left[d_{p10} \left(\frac{3}{2} - \frac{m_p u_p^2}{2kT} \right) + d_{p01} \left(\frac{\varepsilon_{i_rp} - \overline{E}_{Rp}}{kT} \right) \right] \frac{\partial \cdot \mathbf{V}}{\partial \mathbf{r}} + l_{p0} \mathbf{u}_p \cdot \mathbf{t}_p \quad (4.32)$$

where the coefficients a_p, b_p, d_p, l_p are defined in Appendix 4.1, and by $a_{p00} = d_{p00} = 0$.

A similar expression holds for φ_{iq}.

As before, we can write the transport terms as functions of these coefficients. For the stress tensor, we have

$$\mathbf{P} = p\mathbf{I} - 2\mu \overset{0}{\frac{\partial \mathbf{V}}{\partial \mathbf{r}}} - \eta \frac{\partial \cdot \mathbf{V}}{\partial \mathbf{r}} \mathbf{I} \quad (4.33)$$

with

$$p = nkT = \sum_p p_p \quad \text{(static pressure)} \quad (4.34)$$

$$\mu = -(kT)^2 \sum_p \frac{n_p}{m_p} b_{p0} = -(RT)^2 \sum_p \rho_p b_{p00} \quad \text{(dynamic viscosity coefficient)} \quad (4.35)$$

$$\eta = kT \sum_p n_p d_{p10} = p \sum_p \xi_p d_{p10} \quad \text{(bulk viscosity coefficient)} \quad (4.36)$$

The diffusion velocity is equal to

$$\mathbf{U}_p = \frac{kT}{m_p} l_{p0} \mathbf{t}_p \simeq \frac{kT}{m_p} l_{p0} \frac{\partial \xi_p}{\partial \mathbf{r}} \quad (4.37)$$

Despite the similarity of these terms with those of the monatomic case, the corresponding collisional integrals include terms related to inelastic collisions.

As in the case for pure gases, the heat flux term is the sum of a translational flux and a rotational flux:

$$\mathbf{q} = \mathbf{q}_T + \mathbf{q}_R$$

108 CHAPTER 4 QUASI-EQUILIBRIUM REGIMES: GAS MIXTURES

with

$$q_T = -\frac{5}{2}(kT)^2 \sum_p \frac{n_p}{m_p}\left(a_{p10}\frac{1}{T}\frac{\partial T}{\partial r} - l_{p0}\frac{\partial \xi_p}{\partial r}\right)$$

$$q_R = (kT)^2 \sum_p \frac{n_p}{m_p}\left(a_{p01}\frac{1}{T}\frac{\partial T}{\partial r} + l_{p0}\frac{\partial \xi_p}{\partial r}\right) \quad (4.38)$$

Each type of flux is the sum of a conduction flux and a diffusion flux, that is:

$$q = -\lambda\frac{\partial T}{\partial r} - \rho D\sum_p h_p\frac{\partial c_p}{\partial r} \quad (4.39)$$

where

$$\lambda = \lambda_T + \lambda_R$$

and

$$\lambda_T = (kT)^2 \sum_p \frac{n_p}{m_p}\frac{5}{2T}a_{p10}$$

$$\lambda_R = (kT)^2 \sum_p \frac{n_p}{m_p}\frac{1}{T}a_{p01}$$

$$D = -\frac{\rho\, n_p l_{p0} kT}{n^2\, m_p m_q} \quad (4.40)$$

Navier–Stokes equations

Now we can write the Navier–Stokes equations for this type of mixture: the first two equations for mass and momentum conservation are formally identical to the equations for pure diatomic gases (3.50). The species conservation equation is also formally identical to that for monatomic gas mixtures (1.29). As for the energy equation, it may be written in the following form:

$$\rho\frac{de}{dt} = \frac{\partial \cdot}{\partial r}\left(\lambda\frac{\partial T}{\partial r} + \rho D\sum_p h_p\frac{\partial c_p}{\partial r}\right) - p\frac{\partial \cdot V}{\partial r} + 2\mu\overline{\frac{\partial V}{\partial r}:\frac{\partial V}{\partial r}}^{\,0} + \eta\left(\frac{\partial \cdot V}{\partial r}\right)^2$$

(4.41)

with

$$e = \sum_p c_p e_p = \sum_p c_p\frac{\overline{E_p}}{m_p} = \frac{5}{2}RT, \quad h_p = e_p + \frac{p_p}{\rho_p} = \frac{7\,kT}{2\,m_p}, \quad \text{and}$$

$$h = \sum_p c_p h_p = \frac{7}{2}RT.$$

Here, $p = \rho RT$ and μ, η, λ, D are given by the relations (4.35–4.40).

Rotational non-equilibrium

As shown previously for pure gases, we have

$$E_{Rp} = \overline{E}_{Rp}\left(1 + d_{p01}\frac{\partial \cdot V}{\partial r}\right) \qquad (4.42)$$

The corresponding calculations are not developed here. They are practically identical to those developed below for the vibrational mode, which is of greater interest because of a stronger non-equilibrium. Thus, d_{p01} may be connected to various relaxation times ($\tau_{pp}^{TR}, \tau_{pq}^{TR}, \tau_{pq}^{RR} \ldots$). As these times are short compared to the reference time, we generally retain only three global times, firstly and secondly τ_{Rp} and τ_{Rq}, characterizing the TR exchanges of the species p and the species q respectively, retained independently of the collisional partner; and thirdly $\tau_{Rpq} = \tau_{Rqp}$, characterizing the RR exchanges between species. This simplification is used to obtain the transport term expressions presented below (Appendix 4.3).

Simplification of transport terms

Assuming, as in the case for pure gases, that the collisional balance of internal energy is small compared to the kinetic energy balance, the collisional integrals $\alpha, \beta, \delta, \lambda$ may be simplified and written as functions of macroscopic physical quantities (Appendix 4.2). In the same way, the transport terms may be expressed as functions of these quantities[26] (Appendix 4.3).

4.3.2 Two internal modes

The results concerning the Navier–Stokes equations, the transport terms, and their simplified expressions can be deduced from the preceding results, but their formal complexity renders them quite unpresentable in the framework of this book. However, the expression for φ_{ip} is given in Appendix 4.1. Furthermore, the expressions for the transport terms may be deduced from Eqns (4.33)–(4.38), written for one mode, by replacing, respectively, b_{p00} by b_{p000}, d_{p10} by d_{p100}, a_{p10} by a_{p100}, and a_{p01} by a_{p010} in $\mu, \eta, \boldsymbol{q}_T$, and $\boldsymbol{q}_R$. The term for the vibrational heat flux must also be added, that is:

$$\boldsymbol{q}_V = kT\sum_p \frac{n_p}{m_p}\overline{E}_{Vp}\left(a_{p001}\frac{1}{T}\frac{\partial T}{\partial r} + l_{p0}\frac{\partial \xi_p}{\partial r}\right)$$

Only the development for the first-order vibrational non-equilibrium is therefore presented below.

As previously discussed, the presence of two or more species induces the definition of energies specific to each species. Thus, for example, for the vibrational energy of the species p in equilibrium at zero order, we have $E_{Vp} = \overline{E}_{Vp}(T)$, and at this order no definition of a specific vibrational temperature is necessary, in contrast to the first order, for which we have

$$E_{Vp} = \overline{E}_{Vp}\left(1 + d_{p001}\frac{\partial \cdot \mathbf{V}}{\partial \mathbf{r}}\right) \qquad (4.43)$$

We may also define a vibrational temperature specific to the component p, if we assume a Boltzmann distribution for the vibrational population. Thus,

$$E_{Vp} = \overline{E}_{Vp}(T_{Vp}) \qquad (4.44)$$

This is, however, a pure hypothesis, or a convenience.

In Eqn. (4.43), the term d_{p001} is part of a 6×6 Cramer system and, with usual simplifications, may be written as a function of the following five collisional integrals:[27]

$$\langle \Delta\varepsilon_{vpp}^2\rangle_{pp}, \; \langle \Delta\varepsilon_{vqq}^2\rangle_{qq}, \; \langle \Delta\varepsilon_{vp}\Delta\varepsilon_{vq}\rangle_{pq}, \; \langle \Delta\varepsilon_{vp}\Delta\varepsilon_{vpq}\rangle_{pq},$$
$$\text{and } \langle \Delta\varepsilon_{vq}\Delta\varepsilon_{vpq}\rangle_{qp} \qquad (4.45)$$

These integrals are defined and developed in Appendix 4.1. If I_1, I_2, I_3 are combinations of these integrals, also defined in Appendix 4.1, we find from Eqn. (4.43):

$$\frac{E_{Vp} - \overline{E}_{Vp}}{\overline{E}_{Vp}} = \frac{kC_{TR}C_{Vp}}{C_{TRV}^2}I_1\left(1 + \xi_q \frac{I_2 - I_1}{\frac{C_{Vp}}{k}I_3 + \xi_p\frac{C_{Vp}}{C_{Vq}}I_2 + \xi_q I_1}\right)\frac{\partial \cdot \mathbf{V}}{\partial \mathbf{r}} \qquad (4.46)$$

and a similar expression for $E_{Vq} - \overline{E}_{Vq}$.

With the same procedure used for pure gases, we can relate the collisional integrals to phenomenological relaxation times. To do that, we may linearize the general relaxation equation for mixtures (2.39), or we may develop an approximate relaxation equation by means of the usual Chapman–Enskog expansion. The result is identical, and we find:[27]

$$\frac{dE_{Vp}}{dt} = \left[2n\frac{k}{C_{Vp}}\xi_p\langle\Delta\varepsilon_{vpp}^2\rangle_{pp} + 4n\frac{k}{C_{Vp}}\xi_q\left(\begin{array}{c}\langle\Delta\varepsilon_{vpq}\Delta\varepsilon_{vp}\rangle_{pq}\\-\langle\Delta\varepsilon_{vp}\Delta\varepsilon_{vq}\rangle_{pq}\end{array}\right)\right](\overline{E}_{Vp} - E_{Vp})$$
$$+ 4n\frac{k}{C_{Vq}}\xi_q\langle\Delta\varepsilon_{vp}\Delta\varepsilon_{vq}\rangle_{pq}(\overline{E}_{Vq} - E_{Vq}) \qquad (4.47)$$

From the structure of Eqn. (4.47) we can identify the integrals $\langle \cdots \rangle$ with collision frequencies and therefore their inverses with characteristic relaxation times. Thus, as in the case for pure gases, the quantity $2n\frac{k}{C_{Vp}}\langle\Delta\varepsilon_{vpp}^2\rangle_{pp}$ represents the inverse of a relaxation time connected to vibrational exchanges (essentially TV) between the molecules p. Therefore, we define:

$$\left(\tau_{pp}^{TV}\right)^{-1} = 2n\frac{k}{C_{Vp}}\langle\Delta\varepsilon_{vpp}^2\rangle_{pp}$$

Similarly, for the molecules q:

$$\left(\tau_{qq}^{TV}\right)^{-1} = 2n\frac{k}{C_{Vq}}\langle\Delta\varepsilon_{vqq}^2\rangle_{qq}$$

The quantity $4n\frac{k}{C_{Vp}}\langle\Delta\varepsilon_{vpq}\Delta\varepsilon_{vp}\rangle_{pq}$ is associated with the relaxation of $\overline{E}_{Vp} - E_{Vp}$ but also with that of $\overline{E}_{Vq} - E_{Vq}$. Therefore, it participates in the coupling between the vibrational energy relaxations of the components p and q. We can thus relate this quantity to the characteristic time corresponding to the VV exchanges between the molecules p and q, that is:

$$\left(\tau_{pq}^{VV}\right)^{-1} = 4n\langle-\Delta\varepsilon_{vp}\Delta\varepsilon_{vq}\rangle_{pq} \qquad (4.48)$$

This definition of $\left(\tau_{pq}^{VV}\right)^{-1}$ without the term $\frac{k}{C_{Vp}}$ is necessary in order to obtain a symmetrical expression for the equation, giving $\frac{dE_{Vq}}{dt}$, with $\tau_{pq}^{VV} = \tau_{qp}^{VV}$.

Thus, Eqn. (2.39) becomes

$$\frac{dE_{Vp}}{dt} = \left(\frac{\xi_p}{\tau_{pp}^{TV}} + \frac{\xi_q}{\tau_{pq}^{TV}} + \frac{k}{C_{Vp}}\frac{\xi_q}{\tau_{pq}^{VV}}\right)\left(\overline{E}_{Vp} - E_{Vp}\right) - \frac{k}{C_{Vq}}\frac{\xi_q}{\tau_{pq}^{VV}}\left(\overline{E}_{Vq} - E_{Vq}\right)$$

$$(4.49)$$

Of course, a similar equation for $\frac{dE_{Vq}}{dt}$ may be written.

These relaxation equations have a structure analogous to (2.70), obtained for a mixture of gases with molecules considered as harmonic oscillators.

Now the vibrational non-equilibrium equation (4.46) may be written as functions of the phenomenological relaxation times defined above. Setting

$$\tau_{Vp}^{-1} = \frac{\xi_p}{\tau_{pp}^{TV}} + \frac{\xi_q}{\tau_{pq}^{TV}} \quad \text{and} \quad \tau_{Vq}^{-1} = \frac{\xi_q}{\tau_{qq}^{TV}} + \frac{\xi_p}{\tau_{qp}^{TV}} \qquad (4.50)$$

we have

$$\frac{E_{Vp} - \overline{E}_{Vp}}{\overline{E}_{Vp}} = \frac{kC_{TR}}{C_{TRV}^2}\tau_{Vp}\left(1 + \xi_q\frac{\tau_{Vq} - \tau_{Vp}}{\frac{C_{Vp}}{k}\tau_{pq}^{VV} + \xi_p\frac{C_{Vp}}{C_{Vq}}\tau_{Vq} + \xi_q\tau_{Vp}}\right)\frac{\partial\cdot V}{\partial r} \qquad (4.51)$$

$$\frac{E_{Vq} - \overline{E}_{Vq}}{\overline{E}_{Vq}} = \frac{kC_{TR}}{C_{TRV}^2}\tau_{Vq}\left(1 + \xi_p\frac{\tau_{Vp} - \tau_{Vq}}{\frac{C_{Vq}}{k}\tau_{pq}^{VV} + \xi_q\frac{C_{Vq}}{C_{Vp}}\tau_{Vp} + \xi_p\tau_{Vq}}\right)\frac{\partial\cdot V}{\partial r} \qquad (4.52)$$

where τ_p and τ_q represent global relaxation times for the components p and q, respectively, corresponding to the TV exchanges, whatever the collisional partner may be. The first term of the right-hand side of Eqns (4.51) and (4.52) represents the TV relaxation of each component, as in the case for a pure gas, and the second term represents a VV relaxation coupling between the components. We can again find the two extreme cases already discussed in Chapter 2 concerning the order of magnitude of τ_{pq}^{VV}: if this time has a sufficiently large value, both gases relax quasi-independently. On the other hand, if τ_{pq}^{VV} is small, then $T_{Vp} \simeq T_{Vq}$ and the gases relax quasi-simultaneously ('resonant' gases).

4.4 Mixtures of reactive gases

As discussed above, starting from a zero-order equilibrium solution, we have the following system for the distribution function of the species p:

$$(J_{TRVC})_p^0 = 0$$

$$\frac{df_{ip}^0}{dt} = (J_{TRVC})_p^1 \qquad (4.53)$$

The zero-order solution f_{ip}^0 is difficult to obtain from the first equation of the system (4.53), because of the coupling between the various types of collisions. However, we can easily obtain a particular solution resulting from the solution of the following equations written instead of $(J_{TRVC})_p^0 = 0$, i.e.

$$(J_{TRV})_p^0 = 0 \quad \text{and} \quad J_{Cp}^0 = 0 \qquad (4.54)$$

This solution corresponds to an equilibrium Maxwell–Boltzmann distribution for the modes T, R, and V (Eqn. (2.38)) and an equilibrium chemical distribution (Eqn. (2.45)). At the macroscopic level, the solution corresponds to the Euler equations for n, V, T and to the equation $\dot{w}_p = 0$ for the species p (examples given in Chapter 2).

Then, at first order, we can have an approximate solution corresponding to realistic situations widely considered (examples given in Chapter 10). Thus, for φ_{ip}, we consider that the solutions resulting from the equation $\frac{df_{ip}^0}{dt} = (J_{TRV})_p^1$ are the same as those obtained above, with the same Navier–Stokes equations and the same transport terms. Furthermore, we assume that the p equations $\dot{w}_p = 0$ govern the composition of the mixture at zero order. This leads to the conclusion that the chemical production of species is the dominant process for the balance of these species, so that convection and diffusion are negligible. Physical and mathematical interpretations are given in Chapter 7.

Appendix 4.1 Systems of equations for the coefficients a, b, l, d

Binary mixtures of monatomic gases

System for A_p

$$\sum_{m=0}^{1} a_{pm}\left[\left(\alpha_{pm}^{pm}\right)_{pp} + \left(\alpha_{pm}^{pm}\right)_{pq}\right] + \sum_{m=0}^{1} a_{qm} \left(\alpha_{qm}^{pm}\right)_{pq}$$

$$= \int_{v_p} f_p^0 \left(\frac{m_p u_p^2}{2kT} - \frac{5}{2}\right) \mathbf{u}_p \cdot S_{p3/2}^m \mathbf{u}_p d\mathbf{v}_p$$

with

$a_{p0} = a_{q0} = 0$, a_{p1}, and a_{q1}, given by Eqn. (4.9)

$$\left(\alpha_{p1}^{p1}\right)_{pp} = \int_{v_p} J_{pp}^1 \left[S_{p3/2}^1 \mathbf{u}_p\right] \cdot S_{p3/2}^1 \mathbf{u}_p d\mathbf{v}_p$$

$$\left(\alpha_{p1}^{p1}\right)_{pq} = \int_{v_p} J_{pq}^1 \left[S_{p3/2}^1 \mathbf{u}_p\right] \cdot S_{p3/2}^1 \mathbf{u}_p d\mathbf{v}_p$$

$$\left(\alpha_{q1}^{p1}\right)_{pq} = \int_{v_p} J_{pq}^1 \left[S_{p3/2}^1 \mathbf{u}_p\right] \cdot \mathbf{u}_p d\mathbf{v}_p \quad (4.55)$$

System for B_p

The coefficients b_{p0} and b_{q0} are given by the system (4.10), with the right-hand side corresponding to the quantity $\int_{v_p} f_p^0 \frac{m_p}{kT} \overset{0}{\mathbf{u}_p \mathbf{u}_p} : \overset{0}{\mathbf{u}_p \mathbf{u}_p} d\mathbf{v}_p$ and to the symmetrical quantity integrated over v_q. We also have

$$\left(\beta_{p0}^{p0}\right)_{pp} = \int_{v_p} J_{pp}^1 \left[\overset{0}{\mathbf{u}_p \mathbf{u}_p}\right] : \overset{0}{\mathbf{u}_p \mathbf{u}_p} d\mathbf{v}_p$$

$$\left(\beta_{p0}^{p0}\right)_{pq} = \int_{v_p} J_{pq}^1 \left[\overset{0}{\mathbf{u}_p \mathbf{u}_p}\right] : \overset{0}{\mathbf{u}_p \mathbf{u}_p} d\mathbf{v}_p$$

$$\left(\beta_{q0}^{p0}\right)_{pq} = \int_{v_p} J_{pq}^1 \left[\overset{0}{\mathbf{u}_q \mathbf{u}_q}\right] : \overset{0}{\mathbf{u}_p \mathbf{u}_p} d\mathbf{v}_p \quad (4.56)$$

System for L_p

The coefficients l_{p0} and l_{q0} are given by the system (4.11), with right-hand sides corresponding to the quantity $\int_{v_p} f_p^0 \frac{n}{n_p} \boldsymbol{u}_p \cdot \boldsymbol{u}_p d\boldsymbol{v}_p$ and to the symmetrical quantity integrated over $\boldsymbol{v}_q$.

However, the two equations are not independent, and another equation is necessary. This may be obtained by using the condition $\rho_p \boldsymbol{U}_p + \rho_q \boldsymbol{U}_q = 0$, which can be written simply as

$$n_p l_{p0} = n_q l_{q0}$$

Thus

$$l_{p0} \left(\lambda_{p0}^{q0} \right)_{pq} + l_{q0} \left(\lambda_{q0}^{p0} \right)_{pq} = \int_{v_p} f_p^0 \frac{n}{n_p} \boldsymbol{u}_p \cdot \boldsymbol{u}_p d\boldsymbol{v}_p$$

with

$$\left(\lambda_{p0}^{p0} \right)_{pq} = \int_{v_p} J_{pq}^1 [\boldsymbol{u}_p] \cdot \boldsymbol{u}_p d\boldsymbol{v}_p$$

$$\left(\lambda_{q0}^{p0} \right)_{pq} = \int_{v_p} J_{pq}^1 [\boldsymbol{u}_q] \cdot \boldsymbol{u}_p d\boldsymbol{v}_p \tag{4.57}$$

and with the result

$$l_{p0} = -\frac{3}{16} \frac{m_p + m_q}{n_p m} \langle \gamma^2 - \boldsymbol{\gamma} \cdot \boldsymbol{\gamma}' \rangle^{-1} \tag{4.58}$$

Binary mixtures of diatomic gases (one internal mode)

System for A_{ip}

$$A_{ip} = \sum_{mn=0}^{1} a_{pmn} \psi_{pmn}^{3/2} \quad \text{with} \quad \psi_{pmn}^{3/2} = S_{p3/2}^m \left(\frac{m_p u_p^2}{2kT} \right) P_{ip}^n \left(\frac{\varepsilon_{ip}}{kT} \right)$$

Keeping only the first terms, and remembering that $a_{p00} = a_{q00} = 0$, we obtain the following system for the coefficients $a_{p10}, a_{p01}, a_{q10}, a_{q01}$:

$$\sum_{m+n=0}^{1} a_{pmn} \left[\left(\alpha_{pmn}^{p10} \right)_{pp} + \left(\alpha_{pmn}^{p10} \right)_{pq} \right] + a_{qmn} \left(\alpha_{qmn}^{p10} \right)_{pq} = -\frac{15}{2} \frac{n_p kT}{m_p}$$

$$\sum_{m+n=0}^{1} a_{pmn} \left[\left(\alpha_{pmn}^{p01} \right)_{pp} + \left(\alpha_{pmn}^{p01} \right)_{pq} \right] + a_{qmn} \left(\alpha_{qmn}^{p01} \right)_{pq} = 3 \frac{n_p kT}{m_p} \frac{C_{Rp}}{k}$$

APPENDIX 4.1 SYSTEMS OF EQUATIONS FOR a, b, l, d COEFFICIENTS

$$\sum_{m+n=0}^{1} a_{qmn} \left[\left(\alpha_{qmn}^{q10} \right)_{qq} + \left(\alpha_{qmn}^{q10} \right)_{pq} \right] + a_{pmn} \left(\alpha_{pmn}^{q10} \right)_{pq} = -\frac{15}{2} \frac{n_q kT}{m_q}$$

$$\sum_{m+n=0}^{1} a_{qmn} \left[\left(\alpha_{qmn}^{q01} \right)_{qq} + \left(\alpha_{qmn}^{q01} \right)_{pq} \right] + a_{pmn} \left(\alpha_{pmn}^{q01} \right)_{pq} = 3 \frac{n_q kT}{m_q} \frac{C_{Rq}}{k} \quad (4.59)$$

with

$$\left(\alpha_{pmn}^{pm'n'} \right)_{pp} = \sum_i \int_{v_p} J_{pp}^1 \left[\Psi_{pmn}^{3/2} u_p \right] \cdot \Psi_{pm'n'}^{3/2} u_p \, dv_p$$

$$\left(\alpha_{pmn}^{pm'n'} \right)_{pq} = \sum_i \int_{v_p} J_{pq}^1 \left[\Psi_{pmn}^{3/2} u_p \right]_i^k \cdot \Psi_{pm'n'}^{3/2} u_p \, dv_p$$

$$\left(\alpha_{qmn}^{pm'n'} \right)_{pq} = \sum_i \int_{v_p} J_{pq}^1 \left[\Psi_{qmn}^{3/2} u_p \right]_j^l \cdot \Psi_{pm'n'}^{3/2} u_p \, dv_p \quad (4.60)$$

System for B_{ip}

$$B_{ip} = b_{p00}$$

This system is formally identical to the monatomic case (Eqn. (4.10)), with the addition of an index of 0 to the coefficients b and to the integrals β, with

$$\left(\beta_{p00}^{p00} \right)_{pp} = \sum_i \int_{v_p} J_{pp}^1 \left[\overset{0}{u}_p \overset{0}{u}_p \right] : \overset{0}{u}_p \overset{0}{u}_p \, dv_p$$

$$\left(\beta_{p00}^{p00} \right)_{pq} = \sum_i \int_{v_p} J_{pq}^1 \left[\overset{0}{u}_p \overset{0}{u}_p \right]_i^k : \overset{0}{u}_p \overset{0}{u}_p \, dv_p$$

$$\left(\beta_{p00}^{q00} \right)_{pq} = \sum_i \int_{v_p} J_{pp}^1 \left[\overset{0}{u}_p \overset{0}{u}_p \right]_j^l : \overset{0}{u}_p \overset{0}{u}_p \, dv_p \quad (4.61)$$

System for D_{ip}

$$D_{ip} = \sum_{mn=0}^{1} d_{pmn} \Psi_{pmn}^{1/2}$$

Here, we have $d_{p00} = 0$, and the coefficients d_{p10}, d_{p01}, d_{q10}, d_{q01} are given by a system identical to that giving the coefficients a (Eqn. (4.59)), provided we replace the coefficients a with d, the integrals α (Eqn. (4.60)) with the integrals δ (Eqn. (4.62)), and the right-hand sides respectively with $-n_p \frac{C_R}{C_{TR}}$, $-n_p \frac{C_{Rp}}{C_{TR}}$,

$-n_q \frac{C_R}{C_{TR}}, -\frac{C_{Rq}}{C_{TR}}$. The δ integrals are given by the following expressions:

$$(\delta_{pmn}^{pm'n'})_{pp} = \sum_i \int_{v_p} J_{pp}^1 [\Psi_{pmn}^{1/2}] \Psi_{pm'n'}^{1.2} dv_p$$

$$(\delta_{pmn}^{pm'n'})_{pq} = \sum_i \int_{v_p} J_{pq}^1 [\Psi_{pmn}^{1/2}]_i^k \Psi_{pm'n'}^{1.2} dv_p$$

$$(\delta_{pmn}^{pm'n'})_{pq} = \sum_i \int_{v_p} J_{pq}^1 [\Psi_{pmn}^{1/2}]_j^l \Psi_{pm'n'}^{1.2} dv_p \qquad (4.62)$$

System for L_{ip}

This system is formally identical to the monatomic case (Eqn. (4.11)), with the addition of an index of 0 to the coefficients l and to the λ integrals, with

$$\left(\lambda_{p00}^{p00}\right)_{pp} = \sum_i \int_{v_p} J_{pp}^1 [u_p] \cdot u_p dv_p = 0$$

$$\left(\lambda_{p00}^{p00}\right)_{pq} - \sum_i \int_{v_p} J_{pq}^1 [u_p]_i^k \cdot u_p dv_p$$

$$\left(\lambda_{q00}^{p00}\right)_{pq} = \sum_i \int_{v_p} J_{pq}^1 [u_q]_j^l \cdot u_p dv_p = 0 \qquad (4.63)$$

Binary mixtures of diatomic gases (two internal modes)

The expression for φ_{ip} is

$$\varphi_{ip} = \left[a_{p100} \left(\frac{5}{2} - \frac{m_p u_p^2}{2kT} \right) + a_{p010} \left(\frac{\varepsilon_{i_r p} - \overline{E}_{Rp}}{kT} \right) + a_{p001} \left(\frac{\varepsilon_{i_v p} - \overline{E}_{Vp}}{kT} \right) \right]$$

$$\times \frac{1}{T} \frac{\partial T}{\partial r} \cdot u_p + b_{p000} \frac{\partial V}{\partial r} : u_p^0 u_p$$

$$+ \left[d_{p100} \left(\frac{3}{2} - \frac{m_p u_p^2}{2kT} \right) + d_{p010} \left(\frac{\varepsilon_{i_r p} - \overline{E}_{Rp}}{kT} \right) + d_{p001} \left(\frac{\varepsilon_{i_v p} - \overline{E}_{Vp}}{kT} \right) \right]$$

$$\times \frac{\partial \cdot V}{\partial r} + l_{p0} t_p \cdot u_p \qquad (4.64)$$

The formulae giving the coefficients become intricate and therefore are not described in detail here. For example, the systems giving the coefficients a and d become 6×6 systems including 6×9 collisional integrals. This calculation is left to the reader, as the computations represent an extension of the case with one single mode.

Only a few results for the coefficient $D_{ip} = \sum_{mns}^{\infty} d_{pmns} \Psi_{pmns}^{1/2}$ are given here, as well as the collisional integrals necessary for the determination of the linearized vibrational non-equilibrium. Thus, using approximations discussed in Chapter 3, the 6×6 system giving the coefficients $d_{p100}, d_{p010}, d_{p001}, d_{q100}, d_{q010}, d_{q001}$ may be simplified. For d_{p001}, we find the relation appearing in Eqns (4.43) and (4.46), with

$$\Delta \varepsilon_{vpp} = (\varepsilon_{k_v p} + \varepsilon_{l_v p} - \varepsilon_{i_v p} - \varepsilon_{j_v p})/kT$$

$$\Delta \varepsilon_{vp} = (\varepsilon_{k_v p} - \varepsilon_{i_v p})/kT$$

$$\Delta \varepsilon_{vpq} = (\varepsilon_{k_v p} + \varepsilon_{l_v q} - \varepsilon_{i_v p} - \varepsilon_{j_v q})/kT \quad \text{and} \quad \Delta \varepsilon_{vq} = (\varepsilon_{l_v q} - \varepsilon_{j_v q})/kT$$

and with

$$I_1^{-1} = 2n_p \frac{k}{C_{Vp}} \langle (\Delta \varepsilon)_{vpp}^2 \rangle_{pp} + 4n_q \frac{k}{C_{Vp}} \langle \Delta \varepsilon_{vp} \Delta \varepsilon_{vpq} \rangle_{pq}$$

$$I_2^{-1} = 2n_q \frac{k}{C_{Vq}} \langle (\Delta \varepsilon)_{qq}^2 \rangle_{qq} + 4n_p \frac{k}{C_{Vq}} \langle \Delta \varepsilon_{vq} \Delta \varepsilon_{vpq} \rangle_{qp}$$

$$I_3^{-1} = -4n \langle \Delta \varepsilon_{vp} \Delta \varepsilon_{vq} \rangle_{pq}$$

We also have

$$\langle \cdots \rangle_{pq} = \left(\frac{kT}{2\pi m_r}\right)^{1/2} \sum_{i,j,k,l} \frac{\overline{n}_{i_r p} \overline{n}_{i_v p} \overline{n}_{j_r q} \overline{n}_{j_v q}}{n^4}$$

$$\times \int_{\Omega, \gamma_{pq}} \exp\left(-\gamma_{pq}^2\right) \gamma_{pq}^3 (\cdots) I_{ip,jq}^{kp,lq} d\Omega \, d\gamma_{pq} \quad (4.65)$$

Appendix 4.2 Collisional integrals and simplifications

The integrals below apply to the binary mixtures of diatomic gases with one excited internal mode (rotation). The integrals corresponding to the monatomic case are, of course, obtained by cancelling the terms depending on the internal energy. The integrals corresponding to diatomic gases with two internal modes (rotation, vibration) may be obtained by replacing ε_i or ε_{i_r} with $\varepsilon_{i_r} + \varepsilon_{i_v}$. Then, the simplifications used for pure gases (Chapter 3) may be applied. However, as discussed above, the phenomena related to the vibrational non-equilibrium cannot be correctly developed in this way. It is then preferable to separate the contribution of each energy mode in the expansions themselves (Chapter 3).

118 CHAPTER 4 QUASI-EQUILIBRIUM REGIMES: GAS MIXTURES

The α integrals

These integrals appear in the system corresponding to Eqn. (4.59) giving the coefficients a and therefore also in the heat flux (Eqn. (4.38)). This 4×4 system includes 2×12 integrals symmetrical with respect to p and q. These 12 integrals are developed below, first in their general form as functions of integrals $\langle \cdots \rangle$, then in an approximate form with the usual simplifications. Thus, we have

$$\left(\alpha_{p10}^{p10}\right)_{pp} = -8n_p^2 \frac{kT}{m_p} \left\langle \gamma^4 \sin^2 \chi - \Delta\varepsilon_{rpp} \gamma^2 \sin^2 \chi + \frac{11}{8} \left(\Delta\varepsilon_{rpp}\right)^2 \right\rangle_{pp}$$

$$\simeq -5\xi_p \frac{n_p kT}{m_p} \left(\frac{nkT}{\mu_p} + \frac{5}{6} \frac{C_{Rp}}{k} \left(\tau_{pp}^{RT}\right)^{-1} \right)$$

$$\left(\alpha_{p01}^{p01}\right)_{pp} = 8n_p^2 \frac{kT}{m_p} \left\langle \frac{\varepsilon_{irp} - \overline{E}_{Rp}}{kT} \left(\gamma \cdot \gamma' \frac{\varepsilon_{krp} - \varepsilon_{lrp}}{kT} - \gamma^2 \frac{\varepsilon_{irp} - \varepsilon_{jrp}}{kT} + \frac{3}{2}\Delta\varepsilon_{rpp} \right) \right\rangle_{pp}$$

$$\left(\alpha_{p01}^{p10}\right)_{pp} = \left(\alpha_{p10}^{p01}\right)_{pp} = -5n_p^2 \frac{kT}{m_p} \left\langle (\Delta\varepsilon_{rpp})^2 \right\rangle_{pp} \simeq -\frac{5}{2}\xi_p \frac{n_p kT}{m_p} \frac{C_{Rp}}{k} \left(\tau_{pp}^{RT}\right)^{-1}$$

$$= -3\xi_p \frac{n_p kT}{m_p} \frac{C_{Rp}}{k} \left(\frac{kT}{m_p}(D_{pp})^{-1} + \frac{1}{2} \left(\tau_{pp}^{RT}\right)^{-1} \right) \qquad (4.66)$$

$$\left(\alpha_{p10}^{p10}\right)_{pq} = -16 n_p n_q \frac{kT}{m_p} \left(\frac{m_q}{m_p + m_q} \right)^3$$

$$\times \left[\begin{array}{l} \left(\frac{30}{4} \left(\frac{m_p}{m_q}\right)^2 + \frac{25}{4} \right) \langle \gamma^2 - \gamma \cdot \gamma' \rangle_{pq} - 5 \langle \gamma^2 (\gamma^2 - \gamma \cdot \gamma') \rangle_{pq} \\ + \langle \gamma^3 (\gamma^3 - \gamma'^3 \cos \chi) \rangle_{pq} + 2 \frac{m_p}{m_q} \langle \gamma^2 (\gamma^2 - \gamma'^2 \cos^2 \chi) \rangle \\ - \frac{1}{6} \left(\Delta\varepsilon_{rpq}\right)^2 \rangle_{pq} \end{array} \right]$$

$$\simeq -3 \frac{n_p n_q}{n} \left(\frac{kT}{m_p} \right)^2 \left(\frac{m_q}{m_p + m_q} \right)^2 \left[\left(\frac{15}{2} \left(\frac{m_p}{m_q}\right)^2 + \frac{25}{4} \right) - 3 B_{pq}^* \right] D_{pq}^{-1}$$

$$- 20 n_p n_q \frac{kT}{m_q} \left(\frac{m_q}{m_p + m_q} \right)^3 \frac{kT}{\mu'_{pq}}$$

$$\left(\alpha_{p01}^{p01}\right)_{pq} = 16 n_p n_q \frac{kT}{m_p} \frac{m_q}{m_p + m_q}$$

$$\times \left[\begin{array}{l} \frac{3}{2} \frac{m_p}{m_q} \left\langle \Delta\varepsilon_{rp} \left(\frac{\varepsilon_{irp} - \overline{E}_{Rp}}{kT} \right) \right\rangle_{pq} \\ + \left\langle \frac{\varepsilon_{irp} - \overline{E}_{Rp}}{kT} \left(\frac{\varepsilon_{krp} - \overline{E}_{Rp}}{kT} \gamma \gamma' \cos \chi - \frac{\varepsilon_{irp} - \overline{E}_{Rp}}{kT} \gamma^2 \right) \right\rangle_{pq} \end{array} \right]$$

$$\simeq -3\frac{n_p n_q}{n}\frac{kT}{m_p}$$

$$\times \left[\frac{kT}{m_p}\frac{C_{Rp}}{k}D_{pq}^{-1} + \frac{m_p}{m_p+m_q}\left(\frac{C_{Rp}}{k}\left(\tau_{pq}^{RT}\right)^{-1} + \left(\tau_{pq}^{RR}\right) - 1\right)\right]$$

$$\left(\alpha_{p01}^{p10}\right)_{pq} = \left(\alpha_{p10}^{p01}\right)_{pq} = -20 n_p n_q \frac{kT}{m_q}\left(\frac{m_q}{m_p+m_q}\right)^2 \langle \Delta\varepsilon_{rpq}\Delta\varepsilon_{rp}\rangle_{pq}$$

$$\simeq -5\frac{n_p n_q}{n}\frac{kT}{m_q}\left(\frac{m_q}{m_p+m_q}\right)^2 \frac{C_{Rp}}{k}\left(\tau_{pq}^{RT}\right)^{-1}$$

$$\left(\alpha_{q10}^{p10}\right)_{pq} = 16 n_p n_q kT \frac{m_r}{(m_p+m_q)^2}$$

$$\times \begin{bmatrix} \frac{55}{4}\langle \gamma^2 - \boldsymbol{\gamma}\cdot\boldsymbol{\gamma}'\rangle_{pq} - 5\langle \gamma^2\left(\gamma^2 - \boldsymbol{\gamma}\cdot\boldsymbol{\gamma}'\right)\rangle \\ +\langle \gamma^3\left(\gamma^3 - \gamma'^3 \cos\chi\right)\rangle_{pq} \\ -2\langle \gamma^2\left(\gamma^2 - \gamma'^2 \cos^2\chi\right)\rangle - \frac{1}{6}\langle (\Delta\varepsilon_{rpq})^2\rangle_{pq} \end{bmatrix}$$

$$\simeq 3\frac{n_p n_q}{n}\left(\frac{kT}{m_p+m_q}\right)^2 \left(\frac{55}{4} - 3B_{pq}^*\right)D_{pq}^{-1} - 20 n_p n_q (kT)^2$$

$$\times \frac{m_r}{(m_p+m_q)^2}\left(\mu'_{pq}\right)^{-1} \tag{4.67}$$

$$\left(\alpha_{q01}^{p01}\right)_{pq} = 24 n_p n_q \frac{kT}{m_p+m_q}\langle \Delta\varepsilon_{rp}\Delta\varepsilon_{rq}\rangle_{pq}$$

$$\simeq 3\frac{n_p n_q}{n}\frac{kT}{m_p+m_q}\left(\tau_{pq}^{RR}\right)^{-1}$$

$$\left(\alpha_{q01}^{p10}\right)_{pq} = -20 n_p n_q \frac{kT}{m_q}\left(\frac{m_q}{m_p+m_q}\right)^2 \langle \Delta\varepsilon_{rq}\Delta\varepsilon_{rpq}\rangle_{pq}$$

$$\simeq -5\frac{n_p n_q}{n}\frac{kT}{m_q}\left(\frac{m_q}{m_p+m_q}\right)^2 \frac{C_{Rq}}{k}\left(\tau_{qp}^{RT}\right)^{-1}$$

$$\left(\alpha_{q10}^{p01}\right)_{pq} = -20 n_p n_q \frac{kT}{m_p}\left(\frac{m_p}{m_p+m_q}\right)^2 \langle \Delta\varepsilon_{rp}\Delta\varepsilon_{rpq}\rangle_{pq}$$

$$\simeq -5\frac{n_p n_q}{n}\frac{kT}{m_p}\left(\frac{m_p}{m_p+m_q}\right)^2 \frac{C_{Rp}}{k}\left(\tau_{pq}^{RT}\right)^{-1}$$

with

$$\mu'_{pq} = \frac{5}{8}kT\left[\langle \gamma^2\left(\gamma^2 - \gamma'^2 \cos^2\chi\right)\rangle - \frac{1}{6}\langle (\Delta\varepsilon_{rpq})^2\rangle_{pq}\right]^{-1} \tag{4.68}$$

120 CHAPTER 4 QUASI-EQUILIBRIUM REGIMES: GAS MIXTURES

By analogy with pure gases, the term μ'_{pq} may be considered a 'fictitious viscosity'; we also find that $\mu'_{pq} = \mu_p$ when $p = q$.

The non-dimensional quantity B^*_{pq} is such that

$$B^*_{pq} = \frac{1/3 \left[5 \langle \gamma^2 (\gamma^2 - \boldsymbol{\gamma} \cdot \boldsymbol{\gamma}') \rangle - \langle \gamma^3 (\gamma^3 - \gamma'^3 \cos \chi) \rangle_{pq} \right]}{\langle \gamma^2 - \boldsymbol{\gamma} \cdot \boldsymbol{\gamma}' \rangle_{pq}} \quad (4.69)$$

We also have 12 symmetrical integrals, obtained by exchanging the indices p and q in the preceding integrals.

The β integrals

These integrals appear in the (2×3) system giving b_{p00} and b_{q00} in a symmetrical way. Therefore, they also appear in the coefficient of dynamic viscosity. Thus

$$\left(\beta^{p00}_{p00} \right)_{pp} = -16 \left(\frac{n_p kT}{m_p} \right)^2 \left\langle \gamma^4 \sin^2 \chi - \gamma^2 \sin^2 \chi \left(\Delta \varepsilon_{rpp} \right) + \frac{1}{3} \left(\Delta \varepsilon_{rpp} \right)^2 \right\rangle_{pp}$$

$$\simeq -10 \left(\frac{n_p kT}{m_p} \right)^2 \frac{kT}{\mu_p}$$

$$\left(\beta^{p00}_{p00} \right)_{pq} = -32 n_p n_q \left(\frac{kT}{m_p + m_q} \right)^2 \left(\frac{m_q}{m_p} \right)^2$$

$$\times \left\langle \gamma^2 \left(\gamma^2 - \gamma'^2 \cos^2 \chi \right) - \frac{1}{6} \left(\Delta \varepsilon_{rpq} \right)^2 \right\rangle_{pq}$$

$$- \frac{320}{3} n_p n_q \left(\frac{kT}{m_p + m_q} \right)^2 \frac{m_q}{m_p} \langle \gamma^2 - \boldsymbol{\gamma} \cdot \boldsymbol{\gamma}' \rangle_{pq}$$

$$\simeq -20 n_p n_q \left(\frac{kT}{m_p + m_q} \right)^2 \left(\frac{m_q}{m_p} \right)^2$$

$$\times \left(\frac{kT}{n} \frac{m_p + m_q}{m_q^2} D^{-1}_{pq} + kT \left(\mu'_{pq} \right)^{-1} \right)$$

$$\left(\beta^{p00}_{q00} \right)_{pq} = -32 n_p n_q \left(\frac{kT}{m_p + m_q} \right)^2 \left\langle \gamma^2 \left(\gamma^2 - \gamma'^2 \cos^2 \chi \right) - \frac{1}{6} (\Delta \varepsilon_{rpq})^2 \right\rangle_{pq}$$

$$+ \frac{320}{3} n_p n_q \left(\frac{kT}{m_p + m_q} \right)^2 \langle \gamma^2 - \boldsymbol{\gamma} \cdot \boldsymbol{\gamma}' \rangle_{pq}$$

$$\simeq 20 n_p n_q \left(\frac{kT}{m_p + m_q} \right)^2 \left(\frac{kT}{nm_r} D^{-1}_{pq} - kT \left(\mu'_{pq} \right)^{-1} \right) \quad (4.70)$$

The other three integrals are obtained by exchanging the indices p and q.

The δ integrals

The δ integrals appear in the system giving the coefficients d, and therefore in the bulk viscosity and in the rotational non-equilibrium. There are 2×12 integrals in the system. We have

$$\left(\delta^{p10}_{p10}\right)_{pp} = \left(\delta^{p01}_{p01}\right)_{pp} = \left(\delta^{p10}_{p01}\right)_{pp} = \left(\delta^{p01}_{p10}\right)_{pp} = -2n^2 \langle \Delta\varepsilon^2_{rpp}\rangle_{pp}$$

$$\simeq -\frac{n_p^2}{n}\frac{C_{Rp}}{k}\left(\tau^{RT}_{pp}\right)^{-1}$$

$$\left(\delta^{p10}_{p10}\right)_{pq} = -4n_p n_q \left(\frac{m_q}{m_p + m_q}\right)^2 \langle \Delta\varepsilon^2_{rpq}\rangle_{pq}$$

$$\quad - 16 n_p n_q \frac{m_r}{(m_p + m_q)} \langle \gamma^2 - \boldsymbol{\gamma}\cdot\boldsymbol{\gamma}'\rangle_{pq}$$

$$\simeq -\frac{n_p n_q}{nk}\left(\frac{m_q}{m_p + m_q}\right)^2 \left[C_{Rp}\left(\tau^{RT}_{pq}\right)^{-1} + C_{Rq}\left(\tau^{RT}_{qp}\right)^{-1}\right]$$

$$\quad - 3\frac{n_p n_q}{n}\frac{kT}{m_p + m_q}D^{-1}_{pq}$$

$$\left(\delta^{p01}_{p01}\right)_{pq} = -4n_p n_q \langle \Delta\varepsilon^2_{rpq}\rangle_{pq} \simeq -\frac{n_p n_q}{n}\left[\frac{C_{Rp}}{k}\left(\tau^{RT}_{pq}\right)^{-1} + \left(\tau^{RR}_{pq}\right)^{-1}\right]$$

$$\left(\delta^{p01}_{p10}\right)_{pq} = \left(\delta^{p10}_{p01}\right)_{pq} = -4n_p n_q \frac{m_q}{m_p + m_q}\langle \Delta\varepsilon_{rp}\Delta\varepsilon_{rpq}\rangle$$

$$\simeq -\frac{n_p n_q}{n}\frac{m_q}{m_p + m_q}\frac{C_{Rp}}{k}\left(\tau^{RT}_{pq}\right)^{-1}$$

$$\left(\delta^{p10}_{q10}\right)_{pq} = 16 n_p n_q \frac{m_r}{(m_p + m_q)}\left\langle \gamma^2 - \boldsymbol{\gamma}\cdot\boldsymbol{\gamma}' - \frac{1}{4}(\Delta\varepsilon_{rpq})^2\right\rangle$$

$$\simeq -\frac{n_p n_q}{n(m_p + m_q)}\left[m_r\frac{C_{Rp}}{k}\left(\tau^{RT}_{pq}\right)^{-1} + m_r\frac{C_{Rq}}{k}\left(\tau^{RT}_{qp}\right)^{-1} - 3kTD^{-1}_{pq}\right]$$

$$\left(\delta^{p01}_{q01}\right)_{pq} = -4n_p n_q \langle \Delta\varepsilon_{rp}\Delta\varepsilon_{rq}\rangle \simeq \frac{n_p n_q}{n}\left(\tau^{RR}_{pq}\right)^{-1}$$

$$\left(\delta^{p10}_{q01}\right)_{pq} = -4n_p n_q \frac{m_q}{m_p + m_q}\langle \Delta\varepsilon_{rq}\Delta\varepsilon_{rpq}\rangle_{pq} \simeq -\frac{n_p n_q}{n}\frac{m_q}{m_p + m_q}\frac{C_{Rq}}{k}\left(\tau^{RT}_{qp}\right)^{-1}$$

$$\left(\delta^{p01}_{q10}\right)_{pq} = -4n_p n_q \frac{m_p}{m_p + m_q}\langle \Delta\varepsilon_{rp}\Delta\varepsilon_{rpq}\rangle_{pq} \simeq -\frac{n_p n_q}{n}\frac{m_p}{m_p + m_q}\frac{C_{Rp}}{k}\left(\tau^{RT}_{pq}\right)^{-1}$$

(4.71)

The other 12 integrals are obtained by exchanging the indices p and q.

The λ integrals

See Appendix 4.1.

Appendix 4.3 Simplified transport coefficients

The expressions for these coefficients are given for binary mixtures of diatomic gases with one excited internal mode (rotation). It is easy (but tedious) to deduce the equivalent expressions for monatomic gases and for diatomic gases with two modes.

Dynamic viscosity

From the definition, we have

$$\mu = -\rho_p \left(\frac{kT}{m_p}\right)^2 b_{p00} - \rho_q \left(\frac{kT}{m_q}\right)^2 b_{q00}$$

Thus, we find[28] that

$$\mu = \frac{\left\langle \begin{array}{c} \xi_p \xi_q (\mu_p^{-1} + \mu_q^{-1}) + 2\left[n(m_p + m_q)\right]^{-1} D_{pq}^{-1} \\ +2(\xi_p m_p - \xi_q m_q)^2 (m_p + m_q)^2 \left(\mu'_{pq}\right)^{-1} \end{array} \right\rangle}{\left\langle \begin{array}{c} \xi_p \xi_q \mu_p^{-1} \mu_q^{-1} + 2\left[n(m_p + m_q)\right]^{-1} D_{pq}^{-1} \left(\xi_p^2 \mu_p^{-1} + \xi_q^2 \mu_q^{-1}\right) \\ + 2(m_p + m_q)^{-2} (\mu'_{pq})^{-1} \left(\xi_p^2 m_p^2 \mu_p^{-1} + \xi_q^2 m_q^2 \mu_q^{-1}\right) \\ +4\xi_p \xi_q \left[n(m_p + m_q)\right]^{-1} D_{pq}^{-1} \left(\mu'_{pq}\right)^{-1} \end{array} \right\rangle} \quad (4.72)$$

Thus, the dynamic viscosity of a binary mixture μ $(=\mu_{pq})$ may be expressed as a function of:

- the dynamic viscosity of each component μ_p and μ_q
- the binary diffusion coefficient D_{pq}
- a fictitious viscosity μ'_{pq}, related to a momentum transfer $p \rightleftarrows q$.

Bulk viscosity

From the definition, we have $\eta = (n_p d_{p10} + n_q d_{q10}) kT$. Thus, we find[28] that

$$\eta = \frac{C_R}{C_{TR}} p$$

$$\times \frac{\frac{C_{Rp} C_{Rq}}{k} \left(\frac{\xi_p}{\tau_{Rp}} + \frac{\xi_q}{\tau_{Rq}}\right) + (\xi_p C_{Rp} + \xi_q C_{Rq}) \tau_{Rpq}^{-1}}{(C_T + \xi_p C_{Rp} + \xi_q C_{Rq}) \left(\frac{C_{Rp} C_{Rq}}{k^2} \frac{1}{\tau_{Rp} \tau_{Rq}} + \xi_p \frac{C_{Rp}}{k} \frac{1}{\tau_{Rp} \tau_{Rpq}} + \xi_q \frac{C_{Rq}}{k} \frac{1}{\tau_{Rq} \tau_{Rpq}}\right)} \quad (4.73)$$

The bulk viscosity of a binary mixture $\eta (= \eta_{pq})$ depends therefore on relaxation times previously defined:

- The TR relaxation time of species p, τ_{Rp}.
- The TR relaxation time of species q, τ_{Rq}.
- The RR relaxation time $p \rightleftarrows q$, τ_{Rpq} ($= \tau_{Rqp}$).

Thermal conductivities

Translational thermal conductivity

From the definition, we have

$$\lambda_T = \frac{5}{2}kT\left(\frac{n_p k}{m_p}a_{p10} + \frac{n_q k}{m_q}a_{q10}\right)$$

We find[20] that

$$\lambda_T = \frac{75}{4}k$$

$$\times \frac{\left\langle \begin{array}{c} 5\xi_p\xi_q\left[(m_p\mu_p)^{-1} + (m_q\mu_q)^{-1}\right] + 20\frac{m_r}{m_p+m_q}\frac{(\xi_p-\xi_q)^2}{\mu'_{pq}} \\ +3\left[n(m_p+m_q)^2 D_{pq}\right]^{-1}\left[2\xi_p\xi_q\left(\frac{55}{4} - 3B^*_{pq}\right) + \frac{m_p}{m_q}\xi_p^2 + \frac{m_q}{m_p}\xi_q^2\right] \end{array} \right\rangle}{\frac{25\xi_p\xi_q}{\mu_p\mu_q} + \frac{9m_r}{n^2(m_p+m_q)^3}\left[Q^*_{pq}Q^*_{qp} - \left(\frac{55}{4} - 3B^*_{pq}\right)^2\right]\frac{\xi_p\xi_q}{D^2_{pq}}}$$ (4.74)

$$\left\langle \begin{array}{c} +\frac{100m_r}{(m_p+m_q)^2\mu'_{pq}}\left(\frac{m_p\xi_p^2}{\mu_p} + \frac{m_q\xi_q^2}{\mu_q}\right) \\ +\frac{15}{n}\left\{ \begin{array}{c} \frac{\xi_p^2}{m_q}\left(\frac{m_p}{m_p+m_q}\right)^2\frac{Q^*_{qp}}{\mu_p} + \frac{\xi_q^2}{m_q}\left(\frac{m_q}{m_p+m_q}\right)^2\frac{Q^*_{pq}}{\mu_q} \\ +\frac{4m_r}{(m_p+m_q)^4}\left[m_p^2 Q^*_{qp} + m_q^2 Q^*_{pq} - 2m_p m_q\left(\frac{55}{4} - 3B^*_{pq}\right)\right]\frac{\xi_p\xi_q}{\mu'_{pq}} \end{array} \right\} \end{array} \right\rangle$$

with

$$Q^*_{pq} = \frac{5}{4}\left[6\left(\frac{m_p}{m_q}\right)^2 + 5\right] - 3B^*_{pq}, \quad Q^*_{qp} = \frac{5}{4}\left[6\left(\frac{m_q}{m_p}\right)^2 + 5\right] - 3B^*_{qp}, \text{ and } B^*_{pq} = B^*_{qp}$$

From this formula, we again find the Eucken approximation (Eqn. (3.73)) for pure gases ($\xi_p = 1, \xi_q = 0$), that is, $\lambda_T = \frac{15}{4}\frac{k}{m}\mu$.

Rotational thermal conductivity

From the definition, we have

$$\lambda_R = -\frac{n_p kT}{m_p}C_{Rp}a_{p01} - \frac{n_q kT}{m_q}C_{Rq}a_{q01}$$

We find[28]

$$\lambda_R = \frac{n_p C_{Rp}}{\xi_p D_{pp}^{-1} + \xi_q D_{pq}^{-1}} + \frac{n_q C_{Rq}}{\xi_q D_{qq}^{-1} + \xi_p D_{pq}^{-1}} \quad (4.75)$$

This expression corresponds to the Eucken approximation for pure gases. A similar expression is of course found when the vibrational mode is excited.

Appendix 4.4 Alternative technique: Gross–Jackson method

Among the approximate solutions of the Boltzmann equation relevant to gaseous media close to equilibrium, the Gross–Jackson method is a perturbation method which may be applied to non-collisional regimes and thus is independent of the definition of a 'small parameter'.[29] Thus, starting from a zero-order solution, a first-order solution is derived by successive approximations with a linearized collisional term. At each level of approximation N, the collisional operator $L(\varphi)$ is such that

$$L(\varphi) = K^{(N)}(\varphi) - \alpha^{(N)} \varphi \quad (4.76)$$

where, as in the case for the Chapman–Enskog method, φ is the perturbation of the zero-order distribution function f^0. Here, $K^{(N)}$ is an operator with a limited discrete spectrum, and $\alpha^{(N)}$ is a constant intended to represent terms of order higher than N.

The case of a pure gas with only one internal mode (rotation) is treated here.[30] The case of gases with two internal modes or the case of mixtures do not present any further major difficulty.

Starting from the Wang-Chang–Uhlenbeck equation (1.50) with a symmetrical collisional operator, and considering a zero-order solution f_i^0 in equilibrium (Eqn. (2.10)), we try to find the perturbation φ_i such that $f_i = f_i^0 (1 + \varphi_i)$. For the same reasons as those discussed for the Chapman–Enskog method, φ_i is expanded in the basis of the following functions, without defining a 'small parameter':

$$\varphi_i = \sum_{lmn} A^{lmn} \Psi_i^{lmn} Y^l \quad (4.77)$$

with $A^{lmn} = A^{lmn}(\mathbf{r}, t)$ and Y^l (Waldmann irreducible tensors[6]) such that

$$Y^0 = 1, \quad Y^1 = \mathbf{w}, \quad Y^2 = \mathbf{ww} - \frac{1}{3} w^2 \mathbf{I} = \overset{0}{\mathbf{ww}}$$

and
$$\psi_i^{lmn} = \frac{\pi^{1/2}}{2b^n} \left[\frac{m!}{(1+m+\tfrac{1}{2})!} \right]^{1/2} S_{l+1/2}^m (w^2) P^n (E_i)$$

where
$$w = \left(\frac{m}{2kT}\right)^{1/2} u, \quad E_i = \frac{\varepsilon_i}{kT}, \quad b^0 = 1, \quad b^1 = \frac{C_R}{k}$$

Here, $S_{l+1/2}^m$ and P^n are respectively the Sonine and Wang-Chang–Uhlenbeck polynomials.

The functions Ψ_i^{lmn} satisfy the orthogonality condition:

$$\sum_i \int_w \exp\left(-w^2 - E_i\right) \Psi_i^{lmn} Y^l \Psi_i^{lm'n'} Y^l dw = \frac{1}{4}\pi^{1/2} Q_R \delta_{mm'} \delta_{nn'} \int_{4\pi} \left(\frac{Y^l Y^l}{w^{2l}}\right) d\Omega \quad (4.78)$$

The first few basis functions are

$$\Psi_i^{000} = 1, \quad \Psi_i^{100} = \left(\frac{2}{3}\right)^{1/2}, \quad \Psi_i^{010} = \left(\frac{2}{3}\right)^{1/2} \left(\frac{3}{2} - w^2\right)$$

$$\Psi_i^{110} = \left(\frac{4}{15}\right)^{1/2} \left(\frac{5}{2} - w^2\right), \quad \Psi_i^{001} = \frac{k}{C_R}(E_i - \bar{E}_R)$$

$$\Psi_i^{101} = \left(\frac{2k}{3C_R}\right)^{1/2} (E_i - \bar{E}_R), \quad \Psi_i^{200} = \left(\frac{4}{15}\right)^{1/2}$$

The coefficients A^{lmn} may be written as functions of the perturbations of the macroscopic moments δM. Thus, using the notation of Chapter 3, we have

$$A^{000} = 0$$

$$A^{010} = -\left(\frac{3}{2}\right)^{1/2} \frac{T_T - T}{T} = -\left(\frac{3}{2}\right)^{1/2} \delta T_T$$

$$A^{001} = \left(\frac{k}{C_R}\right)^{1/2} \frac{E_R - \bar{E}_R}{kT} = \left(\frac{k}{C_R}\right)^{1/2} \delta E_R \; (= \delta T_R)$$

Other relations may be obtained from the definition of other quantities, such as the fluxes. Thus, $A^{200} = \left(\frac{15}{4}\right)^{1/2} \delta P$. In the same way, the definition of the population of the levels i gives the following relation:

$$\sum_i A^{001} \Psi_i^{001} Y^0 = \frac{n_i - \bar{n}_i}{\bar{n}_i} = \delta n_i$$

We write the linearized collisional operator $J^1 [\varphi_i]$ in the form

$$J^1 [\varphi_i] = f_i^0 \frac{n}{\pi^{3/2} Q} L [\varphi_i]$$

with

$$L[\varphi_i] = \sum_{j,k,l,\Omega,\boldsymbol{w}_j} \int \exp\left(-w_j^2 - E_j\right)\left(\varphi'_k + \varphi'_l - \varphi_i - \varphi_j\right) g I_{i,j}^{k,l} d\Omega\, d\boldsymbol{w}_j \quad (4.79)$$

Using the expansion of φ_i (Eqn. (4.77)), we expand $L\left[\Psi_i^{lmn} Y^l\right]$ in the basis of the functions $\Psi_i^{rst} Y^r$. Thus, we obtain

$$L[\varphi_i] = \sum_{\substack{rmn \\ rst}} A^{rmn} K_{rst}^{rmn} \Psi_i^{rst} Y^r \quad (4.80)$$

where $r = l$, because of the assumed symmetry of collisions, and with

$$K_{rst}^{rmn} = \sum_{i,j,k,l,\Omega,\boldsymbol{w}_i,\boldsymbol{w}_j} \int \exp\left(-w_i^2 - w_j^2 - E_i - E_j\right) \Psi_i^{rst} Y_k^r$$

$$\times \left(\Psi_k^{\prime rmn} Y_k^{\prime r} + \Psi_l^{\prime rmn} Y_l^{\prime r} - \Psi_i^{rmn} Y_i^r - \Psi_j^{rmn} Y_j^r\right)$$

$$\times \left[\frac{\pi^{1/2} Q}{4} \int_{4\pi} \frac{Y^r Y^r}{w_i^{2r}} d\Omega\right]^{-1} g I_{i,j}^{k,l} d\Omega\, d\boldsymbol{w}_i d\boldsymbol{w}_j$$

We also have $K_{rst}^{rmn} = K_{rmn}^{rst}$. These integrals are algebraically related to the integrals α, β, $\delta\ldots$, or $\langle\ldots\rangle$ appearing in the Chapman–Enskog method.

Equation (4.80) is rearranged according to the sequence $r + 2m + 2n$, $r + 2s + 2t$, and for a chosen value of N, we replace those terms that have an order $r + 2m + 2n$ and $r + 2s + 2t$ higher than N with $\alpha_N \delta_{ms} \delta_{nt}$, where α_N is an arbitrary constant. Thus, we have

$$L^{(N)}[\varphi_i] = \sum_{\substack{r+2m+2n\leq N \\ r+2s+2t\leq N}} A^{rmn} K_{rst}^{rmn} \Psi_i^{rst} Y^r + \sum_{r+2m+2n>N} \alpha_N A^{rmn} \Psi_i^{rmn} Y^r$$

(4.81)

By analogy with the 'Maxwellian' molecules which have a spectrum of eigenfunctions comprising the functions $\Psi^{rm} Y^r$, we may choose α_N so that $\alpha_N = K_{N00}^{N00}$. However, these models give a non-null balance for elastic collisions for populations of the level i, that is, $(\Delta n_i)_{el} = \dfrac{n_i}{\bar{n}_i} \sum_{2t>N} A^{00t} K_{N00el}^{N00} \Psi_i^{00t} Y^0$. Therefore, the diagonalization constant α_N must be written

$$\alpha_N = K_{N00}^{N00} - K_{N00el}^{N00} \delta_{r0} \delta_{m0}$$

Then, with the expansion (Eqn. (4.77)) for φ_i, the model equation for the operator L may be written

$$L^{(N)}[\varphi_i] = \sum_{\substack{r+2m+2n\leq N \\ r+2s+2t\leq N}} \{A^{rmn}[K_{rst}^{rmn} - (K_{N00el}^{N00}\delta_{r0}\delta_{m0})\delta_{ms}\delta_{nt}]\}\Psi_i^{rst}Y^r$$

$$+ K_{N00}^{N00}\varphi_i - K_{N00el}^{N00}\delta n_i \quad (4.82)$$

Now we are able to write models of any order N. Of course, their accuracy increases with N, but so also does their complexity.

The gas properties are obtained by replacing J^1 with the approximation $J^{(N)} = f_i^0 \frac{n}{\pi^{3/2}Q} L^{(N)}[\varphi_i]$ in the first-order equation $\frac{df_i^0}{dt} = J^1$, where $\frac{1}{f_i^0}\frac{df_i^0}{dt}$ is given in Chapter 3.

The first-order model ($N = 1$) includes only one collisional integral and does not satisfy the conservation of energy. The second-order model ($N = 2$) satisfies the three conservation laws; it also gives the correct viscosity coefficients, but it is necessary to use the model $N = 3$ in order to obtain a correct modelling of energy fluxes.

Here we give only the expression for the model with $N = 2$, which may be used in many applications, i.e.

$$L^{(2)}[\varphi_i] = -K_{200el}^{200}\left(\delta n_i - \frac{\varepsilon_{i_r} - \overline{E}_R}{kT}\delta T_R\right) - K_{200}^{200}$$

$$\times \left[\left(\frac{3}{2} - w^2\right)\delta T_T + \frac{\varepsilon_{i_r} - \overline{E}_R}{kT} - \varphi_i\right] + K_{010in}^{010}$$

$$\times \left[\left(\frac{3}{2} - w^2\right)(\delta T_R - \delta T_T) + \frac{3k}{2C_R}(\delta T_R - \delta T_T)\left(\frac{\varepsilon_{i_r} - \overline{E}_R}{kT}\right)\right]$$

(4.83)

This model ($N = 2$), gives in particular a correct description for the rotational non-equilibrium (the model $N = 3$ gives the same result). Thus, multiplying the equation $\frac{df_i^0}{dt} = J^{(2)}$ by $\varepsilon_{i_r} - \overline{E}_R$, integrating over the velocity space, and summing over the levels, we find that the rotational non-equilibrium is given by the formula

$$\frac{E_R - \overline{E}_R}{\overline{E}_R} = \frac{C_R}{C_{TR}^2}\frac{1}{K_{010in}^{010}}\frac{\partial \cdot V}{\partial r} \quad (4.84)$$

This formula presents a clear connection with Eqn. (3.51) deduced from the Chapman–Enskog method. We may thus verify that the integral K_{010in}^{010} is proportional to a collision frequency, as are the other K integrals, and that $K_{010in}^{010} \sim (\tau_R)^{-1}$.

128 CHAPTER 4 QUASI-EQUILIBRIUM REGIMES: GAS MIXTURES

To sum up, the Gross–Jackson method, initially developed for molecules with Mawellian interaction potential[31] has been progressively extended to polyatomic gases[30] and to gas mixtures.[32] For $N \geq 3$, the method gives a detailed description of the dynamics of gases. The method may also be applied when the zero-order solution is out of equilibrium,[14] and may be used for non-collisional regimes. The method is often easier to use than the Chapman–Enskog method, but it becomes increasingly complex when $N \geq 3$. Furthermore, since it is a linearized method, it may describe only those situations that are close to the zero-order solution of the distribution function.

Appendix 4.5 Alternative technique: method of moments

As already discussed, in the case of elastic collisions, the distribution function f may be written in the following form:

$$F(v, M^0, M^1, \ldots M^n \ldots) \tag{4.85}$$

where the moments M^n are functions of r and t.

Then, if Ψ is a function of v, we obtain an infinite system of equations for an infinite array of moments by replacing the expression (4.85) in the equation

$$\int \Psi \frac{df}{dt} dv = \int \Psi J dv \tag{4.86}$$

Thus, the Boltzmann equation is equivalent to an infinite system of macroscopic equations.

An approximate solution consists of representing the distribution function by a finite number of moments composed of macroscopic quantities $A(r, t)$. As these equations are not closed, an approximate form for F in Eqn. (4.85) must be found in order to close the system.

One possible form consists of expanding the distribution function in a series of orthogonal polynomials[8] such as the Hermite polynomials $H_{i,j,\ldots,n}^{(n)}$, that is:

$$f = f^0 \left(a^0 H^{(0)} + a_i^1 H_i^{(1)} + \frac{1}{2} a_{ij}^2 H_{ij}^{(2)} + \cdots + \frac{1}{n!} a_{ij\ldots n}^n H_{ij\ldots n}^{(n)} \right)$$

Setting $w = u\sqrt{m/kT}$ and $g(w) = \frac{1}{n}\left(\frac{kT}{m}\right)^{3/2} f$, these polynomials are orthogonal with respect to the weighting function $g^0(w) =$

$(2\pi)^{-3/2} \exp(-w^2/2)$, and the first four correspond to the following tensors:

$$H^{(0)} = 1, \quad H_i^{(1)} = w, \quad H_{ij}^{(2)} = \overset{0}{ww} = w_i w_j - \delta_{ij}, \quad \text{and}$$

$$H_{ijk}^{(3)} = v_i v_j v_k - (v_i \delta_{jk} + v_j \delta_{ik} + v_k \delta_{ij})$$

The contraction of $H_{ijk}^{(3)}$ is $H_i^{(3)} = w(w^2 - 5)$.

Using only $H^{(0)}$, $H_i^{(1)}$, $H_{ij}^{(2)}$, $H_i^{(3)}$ in the expansion, we can find the corresponding coefficients $a_{ij..n}^n$ from the orthogonality properties of Hermite polynomials. These coefficients may be expressed in terms of the first moments of the distribution function, that is:

$$a^0 = 1, \quad a^1 = 0, \quad a_{ij}^2 = \frac{P'}{p}, \quad \text{and} \quad a_i^3 = \frac{2q}{p}\sqrt{\frac{m}{kT}}$$

Finally, at this level of approximation, the expression of the distribution function is

$$f = f^0 \left[1 + \frac{1}{2p} \left(\frac{m}{kT}\right) \mathbf{P'} : \overset{0}{\mathbf{uu}} + \left(\frac{m}{kT}\right)\left(\frac{m}{kT}\frac{u^2}{5} - 1\right) \mathbf{q} \cdot \mathbf{u} \right] \quad (4.87)$$

From Eqns (4.86) and (4.87), we can thus obtain a set of macroscopic equations including the three usual conservation equations (1.26) and two other equations giving $\mathbf{P'}$ and $\mathbf{q}$. This system comprises the '13 moment equations'.[8] However, closure of the system is ensured only if we consider Maxwellian molecules (Appendix 1.3), because of the collisional terms that appear in the last two equations.[1]

In this case, these last two equations may be written in the following form:

$$\frac{\partial \mathbf{P'}}{\partial t} + \mathbf{A} + \frac{\mathbf{P'}}{\tau} = 0 \quad (4.88)$$

$$\frac{\partial \mathbf{q}}{\partial t} + \mathbf{B} + \frac{2}{3}\frac{\mathbf{q}}{\tau} = 0 \quad (4.89)$$

where $\mathbf{A}$ and $\mathbf{B}$ are terms that include space derivatives, and τ is a quantity related to the collisional terms that have the dimension of time (Maxwellian molecules): τ may be considered as the relaxation time of the corresponding processes.

We may note that the expression of the distribution function (Eqn. (4.87)) does not involve an expansion in a series of a 'small' parameter, as is the case for the Chapman–Enskog expansion. Thus, independently of the approximate character of this expression, it may be applied to any type of flow.

However, for small Knudsen numbers ($\tau \to 0$), the influence of the history of the flow decays rapidly, so that, by integrating Eqns (4.88) and (4.89), we have

CHAPTER 4 QUASI-EQUILIBRIUM REGIMES: GAS MIXTURES

the solutions

$$\mathbf{P}' = -\tau \mathbf{A} + \tau \frac{\partial \tau \mathbf{A}}{\partial t} + \cdots \tag{4.90}$$

$$\mathbf{q} = -\frac{3}{2}\tau \mathbf{B} + \frac{9}{4}\tau \frac{\partial \tau \mathbf{B}}{\partial t} + \cdots \tag{4.91}$$

Retaining only the first terms of the right-hand side of these equations and only the lowest-order moments in $\mathbf{A}$ and $\mathbf{B}$, we obtain

$$\mathbf{P}' = -2\tau p \,\overline{\frac{\partial \mathbf{V}}{\partial \mathbf{r}}}^{0}$$

$$\mathbf{q} = -\frac{15}{4}\frac{k}{m}\tau p \frac{\partial T}{\partial \mathbf{r}}$$

Comparing with the corresponding results obtained from the Chapman–Enskog expansion, we can write

$$\mu = \tau p \quad \text{and} \quad \lambda = \frac{15}{4}\frac{k}{m}\tau p$$

Thus, as already shown (Appendix 3.4), we find a linear dependence of μ and λ with T (Maxwellian molecules). We again find the Navier–Stokes equations by this method, but with an approximate value for the transport coefficients.

Retaining the second-order terms of the right-hand side of Eqns (4.90)–(4.91), we obtain higher-order conservation equations equivalent to those corresponding to the second-order terms of the Chapman–Enskog expansion (Burnett equations[4]).

With the present method, it is possible to retain more moments in the expansion: for example, in the '20 moment equations', all third-order terms are taken into account.[4] Other developments of the method are available.[33]

FIVE
Transport and Relaxation in Non-Equilibrium Regimes

5.1 Introduction

The first part of this chapter is devoted to cases for which the collisional characteristic times may be very different to the previous cases analysed in Chapters 3 and 4 (WNE cases). Thus, the regimes analysed here correspond to case 2 defined in Chapter 3 (Eqn. (3.6)), that is, to SNE regimes. The zero-order distribution function is then out of equilibrium, and examples of this were given in Chapter 2. Here, first-order solutions are developed with the Chapman–Enskog method. These solutions present important differences from the WNE case, and rather than completely developing the solutions, the differences are emphasized primarily for the transport terms.

First, we consider vibrational non-equilibrium regimes at zero order; this is a typical case frequently encountered in high-temperature flow. We analyse the consequences for transport in these regimes, and at the same level of approximation, we observe that the transport terms are generally simpler than in the WNE case because of the similarity between the zero-order solutions of the frozen case and non-equilibrium cases (Chapter 3). The chemical non-equilibrium case $(SNE)_C$ is then examined, distinguishing two cases for the vibrational mode, that is, $(WNE)_V$ or $(SNE)_V$. In both instances, the resulting vibration–chemistry interaction is pointed out, and consequences on the reaction-rate constants and vibrational non-equilibrium are analysed.

5.2 Vibrational non-equilibrium gases: SNE case

5.2.1 Pure diatomic gases

This is a general case: the translational and rotational modes are in equilibrium at zero order (for both modes one single temperature is defined), but the vibrational

132 CHAPTER 5 TRANSPORT AND RELAXATION IN NON-EQUILIBRIUM REGIMES

mode is out of equilibrium ($\tau_V \sim \theta$). Therefore, according to case 3 of Chapter 3, we have

$$J^0_{TR} = 0 \tag{5.1}$$

$$\frac{df^0_i}{dt} = J^1_{TR} + J^0_V \tag{5.2}$$

We may generally include the resonant VV collisions in the TR collisions, since their vibrational collisional balance is null ($\Delta\varepsilon_V = 0$). Thus, a vibrational temperature T_V can be defined, and a Boltzmann distribution at this temperature may be assumed at zero order (Eqn. (2.28)). In the same way, this temperature is given by one single relaxation equation (2.29), coupled with the Euler equations.

Taking into account the expression for $\frac{df^0_i}{dt} - J^0_V$, which is the non-homogeneous part of the linearized equation (5.2) (Appendix 5.1), we may write the perturbation φ_i in the following form:[4,14]

$$\varphi_i = A_i \frac{1}{T} \frac{\partial T}{\partial r} \cdot \boldsymbol{u} + B_i \frac{\partial \boldsymbol{V}}{\partial r} : \overset{0}{\boldsymbol{uu}} + D_i \frac{\overset{0}{\partial \cdot \boldsymbol{V}}}{\partial r} + F_i \frac{1}{T_V} \frac{\partial T_V}{\partial r} \cdot \boldsymbol{u} + G_i \tag{5.3}$$

The difference with the WNE case with two internal modes (Appendices 3.2 and 3.3) lies in the splitting of the corresponding term A_i into two terms. One term, A_i, includes the contribution of the translational and rotational modes, and another term, F_i, includes the contribution of the vibrational mode out of equilibrium at zero order. As A_i and F_i are connected to the heat fluxes, we may anticipate a structural modification of these fluxes with respect to the WNE case. A new term, G_i, also appears because of the vibrational non-equilibrium. The equations to be solved are then the following:

$$J^1_{TR}[A_i \boldsymbol{u}] = f^0_i \left(\frac{mu^2}{2kT} - \frac{5}{2} + \frac{\varepsilon_{i_r} - \overline{E}_R}{kT} \right)$$

$$J^1_{TR}\left[B_i \overset{0}{\boldsymbol{uu}}\right] = f^0_i \frac{m}{kT} \overset{0}{\boldsymbol{uu}}$$

$$J^1_{TR}[D_i] = f^0_i \left[\frac{2}{3} \frac{C_R}{C_{TR}} \left(\frac{mu^2}{2kT} - \frac{3}{2} \right) - \frac{k}{C_{TR}} \frac{\varepsilon_{i_r} - \overline{E}_R}{kT} \right]$$

$$J^1_{TR}[F_i \boldsymbol{u}] = f^0_i \frac{\varepsilon_{i_v} - \overline{E}_V}{kT_V} \boldsymbol{u}$$

5.2 VIBRATIONAL NON-EQUILIBRIUM GASES: SNE CASE

$$J^1_{TR}[G_i] = \frac{f_i^0}{n}\left[\frac{1}{C_V T_V}\frac{\varepsilon_{i_v} - \overline{E}_V}{kT_V} - \frac{1}{C_{TR}T}\left(\frac{mu^2}{2kT} - \frac{3}{2} + \frac{\varepsilon_{i_r} - \overline{E}_R}{kT}\right)\right]$$

$$\times \left(\sum_i \varepsilon_{i_v}\int_v J_V^0\, d\boldsymbol{v}\right) - J_V^0 \tag{5.4}$$

The method of solution is similar to the one described in Chapters 3 and 4. A_i, B_i, D_i, F_i, and G_i are expanded in the Sonine–Laguerre and Wang-Chang–Uhlenbeck polynomials (3.61). Keeping only the first terms of these expansions, and taking into account the constraints and various simplifications previously discussed, we obtain the systems of equations for various coefficients as given in Appendix 5.1 with the corresponding collisional integrals.

First, we observe that the terms in $D_i(d_{100}$ and $d_{010})$ and in $B_i(b_{000})$ are identical to those of the WNE case with one internal mode (rotation), without any contribution of the vibrational mode, including the corresponding collisional integrals δ^{100}_{100} and β^{000}_{000}, as if this mode was frozen. Therefore, b_{000} and d_{100} (with $d_{010} = \frac{C_T}{C_R}d_{100}$) are given by Eqn. (3.40). In contrast, the terms for A_i and F_i depend on the vibration, but in the corresponding collisional integrals (arising from J^1_{TR}), the cross sections do not depend on the vibration (Appendix 5.1). Only the term G_i depends on this mode and on its relaxation. Thus, we find

$$g_{100} = \frac{C_R}{\frac{3}{2}k}g_{010} = -\frac{\frac{C_R}{C_{TR}kT}\sum_i \varepsilon_{i_v}\int_v J_V^0\,d\boldsymbol{v} - \frac{1}{4}\sum_i \Delta\varepsilon_r\int_v J_V^0\,d\boldsymbol{v}}{\frac{C_{TR}}{C_R}\delta^{100}_{100}} \tag{5.5}$$

The order of magnitude of g_{100} is τ_R/τ_V, as expected, since $\delta^{100}_{100} \sim \tau_R^{-1}$.

Transport terms

From the previous computations, we find the following expression for **P**:

$$\mathbf{P} = p\mathbf{I} - 2\rho\left(\frac{kT}{m}\right)^2 b_{000}\overline{\frac{\partial \mathbf{V}}{\partial \boldsymbol{r}}}^0 - p\left(d_{100}\frac{\partial \cdot \mathbf{V}}{\partial \boldsymbol{r}} + g_{100}\right)\mathbf{I} \tag{5.6}$$

Thus, as for the WNE case, there is a term with the dynamic viscosity μ where

$$\mu = nkT\left(\frac{kT}{m}\right)b_{000} \tag{5.7}$$

and a bulk viscosity term

$$\eta = nkT d_{100} \tag{5.8}$$

which are identical to the WNE case with one mode, without any contribution of the vibrational mode.

As usual, in the diagonal terms of **P**, there is the static pressure $p = nkT$, but a new term appears; this term, $p_r = -nkTg_{100}$, called the 'relaxation pressure',[4,14] depends on the vibrational relaxation and has an order of magnitude τ_R/τ_V. An approximate expression[14] for this term is proposed below (Eqn. (5.27)).

The heat flux term q may be written $\boldsymbol{q} = \boldsymbol{q}_T + \boldsymbol{q}_R + \boldsymbol{q}_V$, with

$$\boldsymbol{q}_T = -\frac{5}{2}Rp\left(a_{100}\frac{\partial T}{\partial r} + \frac{T}{T_V}f_{100}\frac{\partial T_V}{\partial r}\right)$$

$$\boldsymbol{q}_R = \frac{C_R}{m}p\left(a_{010}\frac{\partial T}{\partial r} + \frac{T}{T_V}f_{010}\frac{\partial T_V}{\partial r}\right)$$

$$\boldsymbol{q}_V = \frac{C_V}{m}p\left(\frac{T_V}{T}a_{001}\frac{\partial T}{\partial r} + f_{001}\frac{\partial T_V}{\partial r}\right) \quad (5.9)$$

where $C_V = C_V(T_V)$.

Thus, each peculiar heat flux depends on both temperature gradients T and T_V. We may write

$$\boldsymbol{q}_T = -\lambda_T\frac{\partial T}{\partial r} - \lambda_{TV}\frac{\partial T_V}{\partial r}$$

$$\boldsymbol{q}_R = -\lambda_R\frac{\partial T}{\partial r} - \lambda_{RV}\frac{\partial T_V}{\partial r}$$

$$\boldsymbol{q}_V = -\lambda_{VTR}\frac{\partial T}{\partial r} - \lambda_V\frac{\partial T_V}{\partial r} \quad (5.10)$$

The conductivity coefficients λ may be expressed as functions of collisional integrals including only TR collisions (Appendix 5.1).

After insertion of the expressions of **P** and q in the conservation equations (1.26), we obtain a Navier–Stokes system closed with a relaxation equation for e_V, that is:

$$\rho\frac{de_V}{dt} + \frac{\partial \cdot \boldsymbol{q}_V}{\partial r} = \sum_i \varepsilon_{i_v}\int_v \left(J_V^0 + J_V^1\right)dv \quad (5.11)$$

In comparison with the zero-order equation, Eqn. (5.11) contains two supplementary terms: a term of vibrational flux $\frac{\partial \cdot \boldsymbol{q}_V}{\partial r}$ and a 'production' term $\sum_i \varepsilon_{i_v}\int_v J_V^1\, dv$. In principle, these terms are known but depend on the physical model chosen for the vibrational transitions. An example is proposed in Appendix 5.2.

As for the rotational non-equilibrium that occurs at first order, it is of course identical to the WNE case with one mode (rotation) (Chapter 3).

5.2.2 Mixtures of diatomic gases

We consider here only binary mixtures (p and q). This is the case described in Chapter 2 at zero order, and in Chapter 4 as a WNE solution. Thus, we find aspects of the 'strong' vibrational non-equilibrium analogous to pure gases, but with several types of relaxation characterized by two vibrational temperatures, and the presence of diffusion phenomena. Complex expressions are then expected, particularly for the transport terms.

The system of equations to be solved is the following:

$$J^0_{TRp} = 0$$
$$\frac{df^0_{ip}}{dt} = J^1_{TRp} + J^0_{Vp} \tag{5.12}$$

with

$$J_{TRp} = J_{TRpp} + J_{TRpq}$$
$$J_{Vp} = J_{Vpp} + J_{Vpq} \tag{5.13}$$

As in the case for pure gases, the resonant VV collisions (between molecules of each species) are included in the TR collisions in order to define vibrational temperatures T_{Vp} and T_{Vq}. As discussed above, the zero-order solution is given by Eqn. (2.38) and the relaxation equations by Eqn. (2.69) for the general case, and by Eqn. (2.70) for the harmonic oscillator model.

We use the expression of $\frac{df^0_{ip}}{dt} - J^0_{Vp}$ and write the perturbation φ_{ip} in the form:[28]

$$\varphi_{ip} = A_{ip} \frac{1}{T} \frac{\partial T}{\partial r} \cdot \boldsymbol{u}_p + B_{ip} \frac{\partial \boldsymbol{V}}{\partial r} : \overset{0}{\boldsymbol{u}_p \boldsymbol{u}_p} + D_{ip} \frac{\partial \cdot \boldsymbol{V}}{\partial r}$$
$$+ F_{ip} \frac{1}{T_{Vp}} \frac{\partial T_{Vp}}{\partial r} \cdot \boldsymbol{u}_p + G_{ip} + L_{ip} \boldsymbol{u}_p \cdot \boldsymbol{t}_p \tag{5.14}$$

Each of these six terms has a particular physical meaning, and in view of the gradient associated to each term, it is clear that the coefficients A_{ip}, B_{ip}, D_{ip}, F_{ip}, G_{ip}, and L_{ip} are respectively connected to the following transport terms:

- Translational and rotational thermal conductivities
- Dynamic viscosity
- Bulk viscosity
- Vibrational thermal conductivity
- Relaxation pressure
- Diffusion coefficient.

The corresponding complex computations are left to the reader following the procedure of the previous calculations. However, the general structure of the transport terms included in the Navier–Stokes equations is given below as functions of the non-null first coefficients of the usual expansions. These coefficients themselves are functions of collisional integrals obtained from the Cramer systems given in Appendix 5.3. As in the case for pure gases, most of these coefficients are identical to those of the WNE case developed for one internal mode (TR collisions), with the exception of the coefficients F_{ip} and G_{ip}, which are directly connected to the flux and to the production of vibrational energy. Thus, we have

$$U_p = \frac{kT}{m_p} l_{p000} \frac{\partial \xi_p}{\partial r} \tag{5.15}$$

$$P = p\mathbf{I} - 2 \sum_p \rho_p \left(\frac{kT}{m_p}\right)^2 b_{p000} \overline{\frac{\partial V}{\partial r}}^0 - \sum_p n_p kT \left(g_{p100} + d_{p100} \frac{\partial \cdot V}{\partial r}\right) \mathbf{I} \tag{5.16}$$

with

$$\mu = \sum_p \rho_p \left(\frac{kT}{m_p}\right)^2 b_{p000} \quad \text{and} \quad \eta = kT \sum_p n_p d_{p100} \tag{5.17}$$

$$p_r = -\sum_p n_p kT g_{p100}$$

$$\mathbf{q} = \sum_p (\mathbf{q}_{Tp} + \mathbf{q}_{Rp} + \mathbf{q}_{Vp}) \tag{5.18}$$

with

$$\mathbf{q}_{Tp} = -\frac{5}{2} R_p P_p \left(a_{p100} \frac{\partial T}{\partial r} + \frac{T}{T_{Vp}} f_{p100} \frac{\partial T_{Vp}}{\partial r} - l_{p000} \frac{\partial \xi_p}{\partial r}\right)$$

$$\mathbf{q}_{Rp} = \frac{P_p}{m_p} \left[C_{Rp}\left(a_{p010}\frac{\partial T}{\partial r} + \frac{T}{T_{Vp}} f_{p010}\frac{\partial T_{Vp}}{\partial r}\right) + \overline{E}_{Rp} l_{p000}\frac{\partial \xi_p}{\partial r}\right]$$

$$\mathbf{q}_{Vp} = \frac{P_p}{m_p} \left[C_{Vp}\left(\frac{T_{Vp}}{T} a_{p001}\frac{\partial T}{\partial r} + f_{p001}\frac{\partial T_{Vp}}{\partial r}\right) + E_{Vp} l_{p000}\frac{\partial \xi_p}{\partial r}\right] \tag{5.19}$$

We deduce from the above equations (5.19) the expressions of the thermal conductivities λ_{Tp} and λ_{TVp} (in $\mathbf{q}_{Tp}$), λ_{Rp} and λ_{RVp} (in $\mathbf{q}_{Rp}$), λ_{VTRp} and λ_{Vp} (in $\mathbf{q}_{Vp}$), and also the expressions of the diffusion fluxes of the particular energies. Finally, we can write:

$$\mathbf{q} = -\lambda \frac{\partial T}{\partial r} - \sum_p \left(\lambda_{Vp} \frac{\partial T_{Vp}}{\partial r} + \rho D h_p \frac{\partial c_p}{\partial r}\right) \tag{5.20}$$

with

$$h_p = e_{TRp} + \frac{p_p}{\rho_p} + e_{Vp} = \frac{7}{2}\frac{kT}{m_p} + e_{Vp} \tag{5.21}$$

$$\lambda = \sum_p \left(-\frac{5}{2}R_p p_p a_{p100} + \frac{p_p}{m_p}C_{Rp}a_{p010} + \frac{p_p}{m_p}C_{Vp}\frac{T_{Vp}}{T}a_{p001}\right) \tag{5.22}$$

$$\lambda_{Vp} = -\frac{5}{2}R_p p_p \frac{T}{T_{Vp}}f_{p100} + \frac{p_p}{m_p}C_{Rp}\frac{T}{T_{Vp}}f_{p010} + \frac{p_p}{m_p}C_{Vp}f_{p001} \tag{5.23}$$

$$D = -\frac{\rho}{n^2}\frac{kT}{m_p m_q}n_p l_{p0}$$

Here, e_{Vp} is given by a relaxation equation similar to Eqn. (5.11) that includes the VV and TV exchanges between the species.

Finally, the conservation equations include the usual Navier–Stokes equations written with the above transport terms, the species conservation equation (4.31), and the vibrational relaxation equation (5.11), in which the first-order term of production ($\sim \tau_R/\tau_V$) is usually neglected.

5.2.3 Usual approximations: SNE case

The above expressions for the transport terms may be simplified as in Chapter 3 with the assumption that the collisional internal energy balance is small compared to the kinetic energy balance. Moreover, as most collisional integrals involve only TR and VV exchanges with $\Delta\varepsilon_V = 0$, we again find expressions already obtained in Chapters 3 and 4.

Pure gases

We have:

$$\mu = \mu_{TR} \simeq \mu_T$$

$$\eta = \eta_{TR} \simeq \frac{kC_R}{C_{TR}^2}\tau_R p$$

$$\lambda_T = \frac{5}{2}\mu\frac{C_T}{m}\left(1 - \frac{5}{4}\frac{\mu}{C_T p}\frac{C_R}{\tau_R} + \frac{\rho}{2C_T p}\frac{C_R D_R}{\tau_R}\right)$$

$$\lambda_R = \rho D_R \frac{C_R}{m}\left[1 - \frac{\mu}{2\tau_R p}\left(\frac{\rho D_R}{\mu} - \frac{5}{2}\right)\right] \tag{5.24}$$

$$\lambda_V = \rho D_V \frac{C_V}{m} \tag{5.25}$$

Here, λ_V has a simpler form than λ_T and λ_R, because the vibrational energy remains frozen and diffuses only by virtue of the TR collisions (included in D_V). A further and more drastic approximation (Chapter 3) leads to $D_R \simeq D_V \simeq D_{pp}$.

Within the framework of these approximations, λ_{TV}, λ_{RV}, and λ_{VTR} are negligible, which is equivalent to writing the heat fluxes in their traditional forms:

$$q_{TR} = -\lambda_{TR} \frac{\partial T}{\partial r}, \quad q_V = -\lambda_V \frac{\partial T_V}{\partial r} \tag{5.26}$$

With the harmonic oscillator model, assuming that only co-linear collisions are efficient ($\Delta \varepsilon_R = 0$), we obtain from Eqn. (5.5) the relaxation pressure:

$$p_r = -n \frac{k C_R}{C_{TR}^2} \frac{\tau_R}{\tau_V} \left(\overline{E}_V - E_V \right) \tag{5.27}$$

The order of magnitude of this term is of course τ_R/τ_V, and p_r is proportional to the vibrational energy non-equilibrium.

Gas mixtures

We have:

$$\mu = \mu_{TR} \quad \text{(Eqn. (4.72))}$$

$$\eta = \eta_{TR} \quad \text{(Eqn. (4.73))}$$

Equations (4.74) and (4.75) give λ_T and λ_R.

$$\lambda_{Vp} \simeq \frac{n_p C_{Vp}}{\left(\frac{\xi_p}{D_{pp}} + \frac{\xi_q}{D_{pq}} \right)} \quad \text{and} \quad \lambda_{Vq} \simeq \frac{n_q C_{Vq}}{\left(\frac{\xi_q}{D_{qq}} + \frac{\xi_p}{D_{pq}} \right)} \tag{5.28}$$

$$\lambda_{TVp} \simeq \lambda_{RVp} \simeq \lambda_{VTRp} \simeq 0 \tag{5.29}$$

The expression of p_r, including the harmonic oscillator model, is given in Appendix 5.4.

5.3 Mixtures of reactive gases: (SNE)$_C$ case

5.3.1 (SNE)$_C$ + (WNE)$_V$ case

In a non-equilibrium reactive gas mixture, chemical reaction times are generally longer than the times characteristic of the T, R, V collisions (Figs 9 and 10), so that, for a molecular species p, we can write

$$J^0_{TRVp} = 0$$

$$\frac{df^0_{ip}}{dt} = J^1_{TRVp} + J^0_{Cp} \tag{5.30}$$

This corresponds to a WNE case for the vibrational mode and a SNE case for chemistry, that is (WNE)$_V$+(SNE)$_C$. A weak vibrational non-equilibrium therefore appears at first order.

At zero order, we have an equilibrium Maxwell–Boltzmann distribution at temperature T and the corresponding Euler equations including the species conservation equations (2.44).

At first order, the perturbation φ_{ip} may be written in the following form:

$$\phi_{ip} = A_{ip}\frac{1}{T}\frac{\partial T}{\partial r}\cdot\boldsymbol{u}_p + B_{ip}\frac{\partial\boldsymbol{V}}{\partial r}:\overset{0}{\boldsymbol{u}_p}\boldsymbol{u}_p + D_{ip}\frac{\partial\cdot\boldsymbol{V}}{\partial r} + G_{ip} + L_{ip}\boldsymbol{u}_p\cdot\boldsymbol{t}_p \tag{5.31}$$

As before, we again find coefficients with the same physical meaning. Here, however, A_{ip}, B_{ip}, D_{ip}, and L_{ip} are functions of collisional integrals including T, R, V collisions, as in the WNE case with two internal modes; moreover, the relaxation of the chemical term is included in G_{ip}. Therefore, the transport terms in the Navier–Stokes equations are identical to the WNE case with two internal modes (Chapter 4), with the exception of a relaxation pressure term connected to G_{ip}. Like D_{ip}, this term is expanded in the polynomial basis $\Psi^{1/2}_{mnq}$, that is, $G_{ip} = \sum_{mnq} g_{mnq}\Psi^{1/2}_{mnq}$, and is given by the following equation:

$$J^1_{TRVp}[G_{ip}] = f^0_{ip}\left[\begin{array}{c}\frac{\dot{w}_p}{\rho_p} - \frac{\dot{w}_p}{nm_p C_{TRV} T}(\overline{E}_{Rp} + \overline{E}_{Vp} - \frac{3}{2}kT) \\ \times\left(\frac{m_p u_p^2}{2kT} - \frac{3}{2} + \frac{\varepsilon_{irp} - \overline{E}_{Rp}}{kT} + \frac{\varepsilon_{ivp} - \overline{E}_{Vp}}{kT}\right)\end{array}\right] - J^0_{Cp}$$

As in the case of the vibrational non-equilibrium, the collisional integrals included in G_{ip} are the same as in D_{ip} (Appendix 5.3), and the relaxation pressure is then given by the relation

$$p_r = -\sum_p n_p kT g_{p100} \tag{5.32}$$

The following species conservation equations have to be added to the Navier–Stokes equations:

$$\rho\frac{dc_p}{dt} - \frac{\partial}{\partial r}\cdot\left(\rho D\frac{\partial c_p}{\partial r}\right) = \dot{w}_p = \sum_{ip}\int_{v_p}\left(J^0_{Cp} + J^1_{Cp}\right)dv_p \tag{5.33}$$

140 CHAPTER 5 TRANSPORT AND RELAXATION IN NON-EQUILIBRIUM REGIMES

And the vibrational non-equilibrium appearing at first order[12] (see Eqn. (3.63)) is equal to

$$E_{Vp} = \overline{E}_{Vp}\left(1 + d_{001}\frac{\partial \cdot V}{\partial r} + g_{001}\right) \tag{5.34}$$

We have similar relations for E_T and E_R.

An important point lies in the development of the production term in Eqn. (5.33) since the expression of this term remains formal in the general case. Thus, typical examples are analysed below.

Dissociation of a pure diatomic gas (dissociation phase)

The first and simplest example is the case of a dissociating pure diatomic gas (Chapter 2), according to the following reaction, written with the neglect of possible recombination and other types of dissociation (Chapter 9):

$$M_2 + M_2 \rightarrow 2M + M_2 \tag{5.35}$$

We have a mixture with two species: molecules p and atoms q. Thus,

$$\frac{\dot{w}_p}{m_p} = \sum_{i_p}\int_{v_p} J_{Dp}d\boldsymbol{v}_p = -k_D n_p^2 \tag{5.36}$$

with

$$J_{Dp} = J_{Dp}^0 + J_{Dp}^1$$

The dissociation-rate constant k_D (Chapter 2) is equal to

$$k_D = \sum_{i_p,j_p}\xi_{i_p}\xi_{j_p}\int_{v_{i_p},v_{j_p}}\frac{f_{i_p}^0 f_{j_p}^0}{n_{i_p}n_{j_p}}I_{i_p,j_p}^q g_{i_p j_p}d\Omega\, d\boldsymbol{v}_{i_p}\boldsymbol{v}_{j_p} = \sum_{i_p,j_p}\xi_{i_p}\xi_{j_p}k_{Di_p j_p} \tag{5.37}$$

where $k_{Di_p j_p}$ represents the dissociation-rate constant per level, independent of the population.

Assuming that the dissociation of a molecule does not depend on the collisional partner and on the rotational levels, we have

$$k_D = \sum_{i_{vp}}\xi_{i_{vp}}k_{Di_{vp}}$$

after summation over the levels j_{v_p}, and where $k_{Di_{vp}} = k_{Di_{vp}}(T)$ does not depend on the vibrational populations.

Moreover:

$$\xi_{i_{vp}} = \frac{1}{n_p}\sum_{i_{rp}}\int_{v_p} f_{ip}^0(1+\varphi_{ip})\,d\boldsymbol{v}_p \tag{5.38}$$

After some calculation,[12] we find

$$k_D = \bar{k}_D \left[1 - \left(g_{001} + d_{001} \frac{\partial \cdot V}{\partial r} \right) \left(\frac{\bar{E}_{Vp} - \bar{E}_{VDp}}{kT} \right) \right] \quad (5.39)$$

where $\bar{k}_D = \bar{k}_D(T)$ represents the dissociation-rate constant at zero order (vibrational equilibrium), and $\bar{E}_{VD}$ is the mean vibrational energy per molecule lost because of the dissociation. With the expressions for g_{001} and d_{001} given in Appendix 5.4 and in Chapter 3, we can calculate k_D to first order.

We observe that the order of magnitude of the non-equilibrium term in Eqn. (5.39) is τ_V/τ_D, because $\tau_D \sim (n_p k_D)^{-1}$. We can eliminate the term $g_{001} + d_{001} \frac{\partial \cdot V}{\partial r}$ with the aid of Eqn. (5.34), so that the dissociation-rate constant may be connected to the vibrational non-equilibrium, that is:[12]

$$k_D = \bar{k}_D \left[1 + \frac{E_V - \bar{E}_V}{\bar{E}_V} \frac{\bar{E}_{VD} - \bar{E}_V}{kT} \right] \quad (5.40)$$

where the indices p have been omitted.

Now, if we define the vibration–dissociation coupling factor $V(T, T_V) = \frac{k_D(T, T_V)}{\bar{k}_D(T)}$, we have

$$V = 1 + \left(\frac{E_V - \bar{E}_V}{\bar{E}_V} \right) \left(\frac{\bar{E}_{VD} - \bar{E}_V}{kT} \right) \quad (5.41)$$

Here, $\bar{E}_{VD}$ can be calculated from its definition. Thus we have

$$\bar{E}_{VD} = \frac{\sum_i \varepsilon_{i_v} \int_v J_D^0 \, dv}{\sum_i \int_v J_D^0 \, dv} = \frac{\sum_{i_v} \varepsilon_{i_v} \bar{n}_{i_v} k_{Di_v}}{\sum_{i_v} \bar{n}_{i_v} k_{Di_v}} \quad (5.42)$$

An oscillator model must now be introduced in the expression of k_{Di_v} (Appendix 5.5). Thus, assuming that the dissociation can occur from any vibrational level with an equal probability[34] ('non-preferential model'), we have

$$k_{Di_v} = \bar{k}_D \frac{\bar{Q}_v}{\exp(-\varepsilon_{i_v}/kT)} \frac{1}{N} \quad (5.43)$$

where N is the number of levels. Of course, other choices are possible,[34] but this model is valid at high temperature. Furthermore, with the harmonic oscillator model, if we assume that the result of the collision does not depend on the nature and the state of the partner, we find (Appendix 5.5) a value for $\bar{E}_{VD}$ close to $0.5 E_D$ and, for the anharmonic oscillator model, a value close to $0.45 \bar{E}_{VD}$, where E_D represents the dissociation energy. Using these data, an example of the computation of $V(T, T_V)$ for N_2 is presented in Fig. 11 for a given temperature T, with T_V varying from 0 to T.

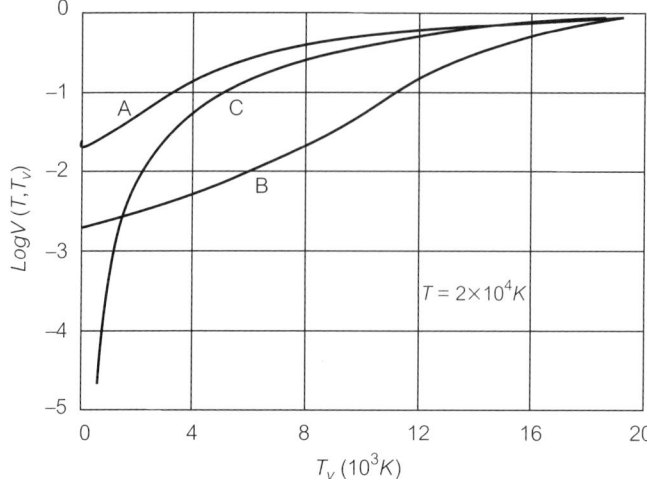

Figure 11. Vibration–dissociation factor V for nitrogen, $T = 2 \times 10^4$ K. A: $(SNE)_C + (WNE)_V$ case: Non-preferential anharmonic oscillator model[34], B: $(SNE)_C + (WNE)_C$ case: Preferential model[35], $-U = E_D/6k$, C: Semi-empirical model[36].

Dissociation of a pure diatomic gas (dissociation–recombination phase)

If we take into account the reverse (backward) reaction of the dissociation reaction (Eqn. (5.34)) (recombination), we have

$$M_2 + M_2 \rightleftarrows 2M + M_2 \tag{5.44}$$

This recombination reaction may be important in expanding flows or in situations close to chemical equilibrium. To zero order, the production term of the molecular component p (Chapter 9) may be written as follows:

$$\sum_{i_p} \int_{v_p} J^0_{Cp} d\boldsymbol{v} = \frac{\dot{w}_p}{m_p} = \bar{k}_R n_p n_q^2 - \bar{k}_D n_p^2 \tag{5.45}$$

For the component q (atoms), we have

$$J^0_{Tq} = 0$$

$$\frac{df^0_q}{dt} = J^1_{Tq} + J^0_{Cq} \tag{5.46}$$

with

$$\int_{v_q} J^0_{Cq} d\boldsymbol{v}_q = -\frac{m_p}{m_q} \sum_{i_p} \int_{v_p} J^0_{Cp} d\boldsymbol{v}_P \tag{5.47}$$

To first order, the structure of φ_{ip} is the same as in the previous case, with the same collisional integrals in $A_{ip}, B_{ip}, D_{ip}, L_{ip}$. Only the coefficient G_{ip} is different because of the term J_{Cp}^0, so that the relaxation pressure is also different (Appendix 5.4); in particular, this pressure disappears close to chemical equilibrium.

The first-order production term may be written in the following simple form:

$$k_R n_p n_q^2 - k_D n_p^2 \qquad (5.48)$$

with k_D and k_R given by formulae similar to Eqn. (5.40), that is:

$$k_D = \bar{k}_D \left[1 + \left(\frac{E_V - \bar{E}_V}{\bar{E}_V} \right) \left(\frac{\bar{E}_{VD} - \bar{E}_V}{kT} \right) \right]$$

$$k_R = \bar{k}_R \left[1 + \left(\frac{E_V - \bar{E}_V}{\bar{E}_V} \right) \left(\frac{\bar{E}_{VR} - \bar{E}_V}{kT} \right) \right] \qquad (5.49)$$

And for E_V, we have the relation given by Eqn. (5.34). Moreover, assuming as before that the result of the collision does not depend on the partner, we have:

$$\bar{E}_{VD} = \bar{E}_{VR} \qquad (5.50)$$

This is true only to zero order (TRV equilibrium).

It is also important to observe that the relation given by Eqn. (2.45) remains valid in the (weak) non-equilibrium region, that is:

$$K_c = \frac{k_D}{k_R} = \frac{\bar{k}_D}{\bar{k}_R} \qquad (5.51)$$

The case of several reactions

For the dissociation of a diatomic gas, we must add the following reaction (reaction 2) to the reaction given by Eqn. (5.44) (reaction 1):

$$M_2 + M \rightleftarrows 3M \qquad (5.52)$$

In the balance of species p, we must therefore add the corresponding term $\dot{w}_p^{(2)}$ to $\dot{w}_p^{(1)}$. The rate constant $k_D^{(2)}$ is modified exactly as $k_D^{(1)}$. However, we may generally assume[34,37] (Chapter 9) that the recombination-rate constant $k_R^{(2)}$ is not influenced by the vibrational non-equilibrium (collisions between atoms); therefore, we have

$$k_R^{(2)} = \bar{k}_R^{(2)}$$

Thus, the relation given by Eqn. (5.50) and therefore that by Eqn. (5.11) are no longer valid.

In the case of several reactions, the balance of each species must take into account the various interactions (vibration–dissociation, vibration–recombination, and vibration–reaction) in the same way. An example is given in Chapter 9 for high-temperature air. Formulae of the same type as Eqn.

(5.49) may therefore be applied to all reactions involving at least one molecular component.[108] Finally, only atom–atom recombination-rate constants are not modified.

5.3.2 (SNE)$_C$ + (SNE)$_V$ case

For very high temperatures, we have $\tau_V \sim \tau_C$. This is particularly true for O_2 at temperatures higher than 10^4 K (Fig. 10). In that case, we must consider a simultaneous non-equilibrium for vibration and chemistry. We thus have the following system for component p:

$$J^0_{TRp} = 0$$

$$\frac{df^0_{ip}}{dt} = J^1_{TRp} + J^0_{Vp} + J^0_{Cp} \qquad (5.53)$$

It seems unnecessary to completely develop the computations, because the transport terms are functions only of TR collisions. Thus, the corresponding results obtained in the WNE case with one internal mode are valid (Chapter 4). We must then take into account the two terms J^0_{Vp} and J^0_{Cp} for the G_{ip} term only (Appendix 5.4).

At first order, the chemical source term interacts with only the rotational mode. Thus, its order of magnitude is τ_R/τ_C, which is practically negligible. Therefore, the vibration–chemistry interaction is present at zero order.

Therefore, the Euler equations should be closed by a vibrational energy conservation equation giving E_{Vp}, or T_{Vp} if we accept the usual assumption of including the resonant VV collisions in the TR collisions. In the same way, a species conservation equation is necessary.

Dissociation of a diatomic gas (dissociation phase)

If we consider the example of a dominant dissociation reaction (Eqn. (5.35)), the species conservation equation is written for zero order in the following form:

$$\frac{\partial n_p}{\partial t} + \frac{\partial \cdot n_p V}{\partial r} = \frac{\dot{w}_p}{m_p} = \sum_{i_p} \int_v J^0_D d\mathbf{v} = -k^0_D n^2_p \qquad (5.54)$$

Thus, neglecting the influence of the collisional partner, summing over the rotational levels, and omitting the index p, we have:

$$k^0_D = \sum_{i_v} \xi^0_{i_v} k_{Di_v} = k^0_D(T, T_V) \qquad (5.55)$$

with
$$\xi_{i_v}^0 = \frac{\exp(-\varepsilon_{i_v}/kT_V)}{Q_V(T_V)}$$

Now a physical model for the dissociation (per level) must be defined: thus, for $k_{Di_{vp}}$, if we adopt the 'non-preferential' model (Eqn. (5.43)), we obtain:[34]

$$\frac{k_D^0(T, T_V)}{\bar{k}_D(T)} = V(T, T_v) = \frac{Q_V(T) Q_V(T_m)}{N Q_V(T_V)} \quad (5.56)$$

with
$$\frac{1}{T_m} = \frac{1}{T_V} - \frac{1}{T}$$

As previously pointed out, there is no significant additional interaction to first order.

The vibrational relaxation equation giving E_{Vp} may also be developed from the Boltzmann equation, i.e.

$$\frac{\partial(n_p E_{Vp})}{\partial t} + \frac{\partial \cdot (n_p E_{Vp} V)}{\partial \boldsymbol{r}} = \sum_{i_{vp}} \varepsilon_{i_{vp}} \int_v J_V^0 \, d\boldsymbol{v} + \sum_{i_{vp}} \varepsilon_{i_{vp}} \int_v J_D^0 \, d\boldsymbol{v} \quad (5.57)$$

The first source term on the right-hand side of Eqn. (5.57) arises from the TRV collisions. Thus, for example, if the harmonic oscillator model is used, this term is equal to $n_p \frac{E_{Vp} - \bar{E}_{Vp}}{\tau_V}$. The second term corresponds to the vibrational energy loss and is therefore equal to $E_{VD}^0 \frac{\dot{w}_p}{m_p} - -E_{VD}^0 k_D^0 n_p^2$. Finally, we find:[12]

$$\frac{dE_{Vp}}{dt} = \frac{E_{Vp} - \bar{E}_{Vp}}{\tau_V} + (E_{Vp} - E_{VD}^0) k_D^0 n_p \quad (5.58)$$

The calculation of E_{VD}^0, presented in Appendix 5.5, gives the result

$$E_{VD}^0 = kT_m \frac{dQ_V(T_m)}{dT_m} \quad (5.59)$$

For fixed T, E_{VD}^0 is shown versus T_V in Fig. 12 for various models.

Dissociation of a diatomic gas (dissociation–recombination phase)

With the reaction given by Eqn. (5.44), we have the species conservation equation

$$\frac{\partial n_p}{\partial t} + \frac{\partial \cdot n_p V}{\partial \boldsymbol{r}} = k_R^0 n_p n_q^2 - k_D^0 n_p^2 \quad (5.60)$$

and the following vibrational relaxation equation:

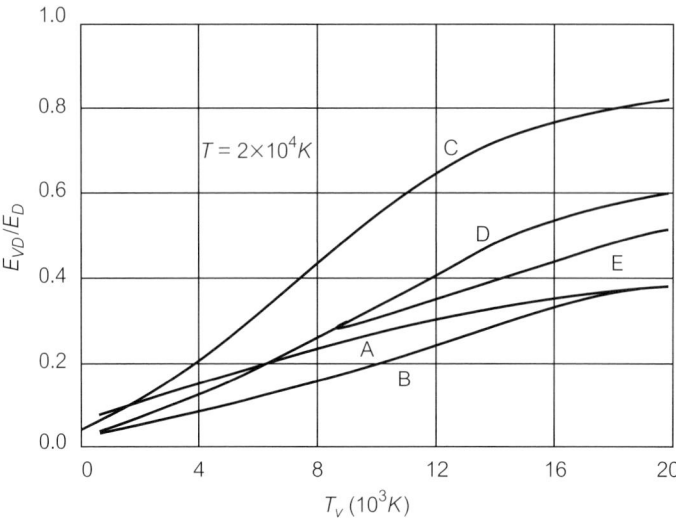

Figure 12. Vibrational energy lost per dissociation for nitrogen (E_{VD}^0/E_D). A: (SNE)$_C$ +(WNE)$_V$ case: Non-preferential model[12], B: (SNE)$_C$ +(SNE)$_V$ case: Non-preferential model[34], C: (SNE)$_C$+(SNE)$_V$ case: Preferential model[35] ($-U = E_D/6k$), D: (SNE)$_C$+(SNE)$_V$ case: Preferential model[35] ($-U = E_D/3k$; E: Physical model[38].

$$\frac{dE_{Vp}}{dt} = \dot{E}_{Vp} + \left(E_{Vp} - E_{VD}^0\right) k_D^0 n_p - \left(E_{Vp} - E_{VR}^0\right) k_R^0 n_q^2 \qquad (5.61)$$

where $\dot{E}_{Vp}$ is the vibrational energy production due to non-reactive collisions.

As before with Eqn. (5.56), we find

$$k_D^0 = V \overline{k}_D$$

If we assume $k_R^0 = \overline{k}_R(T)$ (Chapter 9), we have, in the non-equilibrium zone:

$$\frac{k_D^0}{k_R^0} \neq K_c$$

which is in contrast to the (SNE)$_C$ + (WNE)$_V$ case.

Taking into account the equilibrium conditions, we have

$$E_{VR}^0 = \overline{E}_{VD}^0 = \lim \left(E_{VD}^0\right)_{T_m \to \infty}$$

As before, for the harmonic oscillator model, we obtain

$$\overline{E}_{VD}^0 = \overline{E}_{VD} \simeq 0.5 E_D$$

Of course it is possible to take the reaction given by Eqn. (5.52) into account.

Generalization to a higher number of reactions may be achieved without particular difficulties. For exchange reactions, however, E_{VD} or E_{VR} have to be replaced by a term E_{VC} corresponding to the vibrational energy lost (or gained) by the considered species because of the reaction. This term, similar to E_{VD}, may be written as a function of the activation energy E_C (Chapter 9).

Thus, the vibration–dissociation interaction takes place at zero order and, as already pointed out, the first-order terms are generally negligible ($\sim \tau_R/\tau_V$).

The models $(SNE)_C+(SNE)_V$ and $(SNE)_C+(WNE)_V$ are used in concrete cases (shock waves, nozzle expansions, and so on) in Chapter 9.

Appendix 5.1 Pure gases in vibrational non-equilibrium

Expression for $\frac{df_i^0}{dt}$

$$\frac{df_i^0}{dt} = \left(\frac{mu^2}{2kT} - \frac{5}{2} + \frac{\varepsilon_{i_r} - \overline{E}_R}{kT}\right) \frac{1}{T}\frac{\partial T}{\partial r} \cdot u + \frac{m}{kT}\frac{\partial V}{\partial r} : \overset{0}{uu}$$

$$+ \left[\frac{2}{3}\frac{C_R}{C_{TR}}\left(\frac{mu^2}{2kT} - \frac{3}{2}\right) - \frac{k}{C_{TR}}\frac{\varepsilon_{i_r} - \overline{E}_R}{kT}\right]\frac{\partial \cdot V}{\partial r} + \frac{\varepsilon_{i_v} - \overline{E}_V}{kT_V}\frac{1}{T_V}\frac{\partial T_V}{\partial r} \cdot u$$

$$+ \left[\frac{1}{C_V T_V}\frac{\varepsilon_{i_v} - \overline{E}_V}{kT_V} - \frac{1}{C_{TR}T}\left(\frac{mu^2}{2kT} - \frac{3}{2} + \frac{\varepsilon_{i_r} - \overline{E}_R}{kT}\right)\right]\frac{1}{n}$$

$$\times \sum_i \varepsilon_{i_v} \int_v J_V^0 \, dv$$

Equation systems for the coefficients a, b, d, f, g

Cramer system for the a_{mnq} coefficients

$$a_{100}\alpha_{100}^{100} + a_{010}\alpha_{100}^{010} = -\frac{15}{2}\left(\frac{nkT}{m}\right)$$

$$a_{100}\alpha_{010}^{100} + a_{010}\alpha_{010}^{010} + a_{001}\alpha_{010}^{001} = 3\frac{C_R}{k}\left(\frac{nkT}{m}\right)$$

$$a_{010}\alpha_{001}^{010} + a_{001}\alpha_{001}^{001} = 0 \tag{5.62}$$

The determinant of this system is equal to

$$\Delta = \alpha_{100}^{100}\left(\alpha_{010}^{010}\alpha_{001}^{001} - \alpha_{010}^{001}\alpha_{001}^{010}\right) - \left(\alpha_{100}^{010}\right)^2 \alpha_{001}^{001}$$

CHAPTER 5 TRANSPORT AND RELAXATION IN NON-EQUILIBRIUM REGIMES

Thus, we have

$$a_{100} = -\frac{1}{\Delta}\frac{nkT}{m}\left[\frac{15}{2}\left(\alpha_{010}^{010}\alpha_{001}^{001} - (\alpha_{010}^{001})^2\right) - 3\frac{C_R}{k}\alpha_{010}^{100}\alpha_{001}^{001}\right]$$

$$a_{010} = \frac{1}{\Delta}\frac{nkT}{m}\left(\frac{15}{2}\alpha_{100}^{010}\alpha_{001}^{001} + 3\frac{C_R}{k}\alpha_{100}^{100}\alpha_{001}^{001}\right)$$

$$a_{001} = -\frac{1}{\Delta}\frac{nkT}{m}\left(\frac{15}{2}\alpha_{100}^{010}\alpha_{010}^{001} + 3\frac{C_R}{k}\alpha_{100}^{100}\alpha_{010}^{001}\right) \quad (5.63)$$

with

$$a_{000} = 0 \quad \text{and} \quad \alpha_{100}^{001} = 0 \quad \text{(resonant collisions)}$$

For b_{000}, we have

$$b_{000} = b_{00} = -10\frac{nkT}{m}\left(\beta_{00}^{00}\right)^{-1} \quad \text{(Appendix 3.2)} \quad (5.64)$$

Cramer system for the f_{mnq} coefficients

$$f_{100}\alpha_{100}^{100} + f_{010}\alpha_{100}^{010} = 0$$

$$f_{100}\alpha_{010}^{100} + f_{010}\alpha_{010}^{010} + f_{001}\alpha_{010}^{001} = 0$$

$$f_{010}\alpha_{001}^{010} + f_{001}\alpha_{001}^{001} = 3\frac{C_V}{k}\frac{nkT}{m} \quad (5.65)$$

We find the same collisional integrals as in the system of the a coefficients, with the same determinant. Therefore, we have

$$f_{100} = \frac{1}{\Delta}3\frac{C_V}{k}\frac{nkT}{m}\alpha_{100}^{010}\alpha_{010}^{001}$$

$$f_{010} = -\frac{1}{\Delta}3\frac{C_V}{k}\frac{nkT}{m}\alpha_{100}^{100}\alpha_{010}^{001}$$

$$f_{001} = \frac{1}{\Delta}3\frac{C_V}{k}\frac{nkT}{m}\left(\alpha_{100}^{100}\alpha_{010}^{010} - (\alpha_{010}^{100})^2\right)$$

with

$$f_{000} = 0$$

Cramer system for the d_{mnq} coefficients

We simply have (Appendix 3.2):

$$d_{100} = d_{10} = -n\left(\frac{C_R}{C_{TR}}\right)^2 \delta_{10}^{10}$$

$$d_{010} = d_{01} = -\frac{C_T C_R}{C_{TR}^2}\delta_{10}^{10} \quad (5.66)$$

and
$$d_{000} = 0 \quad \text{and} \quad \delta_{001}^{001} = 0$$

Cramer system for the g_{mnq} coefficients

We have the same integrals as in the system of the d coefficients. However, the two equations of the system are not independent, but another equation is provided by the integrability conditions, and we have

$$g_{100}\delta_{10}^{10} + g_{010}\delta_{10}^{01} = -\frac{1}{kT}\frac{C_R}{C_{TR}}\sum_i \varepsilon_{i_v} \int_v J_V^0\, d\boldsymbol{v} + \frac{1}{4}\sum_i \Delta\varepsilon_r \int_v J_V^0\, d\boldsymbol{v}$$

$$C_T g_{100} = C_R g_{010} \tag{5.67}$$

Finally, we find for g_{100} the relation given by Eqn. (5.5).

Expressions of the collisional integrals

$$\alpha_{100}^{100} = \alpha_{10}^{10}, \quad \alpha_{100}^{010} = \alpha_{010}^{100} = \alpha_{10}^{01} \quad \text{and} \quad \alpha_{010}^{010} = \alpha_{01}^{01} \quad \text{(Appendix 3.3)}$$

$$\alpha_{010}^{001} = \alpha_{001}^{010} = -2n^2 \frac{kT}{m}$$

$$\times \left\langle \left[\frac{\varepsilon_{k_r} - \varepsilon_{l_r}}{kT}\gamma'^2 + \frac{\varepsilon_{i_r} - \varepsilon_{j_r}}{kT}\gamma^2 - \left(\frac{\varepsilon_{i_r} - \varepsilon_{j_r}}{kT} + \frac{\varepsilon_{k_r} - \varepsilon_{l_r}}{kT}\right)\boldsymbol{\gamma}\cdot\boldsymbol{\gamma}'\right] \frac{\varepsilon_{i_v} - \varepsilon_{j_v}}{kT_V}\right\rangle$$

$$\alpha_{001}^{001} = -8n^2 \frac{kT}{m}\left\langle \frac{\varepsilon_{i_v} - E_V}{k'_1 V}\frac{\varepsilon_{i_v} - \varepsilon_{j_v}}{kT_V}(\boldsymbol{\gamma}\cdot\boldsymbol{\gamma}' - \gamma^2)\right\rangle$$

$$\alpha_{100}^{001} = 0$$

Appendix 5.2 First-order expression of the vibrational relaxation equation

The vibrational relaxation equation for a pure gas may be written in the following way after having developed the production terms $\sum_{i,j,k,l} \varepsilon_{i_v}\int_v (J_V^0 + J_V^1)d\boldsymbol{v}$:

$$n\frac{dE_V}{dt} + \frac{\partial\cdot\boldsymbol{q}_V}{\partial\boldsymbol{r}} = \sum_{i,j,k,l}\varepsilon_{i_v}\int_{\Omega,v_i,v_j}\left(f_k^0 f_l^0 - f_i^0 f_j^0\right)g_{ij}I_{i,j}^{k,l}d\Omega\, dv_i dv_j$$

$$+ \sum_{i,j,k,l}\varepsilon_{i_v}\int_{\Omega,v_i,v_j}\left[f_k^0 f_l^0(\varphi_k + \varphi_l) - f_i^0 f_j^0(\varphi_i + \varphi_j)\right]$$

$$\times g_{ij}I_{i,j}^{k,l}d\Omega\, dv_i dv_j \tag{5.68}$$

Here, $I_{i,j}^{k,l}$ includes the vibrational transitions.

The conservation of energy in the collisions gives

$$f_k^0 f_l^0 = f_i^0 f_j^0 \exp\left[\Delta\varepsilon_v \left(1 - \frac{T}{T_V}\right)\right]$$

If we use the harmonic oscillator model, or if we linearize the zero-order production term, we obtain the following result (Eqn. (2.19)):

$$\sum_{i,j,k,l} \varepsilon_{i_v} \int_{v_v} J_V^0 \, d\boldsymbol{v} = n \frac{\overline{E}_V - E_V}{\tau_V}$$

In the first-order production term, only $d_{100}\frac{\partial \cdot \boldsymbol{V}}{\partial \boldsymbol{r}} + g_{100}$ in the expression of φ_i gives non-zero terms which are connected to the rotational relaxation time τ_R (Eqn. (3.53)). If we also take into account the usual approximations ($\Delta\varepsilon \ll \gamma^2$), and if we define a characteristic time τ_V' connected to the diffusion of vibrational energy $\tau_V' \sim D \frac{m}{kT}$, we finally obtain:[39]

$$\frac{dE_V}{dt} + \frac{1}{n}\frac{\partial \cdot \boldsymbol{q}_V}{\partial \boldsymbol{r}} = \frac{\overline{E}_V - E_V}{\tau_V}\left\{1 + \frac{C_R}{C_{TR}^2}\left[\left(C_T - k\frac{\tau_V}{\tau_V'}\right)\tau_R \frac{\partial \cdot \boldsymbol{V}}{\partial \boldsymbol{r}} - C_V \frac{\tau_R}{\tau_V}\right]\right\}$$
$$- \frac{C_R C_V}{C_{TR}^2} kT \frac{\tau_R}{\tau_V} \frac{\partial \cdot \boldsymbol{V}}{\partial \boldsymbol{r}} \qquad (5.69)$$

The correction terms have an order of magnitude τ_R/τ_V or τ_R/θ.

Appendix 5.3 Gas mixtures in vibrational non-equilibrium

Equation systems for the coefficients a, b, d, f, g, l

With the simplifications deduced from the properties of collisional integrals $\left(\alpha_{p100}^{p001} = \alpha_{q100}^{p001} = \delta_{p100}^{p001} = \delta_{q100}^{p001} = 0\right)$, and from the usual approximations $\left(\alpha_{p010}^{p001} = \alpha_{q010}^{p001} = \delta_{p010}^{p001} = \delta_{q010}^{p001} \simeq 0\right)$, we obtain the following systems:

System for the a_{pmnq} coefficients

$$a_{p100}\left[\left(\alpha_{p100}^{p100}\right)_{pp} + \left(\alpha_{p100}^{p100}\right)_{pq}\right] + a_{p010}\left[\left(\alpha_{p010}^{p100}\right)_{pp} + \left(\alpha_{p010}^{p100}\right)_{pq}\right]$$
$$+ a_{q100}\left(\alpha_{q100}^{p100}\right)_{pq} + a_{q010}\left(\alpha_{q010}^{p100}\right)_{pq}$$

APPENDIX 5.3 GAS MIXTURES IN VIBRATIONAL NON-EQUILIBRIUM

$$= -\frac{15}{2} n_p \frac{kT}{m_p}$$

$$a_{p010}\left[\left(\alpha_{p100}^{p010}\right)_{pp} + \left(\alpha_{p100}^{p010}\right)_{pq}\right] + a_{p010}\left[\left(\alpha_{p010}^{p010}\right)_{pp} + \left(\alpha_{p010}^{p010}\right)_{pq}\right]$$

$$+ a_{q100}\left(\alpha_{q100}^{p010}\right)_{pq} + a_{q010}\left(\alpha_{q010}^{p010}\right)_{pq}$$

$$= 3n_p \frac{kT}{m} \frac{C_{rp}}{k}$$

$$a_{p001}\left[\left(\alpha_{p001}^{p001}\right)_{pp} + \left(\alpha_{p001}^{p001}\right)_{pq}\right] + a_{q001}\left(\alpha_{q001}^{p001}\right)_{pq} = 0 \qquad (5.70)$$

We also have, of course, three similar equations for the species q.

System for the f_{pmnq} coefficients

$$f_{p100}\left[\left(\alpha_{p100}^{p100}\right)_{pp} + \left(\alpha_{p100}^{p100}\right)_{pq}\right] + f_{p010}\left[\left(\alpha_{p010}^{p100}\right)_{pp} + \left(\alpha_{p010}^{p100}\right)_{pq}\right]$$

$$+ f_{q100}\left(\alpha_{q100}^{p100}\right)_{pq} + f_{p010}\left(\alpha_{q010}^{p100}\right)_{pq} = 0$$

$$f_{p100}\left[\left(\alpha_{p100}^{p010}\right)_{pp} + \left(\alpha_{p100}^{p010}\right)_{pq}\right] + f_{p010}\left[\left(\alpha_{p010}^{p010}\right)_{pp} + \left(\alpha_{p010}^{p010}\right)_{pq}\right]$$

$$+ f_{q100}\left(\alpha_{q100}^{p010}\right)_{pq} + f_{q010}\left(\alpha_{q010}^{p010}\right)_{pq} = 0$$

$$f_{p001}\left[\left(\alpha_{p001}^{p001}\right)_{pp} + \left(\alpha_{p001}^{p001}\right)_{pq}\right] + f_{q001}\left(\alpha_{q001}^{p001}\right)_{pq} = 3n_p \frac{kT}{m_p} \frac{C_{Vp}}{k} \qquad (5.71)$$

We also have three similar equations for the species q.

System for the b_{pmnq} coefficients

$$b_{p000}\left[\left(\alpha_{p000}^{p000}\right)_{pp} + \left(\alpha_{p000}^{p000}\right)_{pq}\right] + b_{q000}\left(\alpha_{q000}^{p000}\right)_{pq} = 10n_p \frac{kT}{m_p}$$

$$b_{q000}\left[\left(\alpha_{q000}^{q000}\right)_{qq} + \left(\alpha_{q000}^{q000}\right)_{pq}\right] + b_{p000}\left(\alpha_{p000}^{q000}\right)_{pq} = 10n_q \frac{kT}{m_q} \qquad (5.72)$$

System for the d_{pmnq} coefficients

$$d_{p100}\left[\left(\delta_{p100}^{p100}\right)_{pp} + \left(\delta_{p100}^{p100}\right)_{pq}\right] + d_{p010}\left[\left(\delta_{p010}^{p100}\right)_{pp} + \left(\delta_{p010}^{p100}\right)_{pq}\right]$$

$$+ d_{q100}\left(\delta_{q100}^{p100}\right)_{pq} + d_{q010}\left(\delta_{q010}^{p100}\right)_{pq} = -n_p \frac{C_R}{C_{TR}}$$

$$d_{p100}\left[\left(\delta_{p100}^{p010}\right)_{pp} + \left(\delta_{p100}^{p010}\right) pq\right] + d_{p010}\left[\left(\delta_{p010}^{p010}\right)_{pp} + \left(\delta_{p010}^{p010}\right)_{pq}\right]$$

$$+ d_{q100}\left(\delta_{q100}^{p010}\right)_{pq} + d_{q010}\left(\delta_{q010}^{p010}\right)_{pq} = -n_p \frac{C_R}{C_{TR}}$$

We also have two similar equations for the species q, but these four equations are not independent (the sum of the first and the third is equal to the sum of the second and the fourth). Thus, one of these is replaced by an integrability condition involving D_{ip}, that is:

$$\sum_p n_p \left(-C_{Tp} d_{p100} + C_{Rp} d_{p010}\right) = 0 \tag{5.73}$$

System for the g_{pmnq} coefficients

This system is similar to the previous one, with the same collisional integrals, thus:

$$g_{p100}\left[\left(\delta_{p100}^{p100}\right)_{pp} + \left(\delta_{p100}^{p100}\right)_{pq}\right] + g_{p010}\left[\left(\delta_{p010}^{p100}\right)_{pp} + \left(\delta_{p010}^{p100}\right)_{pq}\right]$$

$$+ g_{q100}\left(\delta_{q100}^{p100}\right)_{pq} + g_{q010}\left(\delta_{q010}^{p100}\right)_{pq} = \xi_p \frac{C_{Tp}}{k} \frac{\sum_{p,i} \varepsilon_{i_v p} \int_{v_p} J_{V_p}^0 d\boldsymbol{v}_p}{C_{TR} T}$$

$$g_{p100}\left[\left(\delta_{p100}^{p010}\right)_{pp} + \left(\delta_{p100}^{p010}\right)_{pq}\right] + g_{p010}\left[\left(\delta_{p010}^{p010}\right)_{pp} + \left(\delta_{p010}^{p010}\right)_{pq}\right]$$

$$+ g_{q100}\left(\delta_{q100}^{p010}\right)_{pq} + g_{q010}\left(\delta_{q010}^{p010}\right)_{pq} = -\xi_p \frac{C_{Rp}}{k} \frac{\sum_{p,i} \varepsilon_{i_v p} \int_{v_p} J_{V_p}^0 d\boldsymbol{v}_p}{C_{TR} T}$$

We also have two similar equations for the species q; as before, one of these equations must be replaced by the following integrability condition:

$$\sum_p \left(-C_{Tp} g_{p100} + C_{Rp} g_{p010}\right) = 0 \tag{5.74}$$

Collisional integrals

The α integrals

These include only the collisions TR and VVr; most of them are developed in Appendix 4.2 (one internal mode). Thus, we have

$$\left(\alpha_{p100}^{p100}\right)_{pp} = \left(\alpha_{p10}^{p10}\right)_{pp}, \quad \left(\alpha_{p010}^{p010}\right)_{pp} = \left(\alpha_{p01}^{p01}\right)_{pp}, \quad \text{and} \quad \left(\alpha_{p010}^{p100}\right)_{pp} = \left(\alpha_{p01}^{p10}\right)_{pp}$$

$$\left(\alpha_{p100}^{p100}\right)_{pq} = \left(\alpha_{p10}^{p10}\right)_{pp}, \quad \left(\alpha_{p010}^{p010}\right)_{pq} = \left(\alpha_{p01}^{p01}\right)_{pq}, \quad \text{and} \quad \left(\alpha_{p010}^{p010}\right)_{pq} = \left(\alpha_{p01}^{p01}\right)_{pq}$$

$$\left(\alpha_{q100}^{p100}\right)_{pq} = \left(\alpha_{q10}^{p10}\right)_{pq}, \quad \left(\alpha_{q010}^{p010}\right)_{pq} = \left(\alpha_{q01}^{p01}\right)_{pq}, \quad \text{and} \quad \left(\alpha_{q010}^{p100}\right)_{pq} = \left(\alpha_{q01}^{p10}\right)_{pq}$$

$$\left(\alpha_{q100}^{p010}\right)_{pq} = \left(\alpha_{q10}^{p01}\right)_{pq}, \quad \left(\alpha_{p001}^{p100}\right)_{pp} = 0, \quad \text{and} \quad \left(\alpha_{p001}^{p100}\right)_{pq} = 0$$

With the usual approximations, we also have

$$\left(\alpha_{p001}^{p100}\right)_{pq} = \left(\alpha_{p001}^{p010}\right)_{pq} = \left(\alpha_{q001}^{p001}\right)_{pq} = \left(\alpha_{q001}^{p100}\right)_{pq} = \left(\alpha_{p001}^{q100}\right)_{pq} = 0$$

$$\left(\alpha_{p001}^{p001}\right)_{pp} = 8n^2 \frac{kT}{m_p} \left\langle \frac{\varepsilon_{i_vp} - E_{Vp}}{kT_{Vp}} \left(\frac{\varepsilon_{k_vp} - \varepsilon_{l_vp}}{kT_{Vp}} \gamma_p \cdot \gamma'_p - \frac{\varepsilon_{i_vp} - \varepsilon_{j_vp}}{kT_{Vp}} \gamma^2 \right) \right\rangle$$

$$\simeq -3n_p^2 \left(\frac{kT}{m_p}\right)^2 \frac{C_{Vp}}{k} \frac{1}{nD_{pp}}$$

$$\left(\alpha_{p001}^{p001}\right)_{pq} \simeq -3n_p n_q \left(\frac{kT}{m_p}\right)^2 \frac{C_{Vp}}{k} \frac{1}{nD_{pq}}$$

The β integrals

In the same way, we have

$$\left(\beta_{p000}^{p000}\right)_{pp} = \left(\beta_{p100}^{p100}\right)_{pp}, \quad \left(\beta_{p000}^{p000}\right)_{pq} = \left(\beta_{p00}^{p00}\right)_{pq}, \quad \text{and} \quad \left(\beta_{q000}^{p000}\right)_{pq} = \left(\beta_{q00}^{p00}\right)_{pq}$$

The δ integrals

$$\left(\delta_{p100}^{p100}\right)_{pp} = \left(\delta_{p010}^{p010}\right)_{pp} = \left(\delta_{p010}^{p100}\right)_{pp} = \left(\delta_{p10}^{p10}\right)_{pp}, \quad \text{and} \quad \left(\delta_{p100}^{p100}\right)_{pq} = \left(\delta_{p10}^{p10}\right)_{pq}$$

$$\left(\delta_{p010}^{p010}\right)_{pq} = \left(\delta_{p01}^{p01}\right)_{pq}, \quad \left(\delta_{p010}^{p100}\right)_{pq} = \left(\delta_{p01}^{p01}\right)_{pq}, \quad \text{and} \quad \left(\delta_{q100}^{p100}\right)_{pq} = \left(\delta_{q10}^{p10}\right)_{pq}$$

$$\left(\delta_{q010}^{p010}\right)_{pq} = \left(\delta_{q01}^{01}\right)_{pq}, \quad \left(\delta_{q010}^{p100}\right)_{pq} = \left(\delta_{q01}^{p10}\right)_{pq}, \quad \text{and} \quad \left(\delta_{q100}^{p010}\right)_{pq} = \left(\delta_{q10}^{p01}\right)_{pq}$$

154 CHAPTER 5 TRANSPORT AND RELAXATION IN NON-EQUILIBRIUM REGIMES

$$\left(\delta_{p001}^{p001}\right)_{pp} = \left(\delta_{p001}^{p100}\right)_{pp} = \left(\delta_{p001}^{p010}\right)_{pp} = 0 \text{ and } \left(\delta_{p001}^{p001}\right)_{pq} = \left(\delta_{p001}^{p100}\right)_{pq}$$

$$= \left(\delta_{p001}^{p010}\right)_{pq} = 0$$

$$\left(\delta_{q001}^{p001}\right)_{pq} = \left(\delta_{q001}^{p100}\right)_{pq} = \left(\delta_{q001}^{p010}\right)_{pq} = 0 \text{ and } \left(\delta_{p001}^{q100}\right)_{pq} = \left(\delta_{p001}^{q010}\right)_{pq} = 0$$

Appendix 5.4 Expressions of g coefficients and relaxation pressure

Pure gases in vibrational non-equilibrium

$$G_i = g_{100}\left(\frac{3}{2} - \frac{mu^2}{2kT}\right) + g_{010}\left(\frac{\varepsilon_{i_r} - \overline{E}_R}{kT}\right)$$

$$g_{000} = 0$$

The term g_{001} is not involved ($\delta_{001}^{001} = 0$).
$g_{010} = \frac{C_T}{C_R}g_{100}$; g_{100} is given by Eqn. (5.5), and an approximate value of $p_r/p = -g_{100}$ is given by Eqn. (5.27).

Gas mixtures in vibrational non-equilibrium

$$G_{ip} = g_{p100}\left(\frac{3}{2} - \frac{m_p u_p^2}{2kT}\right) + g_{p010}\left(\frac{\varepsilon_{i_rp} - \overline{E}_{Rp}}{kT}\right)$$

The system involving the g coefficients is given in Appendix 5.3.
The term g_{p001} is not involved, since $\left(\delta_{p001}^{p001}\right)_{pp} = \left(\delta_{p001}^{p001}\right)_{pq} = 0$.
With the usual approximations, $g_{p100} = g_{q100}$, and

$$p_r = -\sum_p n_p g_{p100} kT$$

With the hypotheses of Chapter 4 concerning the rotational relaxation times τ_{Rp}, τ_{Rq}, and τ_{Rpq}, we find:

$$p_r = -\frac{1}{C_{TR}}\sum_{p,i_v}\varepsilon_{i_vp}\int_{\mathbf{v}_p} J_{Vp}^0 d\mathbf{v}_p$$

$$\times \frac{\frac{C_{Rp}C_{Rq}}{k}\left(\frac{\xi_p C_{Rp}}{\tau_{Rq}} + \frac{\xi_q C_{Rq}}{\tau_{Rp}}\right) + \frac{(\xi_p C_{Rp} + \xi_q C_{Rq})^2}{\tau_{Rpq}}}{(C_T + \xi_p C_{Rp} + \xi_q C_{Rq})\left(\frac{C_{Rp}C_{Rq}}{k^2\tau_{Rp}\tau_{Rq}} + \frac{\xi_p C_{Rp}}{k\tau_{Rp}\tau_{Rpq}} + \frac{\xi_q C_{Rq}}{k\tau_{Rq}\tau_{Rpq}}\right)}$$

Here, $\sum_{i_{vp}} \varepsilon_{i_{vp}} \int_{v_p} J^0_{Vp} d\boldsymbol{v}_p$ may be developed in the framework of the harmonic oscillator model (Chapter 2).

Mixtures of reactive gases
(WNE)$_V$ + (SNE)$_C$ case (dissociation)

$$G_{ip} = g_{p100} \left(\frac{3}{2} - \frac{m_p u_p^2}{2kT} \right) + g_{p010} \left(\frac{\varepsilon_{i_{rp}} - \overline{E}_{Rp}}{kT} \right) + g_{p001} \left(\frac{\varepsilon_{i_{vp}} - \overline{E}_{Vp}}{kT} \right)$$

$$G_q = g_{q1} \left(\frac{3}{2} - \frac{m_q u_q^2}{2kT} \right)$$

Here, $g_{q1} \sim \tau_{Tq}/\tau_C$ is negligible.

The system giving the g_p coefficients is the following (omitting the index p):

$$-C_T g_{100} + C_R g_{010} + C_V g_{001} = 0$$
$$(g_{100} + g_{010}) \delta^{010}_{010} = A$$
$$(g_{100} + g_{001}) \delta^{001}_{001} = B \qquad (5.75)$$

Then

$$g_{100} = A \frac{C_R}{C_{TRV}} \left(\delta^{010}_{010} \right)^{-1} + B \frac{C_V}{C_{TRV}} \left(\delta^{001}_{001} \right)^{-1}$$

$$g_{010} = A \frac{C_{TV}}{C_{TRV}} \left(\delta^{010}_{010} \right)^{-1} - B \frac{C_V}{C_{TRV}} \left(\delta^{001}_{001} \right)^{-1}$$

$$g_{001} = -A \frac{C_R}{C_{TRV}} \left(\delta^{010}_{010} \right)^{-1} + B \frac{C_{TR}}{C_{TRV}} \left(\delta^{001}_{001} \right)^{-1}$$

with

$$A = \left[\frac{\overline{E}_{Rp} - \overline{E}_{RD}}{kT} - \xi_p \frac{C_{Rp}}{k} \left(\frac{\overline{E}_{Rp} + \overline{E}_{Vp} - E_T}{C_{TRV} T} \right) \right] \frac{\dot{w}_p}{m_p}$$

$$B = \left[\frac{\overline{E}_{Vp} - \overline{E}_{VD}}{kT} - \xi_p \frac{C_{Vp}}{k} \left(\frac{\overline{E}_{Rp} + \overline{E}_{Vp} - E_T}{C_{TRV} T} \right) \right] \frac{\dot{w}_p}{m_p}$$

$$\delta^{010}_{010} = -n_p \frac{C_{Rp}}{k} \tau_R^{-1}, \quad \delta^{001}_{001} = -n_p \frac{C_{Vp}}{k} \tau_V^{-1}$$

We see that

$$g_{100} + g_{010} \sim \frac{\tau_R}{\tau_D} \quad \text{and} \quad g_{100} + g_{001} \sim \frac{\tau_V}{\tau_D}$$

Then

$$p_r = -n_p kT g_{100} \simeq \frac{k}{C_{TRV}} k\tau_V B \sim \frac{\tau_V}{\tau_D}$$

$(SNE)_V + (SNE)_C$ case

The system for the g coefficients is reduced to

$$-C_T g_{100} + C_R g_{010} = 0$$

$$g_{100}\delta_{100}^{010} + g_{010}\delta_{010}^{010} \simeq \frac{\dot{w}_p}{m_p} \sim \frac{n_p}{\tau_D} \quad (5.76)$$

Since

$$\delta_{100}^{010} \sim \delta_{010}^{010} \sim n_p \frac{C_{Rp}}{k} \tau_R^{-1}$$

we then have

$$g_{100} \sim {\tau_R}/{\tau_D} \sim g_{010} \quad \text{(negligible quantity)}$$

Appendix 5.5 Vibration–dissociation–recombination interaction

Interaction models

Non-preferential model

The dissociation-rate constant per vibrational level, k_{Di_v}, is defined in Chapter 2 and by Eqn. (5.37).

The probability for a dissociating molecule in level i_v is equal to

$$p_{i_v} = \frac{n_{i_v} k_{Di_v}}{n k_D} \quad (5.77)$$

If there is no preferential level, we have $p_{i_v} = \frac{1}{N}$, and

$$k_{Di_v} = \left(\xi_{i_v}\right)^{-1} \frac{k_D}{N} \quad (5.78)$$

If we assume that the rotational mode is in equilibrium, k_{Di_v} depends only on T, and we have

$$k_{Di_v} = \frac{\overline{k}_D}{N} \frac{\overline{Q}_V}{\exp\left(-\varepsilon_{i_v}/kT\right)} \quad (5.79)$$

Preferential model

The dissociation probability for a molecule in level i_v is proportional to $\exp\left(-\frac{E_D - \varepsilon_{i_v}}{kU}\right)$, where $-U$ is a characteristic temperature, a priori unknown (if $-U \to \infty$, we again find the non-preferential model).

We also know that the collision rate of molecules that have an energy higher than $E_D - \varepsilon_{i_v}$ is $\frac{n_{i_v}}{n} \exp\left(-\frac{E_D - \varepsilon_{i_v}}{kT}\right)$.

Then

$$p_{i_v} = C \frac{n_{i_v}}{n} \exp\left[-\frac{E_D - \varepsilon_{i_v}}{k}\left(\frac{1}{U} + \frac{1}{T}\right)\right] \tag{5.80}$$

If the vibrational population is out of equilibrium, we have

$$\frac{n_{i_v}}{n} = \frac{\exp\left(-\frac{\varepsilon_{i_v}}{kT_V}\right)}{Q_V(T_V)}$$

As $\sum_{i_v} p_{i_v} = 1$, we have

$$C = \frac{Q_V(T_V)}{Q_V(T_F)} \exp\left[\frac{E_D}{k}\left(\frac{1}{T} + \frac{1}{U}\right)\right]$$

Thus

$$p_{i_v} = \frac{\exp\left(-\frac{\varepsilon_{i_v}}{kT_F}\right)}{Q_V(T_F)} \tag{5.81}$$

and

$$k_{D i_v} = p_{i_v} k_D \left(\frac{n_{i_v}}{n}\right)^{-1}$$

so that we obtain the non-equilibrium value for k_D:

$$k_D = k_{D i_v} \left(\frac{n_{i_v}}{n}\right) p_{i_v}^{-1}$$

For equilibrium conditions, $T_V = T$, $T_F = -U$, and $p_{i_v \, eq} = \frac{\exp(\varepsilon_{i_v}/kU)}{Q_V(-U)}$; therefore:

$$\overline{k}_D = k_{D i_v} p_{i_v \, eq} \frac{\exp\left(-\varepsilon_{i_v}/kT\right)}{\overline{Q}_V} \tag{5.82}$$

and

$$\frac{k_D}{\overline{k}_D} = V(T, T_V, U) = \frac{Q_V(T) \, Q_V(T_F)}{Q_V(T_V) \, Q_V(-U)} \tag{5.83}$$

When $U \to -\infty$, $Q_V(-U) = N$, and we again find the previous model, with

$$\frac{k_D}{\overline{k_D}} = \frac{Q_V(T) Q_V(T_m)}{Q_V(T_V) N} \quad \text{and} \quad \frac{1}{T_m} = \frac{1}{T_V} - \frac{1}{T} \tag{5.84}$$

The difficulty lies in the choice of the temperature $-U$, which is a state variable ($U = F(T)$). An example of a representation of V is given in Fig. 11 for a value of $-U$ equal to $E_D/6k$. For higher temperatures, however, the non-preferential model is adequate.

Vibrational energy lost per dissociation act

This energy is defined by the following relation, after summation over the rotational levels and the levels j_v:

$$E_{VD} = \frac{\sum_{i_v} \varepsilon_{i_v} \int_v J_D dv}{\sum_{i_v} \int_v J_D dv} = \frac{\sum_{i_v} \varepsilon_{i_v} n_{i_v} k_{D i_v}}{\sum_{i_v} n_{i_v} k_{D i_v}} \tag{5.85}$$

(WNE)$_V$ + (SNE)$_C$ regime

At zero order, if we use the non-preferential model, we have

$$E_{VD} = \overline{E}_{VD} = \frac{1}{N} \sum_{N-1} \varepsilon_{i_v}.$$

And with the harmonic oscillator model, we obtain

$$\overline{E}_{VD} = \frac{1}{N} \sum_{N-1} k\theta_v i_v = \frac{k\theta_v (N-1)}{2} \tag{5.86}$$

For example, for nitrogen: $N = 33$, $\theta_v = 3354$ K, and $E_D = 1.56 \times 10^{-19}$ J, so that

$$\overline{E}_{VD} = 0.47 E_D$$

For oxygen: $N = 27$, $\theta_v = 2239$ K, and $E_D = 8.19 \times 10^{-19}$ J, so that

$$\overline{E}_{VD} = 0.49 E_D$$

With the anharmonic oscillator model, ε_{i_v} is given in Appendix 1.3, with

$$h\nu_v = k\theta_v = 4.69 \times 10^{-20} \text{ J}, \quad h\nu_v x_e = 2.87 \times 10^{-22} \text{ J, and}$$

$$h\nu_v y_e = 1.49 \times 10^{-25} \text{ J (for nitrogen)}$$

$$h\nu_v = 3.14 \times 10^{-20} \text{ J}, \quad h\nu_v x_e = 2.40 \times 10^{-22} \text{ J, and}$$

$$h\nu_v y_e = 1.08 \times 10^{-24} \text{ J (for oxygen)}$$

APPENDIX 5.5 VIBRATION–DISSOCIATION–RECOMBINATION INTERACTION

Then, for nitrogen:

$$\overline{E}_{VD} = 0.46 E_D, \text{ with } N = 45$$

And for oxygen:

$$\overline{E}_{VD} = 0.44 E_D, \text{ with } N = 32$$

Here, E_{VD} may also be calculated at first order, with

$$n_{i_v} = \overline{n}_{i_v}\left[1 + \left(d_{001}\frac{\partial \cdot V}{\partial r} + g_{001}\right)\left(\frac{\varepsilon_{i_v} - \overline{E}_V}{kT}\right)\right]$$

$$= \overline{n}_{i_v}\left[1 + \left(\frac{E_V - \overline{E}_V}{\overline{E}_V}\right)\left(\frac{\varepsilon_{i_v} - \overline{E}_V}{kT}\right)\right]$$

And we find[8]

$$\frac{E_{VD}}{\overline{E}_{VD}} = \frac{\overline{k}_D}{\overline{k}_D}\left\{1 + \left(\frac{E_V - \overline{E}_V}{\overline{E}_V}\right)\left[\frac{\theta_v}{T}\left(\frac{2N-1}{3}\right) - \frac{\overline{E}_V}{kT}\right]\right\} \quad (5.87)$$

An example of this expression (nitrogen) is shown in Fig. 12 (with $\frac{k_D}{\overline{k}_D}$ given by Eqn. (5.40)) and compared with other results.

(SNE)$_V$ + (SNE)$_C$ regime

At zero order and with the non-preferential model, k_D is given by Eqn. (5.84). For E_{VD}, we have from Eqn. (5.85):

$$E_{VD}^0 = \frac{\sum_{i_v}\varepsilon_{i_v}\exp\left(-\frac{\varepsilon_{i_v}}{kT_m}\right)}{\sum_{i_v}\exp\left(-\frac{\varepsilon_{i_v}}{kT_m}\right)} \quad (5.88)$$

This expression may be written in the form of Eqn. (5.59), and with the harmonic oscillator model, we find

$$\frac{E_{VD}^0}{k} = \frac{\theta_v}{\exp\left(-\frac{\theta_v}{T_m}\right) - 1} - \frac{N\theta_v}{\exp\left(-\frac{N\theta_v}{T_m}\right) - 1} \quad (5.89)$$

For nitrogen, this expression is also represented in Fig. 12.

SIX

Generalized Chapman–Enskog Method

6.1 Introduction

This chapter is devoted to the presentation of a generalized Chapman–Enskog method (GCE), applicable to those cases where the degree of non-equilibrium is unknown, and thus GCE can include those WNE and SNE cases which apply only within their range of validity. In practice, a flow is rarely maintained either in an equilibrium or a non-equilibrium state. A classical example is a flow subjected to a strong aerodynamic disturbance, such as a shock wave, and which is initially out of equilibrium but tends ultimately to an equilibrium state. It is thus necessary that the expansion of the distribution function describes the passage from one regime to another. This may be carried out with the GCE method, which is indeed a matching method.

This method is then applied to the cases previously employed: pure gases and gas mixtures in vibrational and chemical non-equilibrium. General methods of the analysis of flows including simultaneous vibrational and chemical non-equilibrium are then proposed. These flows represent significant cases in hypersonic regimes, and examples are given in the following chapters.

6.2 General method

As pointed out in the introduction, in many applications concerning high-speed flows, the more or less progressive passage from non-equilibrium zones to equilibrium zones (or vice versa) cannot always be described by solving the Boltzmann equation as performed in the preceding chapters. These zones of transition are, however, correctly taken into account at the zeroth order of the distribution function, i.e. at the macroscopic level, by the Euler equations, which are closed with the relaxation equations (vibrational or chemical). For example, solutions of the equation $J_I^0 = 0$ may tend in space or time to an equilibrium solution of the equation $J_{II}^0 = 0$, because the collisions of type I are generally particular collisions of type II. The examples previously given in vibrational or chemical non-equilibrium cases illustrate this behaviour ($n_{i_v} \to \overline{n}_{i_v}$, $E_V \to \overline{E}_V$, and so on).

The situation differs at the first order of the distribution function, because the corresponding equations are basically different, since they utilize different types of collision at the different orders. Thus, the non-equilibrium SNE solutions cannot tend to the WNE quasi-equilibrium solutions, since the former include collisions only of type I whereas the latter include collisions of type I and II (Eqns (3.5) and (3.6)). The Navier–Stokes equations are formally the same in both cases, but the transport terms are of course different, since they include different collisions. Moreover, there are relaxation equations in the SNE cases and not in the WNE cases (Table 2, Appendix 6.1). Therefore a matching method is essential to pass from one solution to the other.[40] The frozen case is of course not affected.

This method adds the first-order collisional term of type II in the first-order equation of the SNE system (Eqn. (3.6)). Then, this system becomes[41]

$$J_I^0 = 0$$
$$\frac{df_{ip}^0}{dt} = J_I^1 + J_{II}^0 + J_{II}^1 \tag{6.1}$$

Thus, the supplementary term J_{II}^1, negligible far from the equilibrium, is included in the first-order solution when the medium tends to equilibrium. Thus, we again find the SNE case when $J_{II}^1 \to 0$ and the WNE case when $J_{II}^0 \to 0$.

To sum up, we have a zero-order non-equilibrium situation for type II collisions, that is, of SNE type governed by the Euler and relaxation equations, which also cover equilibrium situations. At the first order, we can also describe situations covering simultaneously equilibrium and non-equilibrium regimes for type II collisions.

Two important examples previously treated either as SNE cases or as WNE cases are examined below. First is the case of a pure gas flow in vibrational non-equilibrium, and the second case is that of a reactive flow in the dissociation–recombination regime. The generalization to more complex cases does not differ in principle.

6.3 Vibrationally excited pure gases

The GCE solution must describe the vibrational non-equilibrium and equilibrium regimes at zero and first order. Thus, we have the following system:

$$J_{TR}^0 = 0$$
$$\frac{df_i^0}{dt} = J_{TR}^1 + J_V^0 + J_V^1 \quad (6.2)$$

At zero order, we have for f_i^0 Eqn. (2.12), with the corresponding Euler equations and the relaxation equation (2.29) if we include resonant VV collisions in the TR collisions. This solution fits the equilibrium solution (Eqn. (2.10)) when $\tau_V \to 0$.

At first order, the difficulty arises from the structure of the term J_V^1, which is strongly different from J_{TR}^1. Thus:

$$J_{TR}^1 = \sum_{j,k,l} \int_{\Omega, \mathbf{v}_j} f_i^0 f_j^0 \left(\varphi_k + \varphi_l - \varphi_i - \varphi_j \right) I_{i,j}^{k,l}(TR) \, g_{ij} \, d\Omega \, d\mathbf{v}_j$$

$$J_V^1 = \sum_{j,k,l} \int_{\Omega, \mathbf{v}_j} \left[f_k^0 f_l^0 \left(\varphi_k + \varphi_l \right) - f_i^0 f_j^0 \left(\varphi_i + \varphi_j \right) \right] I_{i,j}^{k,l}(V) \, g_{ij} \, d\Omega \, d\mathbf{v}_j$$

Here, $I_{i,j}^{k,l}(TR)$ and $I_{i,j}^{k,l}(V)$ represent the cross sections of the TR and V collisions respectively; f_i^0 (the solution of $J_{TR}^0 = 0$) is built with the collisional invariants of TR collisions, which are partially different from those of V collisions.

In contrast to the operator J_{TR}^1, the operator J_V^1 is not self-adjoint, and its eigenvalues are unknown. It is, however, possible to split it into two parts. Thus:

$$J_V^1 = J_V^{A1} + J_V^{NA1}$$

where J_V^{A1} is a self-adjoint operator such as

$$J_V^{A1} = \frac{1}{2} \sum_{j,k,l} \int_{\Omega, \mathbf{v}_j} \left(f_k^0 f_l^0 + f_i^0 f_j^0 \right) \left(\varphi_k + \varphi_l - \varphi_i - \varphi_j \right) I_{i,j}^{k,l}(V) \, g_{ij} \, d\Omega \, d\mathbf{v}_j \quad (6.3)$$

The eigenvalues of J_V^{A1} are the collisional invariants of type II collisions.
The non-self-adjoint operator J_V^{NA1} is

$$J_V^{NA1} = \frac{1}{2} \sum_{j,k,l} \int_{\Omega, \mathbf{v}_j} \left(f_k^0 f_l^0 - f_i^0 f_j^0 \right) \left(\varphi_k + \varphi_l + \varphi_i + \varphi_j \right) I_{i,j}^{k,l}(V) \, g_{ij} \, d\Omega \, d\mathbf{v}_j$$

$$(6.4)$$

6.3 VIBRATIONALLY EXCITED PURE GASES

The order of magnitude of this operator is smaller than J_V^{A1} and therefore may be neglected at the first order of the distribution function.[40]

The system given by Eqn. (6.2) may then be rewritten as follows:

$$J_{TR}^0 = 0$$

$$\frac{df_i^0}{dt} = J_{TR}^1 + J_V^0 + J_V^{A1} \quad (6.5)$$

The expression of φ_i is the same as in the corresponding SNE case (Eqn. (5.3)), and the equations giving the coefficients A_i, B_i, D_i, F_i, and G_i are given in Eqn. system (5.4), in which the operator J_{TR}^1 is replaced by $(J_{TR}^1 + J_V^{A1})$ but the right-hand sides of these equations are identical. For the coefficients a, b, d, f, and g, we therefore have the same systems of equations (Appendix 5.1), but collisional integrals that include J_V^{A1} must be linearly added to those that include J_{TR}^1. Thus, we have integrals specific to TR collisions, $\langle \cdots \rangle_{TR}$, and integrals specific to V collisions, $\langle \cdots \rangle_V$, that is:

$$\langle \cdots \rangle_{TR} = \left(\frac{kT}{\pi m}\right)^{1/2} \sum_{i,j,k,l} \left[\frac{\overline{n}_{i_r} \overline{n}_{j_r}}{n^2} \int_{\Omega,\gamma} \exp\left(-\gamma^2\right) \gamma^3 (\cdots) I_{i,j}^{k,l} (TR) \, d\Omega \, d\gamma \right] \quad (6.6)$$

$$\langle \cdots \rangle_V = \left(\frac{kT}{\pi m}\right)^{1/2}$$

$$\times \sum_{i,j,k,l} \left\{ \frac{\overline{n}_{i_r} \overline{n}_{j_r} n_{i_v} n_{j_v}}{n^4} \frac{1}{2} \left[\begin{array}{c} 1 + \exp\left(\Delta \varepsilon_v \left(1 - \frac{T}{T_V}\right)\right) \\ \times \int_{\Omega,\gamma} \exp\left(-\gamma^2\right) \gamma^3 (\cdots) I_{i,j}^{k,l} (V) \, d\Omega \, d\gamma \end{array} \right] \right\} \quad (6.7)$$

where

$$\frac{\overline{n}_{i_r}}{n} = \frac{\exp\left(-\frac{\varepsilon_{i_r}}{kT}\right)}{Q_R(T)}$$

$$\frac{n_{i_v}}{n} = \frac{\exp\left(-\frac{\varepsilon_{i_v}}{kT_v}\right)}{Q_V(T_V)}$$

and

$$\Delta \varepsilon_v = \frac{\varepsilon_{k_v} + \varepsilon_{l_v} - \varepsilon_{i_v} - \varepsilon_{j_v}}{kT} \neq 0$$

With the harmonic oscillator model, we have $\Delta \varepsilon_v = \frac{h \nu_v}{kT}$.

The collisional integrals α, β, δ involved in the expression of the coefficients a, b, d, f, and g are presented in Appendix 6.1.

6.3.1 Transport terms

Of course, the Navier–Stokes equations are valid, but the expressions for the transport terms are defined below:

The formal expression of the stress tensor **P** is the same as in the SNE case (Eqn. (5.6)), with a dynamic viscosity coefficient μ equal to

$$\mu = nkT\left(\frac{kT}{m}\right) \quad b_{000} = -10\left(\frac{nkT}{m}\right)\left(\beta_{000}^{000}\right)^{-1} \tag{6.8}$$

with b_{000} and β_{000}^{000} given in Appendix 6.1. Thus

$$\mu = \frac{5}{8}kT \left[\begin{array}{l} \langle \gamma^4 \sin^2 \chi - (\Delta\varepsilon_r)\,\gamma^2 \sin^2 \chi + \frac{1}{3}(\Delta\varepsilon_r)^2 \rangle_{TR} \\ +\langle \gamma^4 \sin^2 \chi - (\Delta\varepsilon_r + \Delta\varepsilon_v)\,\gamma^2 \sin^2 \chi + \frac{1}{3}(\Delta\varepsilon_r + \Delta\varepsilon_v)^2 \rangle_V \end{array} \right]^{-1} \tag{6.9}$$

or, in condensed form (with obvious notations):

$$\mu = \frac{5}{8}kT\left[\langle\phi\rangle_{TR} + \langle\phi\rangle_V\right]^{-1}$$

We can also write

$$\mu = \mu_{TR}\frac{1}{1+R} \quad \text{where } R = \frac{\langle\phi\rangle_V}{\langle\phi\rangle_{TR}}.$$

Thus, if the vibration is frozen (no V collision), we again find the SNE case ($R = 0$), and we have

$$\mu = \mu_{TR}$$

If $T_V \to T$, we again find the WNE case, with $\langle\phi\rangle_{TR} + \langle\phi\rangle_V = \langle\phi\rangle_{TRV}$. Thus

$$\mu = \mu_{TRV}$$

For the bulk viscosity η, we have

$$\eta = nkTd_{100} = -nkT\left(\frac{C_R}{C_{TR}}\right)^2 \left(\delta_{100}^{100}\right)^{-1} \tag{6.10}$$

with $\delta_{100}^{100} = -2n^2\left[\langle(\Delta\varepsilon_r)^2\rangle_{TR} + \langle(\Delta\varepsilon)^2\rangle_V\right]$ (Appendix 6.1), and $\Delta\varepsilon = \Delta\varepsilon_r + \Delta\varepsilon_v$.

As before, if we set $R' = \frac{\langle(\Delta\varepsilon)^2\rangle_{TR}}{\langle(\Delta\varepsilon_r)^2\rangle_V}$, we have

$$\eta = \eta_{TR}\frac{1}{1+R'} \tag{6.11}$$

Thus, we find a bulk viscosity term that includes the SNE and WNE solutions. In fact, this term exists because the rotational mode has been assumed in equilibrium at zero order (Chapter 3).

When $T_V \to T$, we have $\eta \to \eta_{TRV}$.

There is also a relaxation pressure term p_r equal to

$$p_r = -nkT g_{100} \tag{6.12}$$

where g_{100} is given by Eqn. (5.5), in which δ_{100}^{100} is given by Eqn. (6.32).
Then

$$p_r = (p_r)_{TR} \frac{1}{1 + R'} \tag{6.13}$$

Thus, in the SNE regime $R' \simeq 0$ and $p_r = (p_r)_{TR}$, whereas close to equilibrium the numerator of g_{100} tends to zero and $p_r \to 0$, which is the result found in the WNE case.

Remarks

If the rotational mode is also assumed out of equilibrium at zero order (including a rotational relaxation equation), no bulk viscosity or relaxation pressure terms appear. The rotational and vibrational non-equilibria are governed by relaxation equations.

As for the heat fluxes, the relations given by Eqn. (5.9) remain valid, but in the coefficients a and f, the integrals $\langle \cdots \rangle$ have to be replaced by $\langle \cdots \rangle_{TR} + \langle \cdots \rangle_V$ (Appendix 6.1).

6.3.2 Approximate expressions of heat fluxes

It is interesting to find the 'matching' between the SNE and WNE expressions of heat fluxes simplified with the hypotheses already used ($\Delta\varepsilon \ll \gamma^2$, harmonic oscillator model). Thus, we find[41]

$$\lambda_T = \frac{5}{2} \mu \frac{C_T}{m} \left\{ 1 - \frac{5}{4} \frac{\mu}{C_R p} \left[\frac{C_R}{\tau_R} + \frac{C_V}{\tau_V} \frac{T_V}{T} \frac{\overline{Q}_V}{Q_V} \frac{1}{2} \right. \right.$$
$$\left. \left. \times \left[1 + \exp\left(\frac{h\nu_v}{kT} \left(1 - \frac{T}{T_V} \right) \right) \right] \right] + \frac{\rho}{2 C_T p} \left(\frac{C_R D_R}{\tau_R} \right) \right\}$$

with

$$\overline{Q}_V = \left[1 - \exp\left(-\frac{h\nu_v}{kT} \right) \right]^{-1} \quad \text{and} \quad Q_V = \left[1 - \exp\left(-\frac{h\nu_v}{kT_V} \right) \right]^{-1} \tag{6.14}$$

$$\lambda_{TV} = \frac{5}{4} \frac{\rho\mu}{mp} \frac{C_V D_V}{\tau_V} \frac{T_V}{T} \frac{\overline{Q}_V}{Q_V} \frac{1}{2} \left\{ 1 + \exp\left[\frac{h\nu_v}{kT} \left(1 - \frac{T}{T_V} \right) \right] \right\} \tag{6.15}$$

$$\lambda_R = \rho D_R \frac{C_R}{m} \left[1 - \frac{\mu}{2\tau_R p} \left(\frac{\rho D_R}{\mu} - \frac{5}{2} \right) \right] \tag{6.16}$$

$$\lambda_{RV} \simeq 0 \tag{6.17}$$

$$\lambda_V = \rho D_V \frac{C_V}{m} \left\{ 1 - \frac{\rho D_V}{2\tau_V p} \frac{\overline{Q}_V}{Q_V} \frac{1}{2} \left[1 + \exp\left(\frac{h\nu_v}{kT}\left(1 - \frac{T}{T_V}\right)\right) \right] \right\} \tag{6.18}$$

$$\lambda_{VTR} = \frac{5}{4} \frac{\rho \mu}{mp} \frac{C_V D_V}{\tau_V} \frac{T_V}{T} \frac{\overline{Q}_V}{Q_V} \frac{1}{2} \left\{ 1 + \exp\left[\frac{h\nu_v}{kT}\left(1 - \frac{T}{T_V}\right)\right] \right\} = \lambda_{TV} \tag{6.19}$$

When $T_V \to T$, we again find the relations of the WNE case (Eqn. (3.69)), and when $\tau_V \to \infty$, we again find the relations of the SNE case, equivalent to the WNE case with one internal mode (rotation), itself deduced from the relations given in Eqn. (3.69).

We can see that the term which includes the vibrational non-equilibrium, $\frac{T_V}{T} \frac{\overline{Q}_V}{Q_V} \frac{1}{2} \left\{ 1 + \exp\left[\frac{h\nu_v}{kT}\left(1 - \frac{T}{T_V}\right)\right] \right\}$, has a weak influence on the values of λ_T and λ_R which practically depend only on T. Similarly, the values of λ_{TV} and λ_{RV} are small compared to λ_T and λ_R ($\sim \tau_V^{-1}$). In contrast, the vibrational conductivity λ_V strongly depends on the non-equilibrium, essentially with the term $C_V = C_V(T_V)$, whereas the term λ_{VTR} ($\sim \tau_V^{-1}$) is small compared to λ_V. Examples of curves ($\lambda_V = f(T, T_V)$) are represented in Fig. 16 (a) and (b).

6.4 Extension to mixtures of vibrational non-equilibrium gases

The SNE solutions have been more or less developed in the previous chapters. Here, then, we merely give the general outline for the GCE solutions. Thus, for a binary mixture in vibrational non-equilibrium at zero order,[28] we have the following system for the component p:

$$J^0_{TRp} = 0$$

$$\frac{df^0_{ip}}{dt} = J^1_{TRp} + J^0_{Vp} + J^{A1}_{Vp} \tag{6.20}$$

with

$$J^{A1}_{Vp} = \frac{1}{2} \sum_{j,k,l} \int_{\Omega, v_{j_p}} \left(f^0_{k_p} f^0_{l_p} + f^0_{i_p} f^0_{j_p} \right) \left(\varphi_{k_p} + \varphi_{l_p} - \varphi_{i_p} - \varphi_{j_p} \right) I^{k_p, l_p}_{i_p, j_p}(V) g_{i_p j_p} d\Omega \, dv_{j_p}$$

$$+ \frac{1}{2} \sum_{j,k,l} \int_{\Omega, v_{j_q}} \left(f^0_{k_p} f^0_{l_q} - f^0_{i_p} f^0_{j_q} \right) \left(\varphi_{k_p} + \varphi_{l_q} - \varphi_{i_p} - \varphi_{j_q} \right) I^{k_p, l_q}_{i_p, j_q}(V) g_{i_p j_q} d\Omega \, dv_{j_q}$$

$$\tag{6.21}$$

The complete expressions of the transport terms included in the Navier–Stokes equations cannot be given here, but their approximate expressions may be deduced from the preceding formulations. Thus, the vibrational conductivity for the component p has the following expression:

$$\lambda_{Vp} = \frac{nD_{pp}C_{Vp}}{1 + \frac{\xi_q}{\xi_p}\frac{D_{pp}}{D_{pq}} + \frac{m_p}{kT}\frac{\overline{Q}_{Vp}}{Q_{Vp}}\left(\frac{1}{2}\frac{D_{pp}}{\tau_{TVpp}} + \frac{\xi_q}{\xi_p}\frac{m_p}{m_p+m_q}\frac{D_{pp}}{\tau_{TVpq}}\right)\frac{1}{2}\left\{1 + \exp\left[-\frac{hv_{vp}}{kT}\left(1 - \frac{T}{T_{Vp}}\right)\right]\right\}}$$

$$+ \frac{\xi_q}{\xi_p}\frac{m_p}{kT}\frac{m_p}{m_p+m_q}\frac{Q_{Vq}}{Q_{Vp}}\frac{D_{pp}}{\tau_{VVpq}}\frac{1}{2}\left\{1 + \exp\left[-\frac{hv_{vp}}{kT}\left(1 - \frac{T}{T_{Vp}}\right)\right]\exp\left[-\frac{hv_q}{kT}\left(1 - \frac{T}{T_{Vq}}\right)\right]\right\} \quad (6.22)$$

with τ_{TVpp}, τ_{TVpq}, τ_{VVpq} given by Eqns (2.71) and (2.72).

Examples are represented in Figs 17 and 18.

6.5 Reactive gases

The GCE method is applied to reactive gas mixtures in the case where we may assume that, for the molecular components, the vibrational mode is weakly disequilibrated. This is a $(GCE)_C + (WNE)_V$ case that includes all chemical regimes (equilibrium and non-equilibrium) and that represents a general case in hypersonic flow, extending the $(SNE)_C + (WNE)_V$ case of Chapter 5.

For the component p, the relations of Eqn. (5.29) become

$$J^0_{TRVp} = 0$$

$$\frac{df^0_{ip}}{dt} = J^1_{TRVp} + J^0_{Cp} + J^{A1}_{Cp} \quad (6.23)$$

The zero-order solution, the structure of φ_{ip}, and the systems of equations giving the coefficients a, b, d, g, and l are the same as in the corresponding $(SNE)_C$ case. However, the operator J^1_{TRVp} must be replaced by $J^1_{TRVp} + J^{A1}_{Cp}$, so that the collisional integrals depending on TRV collisions, that is α, β, and δ, must be replaced by the integrals $\alpha_{TRV} + \alpha_C$, $\beta_{TRV} + \beta_C$, and $\delta_{TRV} + \delta_C$ respectively.

Considering the case of a dissociating pure gas, as in Chapter 5, we easily find the first-order expression for the dissociation-rate constant k_D, i.e.

$$k_D = k^0_D\left[1 - \left(g_{001} + d_{001}\frac{\partial \cdot V}{\partial r}\right)\left(\frac{\overline{E}_{Vp} - \overline{E}_{VDp}}{kT}\right)\right] \quad (6.24)$$

with

$$k^0_D = \overline{k}_D$$

168 CHAPTER 6 GENERALIZED CHAPMAN–ENSKOG METHOD

Equation (6.24) is formally identical to Eqn. (5.39), but the coefficients g_{001} and d_{001} include the integrals $\left(\delta_{001}^{001}\right)_{TRV} + \left(\delta_{001}^{001}\right)_C$, with

$$\left(\delta_{001}^{001}\right)_{TRV} = -2n^2 \langle (\Delta\varepsilon_{vp})^2 \rangle \tag{6.25}$$

Taking into account the expression of J_{Cp}^{A1}, we also find

$$\left(\delta_{001}^{001}\right)_C = \left(\frac{\overline{E}_{Vp} - \overline{E}_{VD}}{kT}\right)^2 \dot{w}_p \tag{6.26}$$

Thus, setting $R'' = \dfrac{\left(\delta_{001}^{001}\right)_C}{\left(\delta_{001}^{001}\right)_{TRV}}$ and $(\sim \frac{\tau_V}{\tau_C})$, we obtain

$$d_{001} = (d_{001})_{TRV} \frac{1}{1+R''} \quad \text{and} \quad g_{001} = (g_{001})_{TRV} \frac{1}{1+R''} \tag{6.27}$$

We find similar relations for the other collisional integrals; thus, for example, we have

$$\left(\delta_{010}^{010}\right)_{TRV} = -2n_p^2 \langle (\Delta\varepsilon_{rp})^2 \rangle \quad \text{and} \quad \left(\delta_{010}^{010}\right)_C = \left(\frac{\overline{E}_{Rp} - \overline{E}_{RD}}{kT}\right)^2 \dot{w}_p \tag{6.28}$$

For vibrational energy E_{Vp}, we obtain the same formal expression as Eqn. (5.34). Thus, we arrive at the conclusion that the relation between the dissociation-rate constant k_D and the vibrational energy E_V is the same as in the (WNE)$_V$+(SNE)$_C$ case, that is:[12]

$$k_D = \overline{k}_D \left[1 - \left(\frac{E_V - \overline{E}_V}{\overline{E}_V}\right)\left(\frac{\overline{E}_V - \overline{E}_{VD}}{kT}\right)\right] \tag{6.29}$$

Far from chemical equilibrium, we again find the (WNE)$_V$+(SNE)$_C$ case of Chapter 5 and, in particular, the same transport coefficients.

Close to equilibrium, $\dot{w}_p \to 0$ and $p_r \to 0$. The other transport terms remain close to the corresponding terms without chemical reaction, i.e.

$$\mu = (\mu)_{TRV}, \quad \eta = (\eta)_{TRV}, \quad \lambda = (\lambda)_{TRV}$$

The preceding simplifications concerning these terms may eventually be used. In particular, whether or not a sample is in the equilibrium regime, we know that $\mu_{TRV} \simeq \mu_{TR} \simeq \mu_T$. The value of η depends on the degree of excitation of internal modes and is proportional to the corresponding relaxation times. Finally, for heat fluxes, various expressions have been given, in particular in the non-equilibrium regime (Chapter 5).

6.6 Conclusions on non-equilibrium flows

At the end of this chapter, a general problem is posed: which equations must be used in order to describe as accurately as possible the flows in vibrational and chemical non-equilibrium?

The $(WNE)_V + (GCE)_C$ case examined in the present chapter seems to be a valid answer to this question, provided τ_V is smaller than τ_C, or as previously stated, if they are of the same order of magnitude. The regions where strong vibrational non-equilibrium prevails are excluded, for example the zones just behind the shock waves.

The $(SNE)_V + (SNE)_C$ case, described in Chapter 5, would cover these zones at the Euler level while taking into account the vibration–dissociation interaction, including various models, preferential or not. However, at the Navier-Stokes level, the transport terms depend only on the TR collisions! A $(GCE)_V + (GCE)_C$ method should be used, but would raise many difficulties.

An alternative solution,[42] called the 'mixed solution' (or MS model), consists of using the Navier-Stokes equations closed using species conservation equations and vibrational relaxation equations ((5.60) and (5.61)). Thus, this solution starts from an SNE model but keeps the hierarchy $\tau_V \leq \tau_C$ by using the reaction-rate constants of the type of Eqn. (5.40), which take into account the vibration–reaction interaction. In the same way, in the relaxation equations, the source terms E_{VD} ... are calculated at first order (Eqn. (5.87)), and for the transport terms, the relations given by Eqns (6.9)–(6.11) and (6.14)–(6.19) are used. The regions in very strong non-equilibrium are of course excluded, as are the Navier-Stokes equations themselves.

In the applications treated in the second part of this book, this last solution is generally used (Chapter 7, Appendix 7.1), but it is of course not always necessary to take into account vibrational and chemical non-equilibrium simultaneously (Chapters 9 and 10).

Appendix 6.1 Vibrationally excited pure gases

Comparison of the WNE and SNE methods (summary)

A summary of the comparison is presented in Table 2, including equations, corresponding regimes, and respective transport terms.

Table 2. Comparison of the WNE and SNE methods for non-dissociated pure gases.

Method	WNE	SNE
Zero order	$J^0_{TRV} = 0$ TRV equilibrium, with single temperature T. Euler equations.	$J^0_{TR} = 0$ TR equilibrium, with common temperature T. Vibrational non-equilibrium, with temperature T_V. Euler equations, with vib. relaxation equation.
First order	$\frac{df^0_i}{dt} = J^1_{TRV}$ Weak RV non-equilibrium. Navier–Stokes equations.	$\frac{df^0_i}{dt} = J^1_{TR} + J^0_V$ Weak R non-equilibrium. Navier–Stokes equations, with vib. relaxation equation.
Transport	Transport terms μ, η, λ depending on TRV collisions. Heat fluxes depending on the gradient of T.	Transport terms $\mu, \eta, \lambda_T, \lambda_R, \lambda_V$ depending on TR collisions. Heat fluxes depending on the gradients of T and T_V.

GCE method: equation systems and collisional integrals

The equation system (6.5) is equivalent to an SNE regime, with the coefficients $a, b, d, f,$ and g given in Appendix 5.1. However, the collisional integrals α, β, δ are different, since they represent here sums of integrals $\langle \cdots \rangle_{TR}$ and $\langle \cdots \rangle_V$ (Eqns (6.6) and (6.7)). Thus, in the systems given by Eqns (5.63), (5.64), (5.65), and (5.66), we have the following expressions for these integrals:

The α collisional integrals

$$\alpha^{100}_{100} = -8n^2 \frac{kT}{m}$$
$$\times \left[\begin{array}{l} \langle \gamma^4 \sin^2 \chi - (\Delta\varepsilon_r) \gamma^2 \sin^2 \chi + \frac{11}{8} (\Delta\varepsilon_r)^2 \rangle_{TR} \\ + \langle \gamma^4 \sin^2 \chi - (\Delta\varepsilon_r + \Delta\varepsilon_v) \gamma^2 \sin^2 \chi + \frac{11}{8} (\Delta\varepsilon_r + \Delta\varepsilon_v)^2 \rangle_V \end{array} \right]$$

$$\alpha^{010}_{100} = -5n^2 \frac{kT}{m} \left[\langle (\Delta\varepsilon_r)^2 \rangle_{TR} + \langle (\Delta\varepsilon_r)(\Delta\varepsilon_r + \Delta\varepsilon_v) \rangle_V \right]$$

$$\alpha^{001}_{100} = -5n^2 \frac{kT}{m} \left[\langle (\Delta\varepsilon_v)(\Delta\varepsilon_r + \Delta\varepsilon_v) \rangle_V \right] \text{ because } \langle (\Delta\varepsilon_v) \rangle_{TR} = 0$$

$$\alpha^{010}_{010} = 8n^2 \frac{kT}{m} \left[\begin{array}{l} \left\langle \left(\frac{\varepsilon_{ir}-\bar{E}_R}{kT}\right) \left(\frac{3}{2}\Delta\varepsilon_r + \frac{\varepsilon_{kr}-\varepsilon_{lr}}{kT}\boldsymbol{\gamma}\cdot\boldsymbol{\gamma}' - \frac{\varepsilon_{ir}-\varepsilon_{jr}}{kT}\gamma^2\right) \right\rangle_{TR} \\ + \left\langle \left(\frac{\varepsilon_{ir}-\bar{E}_R}{kT}\right) \left(\frac{3}{2}\Delta\varepsilon_r + \frac{\varepsilon_{kr}-\varepsilon_{lr}}{kT}\boldsymbol{\gamma}\cdot\boldsymbol{\gamma}' - \frac{\varepsilon_{ir}-\varepsilon_{jr}}{kT}\gamma^2\right) \right\rangle_V \end{array} \right]$$

(6.30)

$$\alpha_{010}^{001} = -2n^2 \frac{kT}{m}$$

$$\times \begin{bmatrix} \left\langle \frac{\varepsilon_{kr}-\varepsilon_{lr}}{kT}\gamma'^2 + \frac{\varepsilon_{ir}-\varepsilon_{jr}}{kT}\gamma^2 - \left(\frac{\varepsilon_{ir}-\varepsilon_{jr}}{kT} + \frac{\varepsilon_{kr}-\varepsilon_{lr}}{kT}\right)\frac{\varepsilon_{iv}-\varepsilon_{jv}}{kT_V}\boldsymbol{\gamma}\cdot\boldsymbol{\gamma}'\right\rangle_{TR} \\ +\langle \frac{3}{2}(\Delta\varepsilon_r)(\Delta\varepsilon_v) + \frac{\varepsilon_{kr}-\varepsilon_{lr}}{kT}\frac{\varepsilon_{ky}-\varepsilon_{lv}}{kT_V}\gamma'^2 + \frac{\varepsilon_{ir}-\varepsilon_{jr}}{kT}\frac{\varepsilon_{iv}-\varepsilon_{jv}}{kT_V}\gamma^2 \\ -\left(\frac{\varepsilon_{ir}-\varepsilon_{jr}}{kT}\frac{\varepsilon_{ky}-\varepsilon_{lv}}{kT_V} + \frac{\varepsilon_{kr}-\varepsilon_{lr}}{kT}\frac{\varepsilon_{iv}-\varepsilon_{jv}}{kT_V}\right)\boldsymbol{\gamma}\cdot\boldsymbol{\gamma}'\rangle_V \end{bmatrix}$$

$$\alpha_{001}^{001} = 8n^2 \frac{kT}{m} \begin{bmatrix} \left\langle \frac{\varepsilon_{iv}-E_V}{kT_V}\frac{\varepsilon_{iv}-\varepsilon_{jv}}{kT_V}(\boldsymbol{\gamma}\cdot\boldsymbol{\gamma}'-\gamma^2)\right\rangle_{TR} \\ +\left\langle \frac{\varepsilon_{iv}-E_V}{kT_V}\left(\frac{3}{2}\Delta\varepsilon_v + \frac{\varepsilon_{ky}-\varepsilon_{lv}}{kT_V}\boldsymbol{\gamma}\cdot\boldsymbol{\gamma}' - \frac{\varepsilon_{iv}-\varepsilon_{jv}}{kT_V}\gamma^2\right)\right\rangle_V \end{bmatrix}$$

The β collisional integral

$$\beta_{000}^{000} = -16n^2 \left(\frac{kT}{m}\right)^2$$

$$\times \begin{bmatrix} \langle \gamma^4 \sin^2\chi - (\Delta\varepsilon_r)\gamma^2 \sin^2\chi + \frac{1}{3}(\Delta\varepsilon_r)^2\rangle_{TR} \\ +\langle \gamma^4 \sin^2\chi - (\Delta\varepsilon_r + \Delta\varepsilon_v)\gamma^2 \sin^2\chi + \frac{1}{3}(\Delta\varepsilon_r + \Delta\varepsilon_v)^2\rangle_V \end{bmatrix}$$

(6.31)

The δ collisional integrals

$$\delta_{100}^{100} = -2n^2 \left[\langle(\Delta\varepsilon_r)^2\rangle_{TR} + \langle(\Delta\varepsilon_r + \Delta\varepsilon_v)^2\rangle_V\right]$$

$$\delta_{100}^{010} = -2n^2 \left[\langle(\Delta\varepsilon_r)^2\rangle_{TR} + \langle\Delta\varepsilon_r(\Delta\varepsilon_r + \Delta\varepsilon_v)\rangle_V\right]$$

$$\delta_{100}^{001} = -2n^2 \left[\langle\Delta\varepsilon_v(\Delta\varepsilon_r + \Delta\varepsilon_v)\rangle_V\right]$$

$$\delta_{010}^{010} = -2n^2 \left[\langle(\Delta\varepsilon_r)^2\rangle_{TR} + \langle(\Delta\varepsilon_r)^2\rangle_V\right]$$

$$\delta_{010}^{001} = -2n^2 \left[\langle(\Delta\varepsilon_r)(\Delta\varepsilon_v)\rangle_V\right]$$

$$\delta_{001}^{001} = -2n^2 \left[\langle(\Delta\varepsilon_v)^2\rangle_V\right]$$

(6.32)

Appendix 6.2 Transport terms in non-dissociated media

Viscosities, conductivities (equilibrium case)

As an example, results of the calculations for μ and η are represented as functions of temperature in Fig. 13 (a) and (b), for nitrogen and oxygen respectively, below the threshold of significant dissociation.[28]

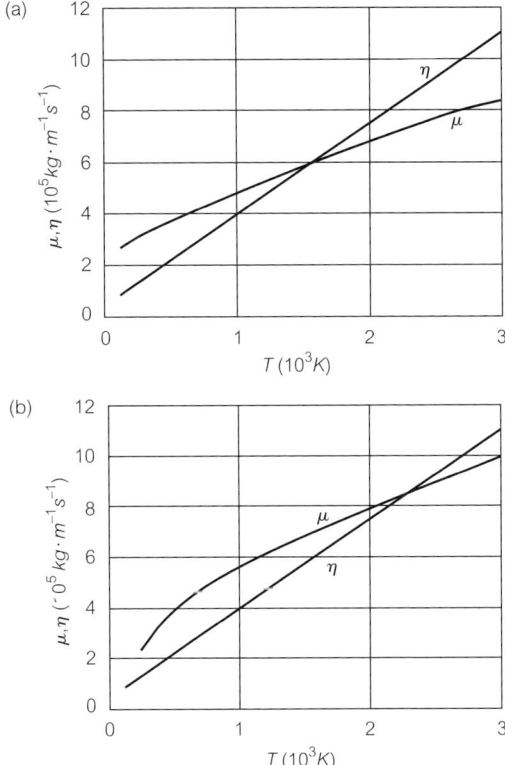

Figure 13. (a) Viscosity coefficients (nitrogen); (b) Viscosity coefficients (oxygen).

We observe that the coefficients μ and η are close to each other, but the associated gradients may be very different (Chapter 8), and of course the same is true for the corresponding terms in the conservation equations.

An example[28] of the calculation of μ for a N_2/H_2 mixture from Eqn. (4.72) is represented in Fig. 14 and compared to the values given by a barycentric formula often used for mixtures[43], $\left(\frac{1}{\mu} = \frac{\xi_p}{\mu_p} + \frac{\xi_q}{\mu_q}\right)$. The thermal conductivity λ of the same mixture in equilibrium, calculated from Eqns (4.74) and (4.75), is also represented in Fig. 15.

Thermal conductivities (non-equilibrium GCE case)

As pointed out above, vibrational conductivity is strongly influenced by the non-equilibrium: examples of the calculation of λ_{Vp} in a mixture (air)[28] for nitrogen and oxygen (Eqn. (6.22)) are represented in Fig. 16.

Note that the collisional integrals necessary for the preceding calculations have been dealt with in many references (see, for example, Refs. 3, 44, and 45).

APPENDIX 6.3 EXAMPLE OF GASES WITH DOMINANT VV COLLISIONS

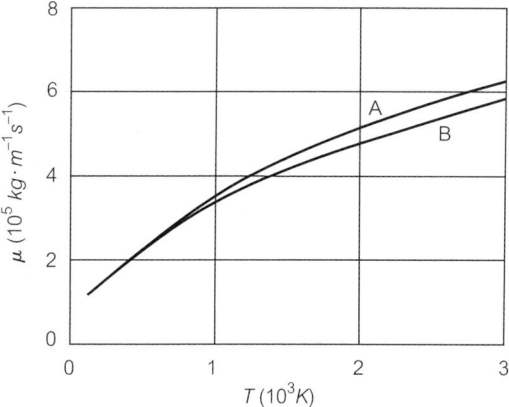

Figure 14. Viscosity of the mixture N_2/H_2 ($\xi_{H_2} = 0.8$). A: Ref. 28; B: Ref. 43.

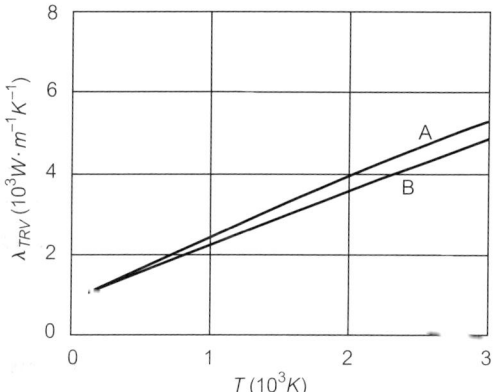

Figure 15. Thermal conductivity of the mixture N_2/H_2 ($\xi_{H_2} = 0.65$). A: Ref. 28; B: Ref. 43.

For a given collisional model, these integrals depend on temperatures T and T_V only.

Appendix 6.3 Example of gases with dominant VV collisions

In this case, the system successively giving f_i^0 and φ_i is the following:

$$J_{VV}^0 = 0$$

$$\frac{df_i^0}{dt} = J_{VV}^1 + J_{TV}^0 \tag{6.33}$$

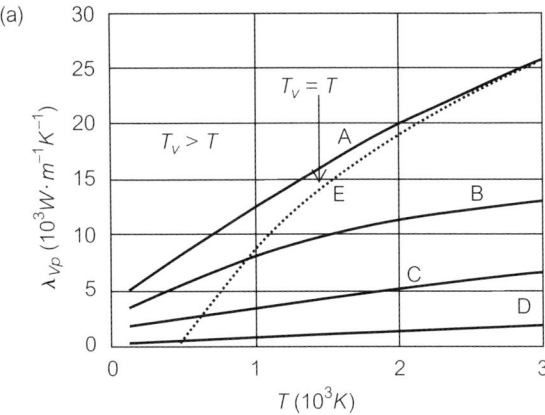
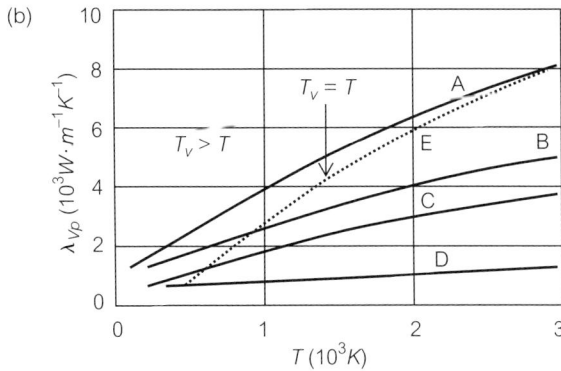

Figure 16. (a) Vibrational conductivity of nitrogen in a mixture (air). A: $T_V = 3000$ K, B: $T_V = 1000$ K, C: $T_V = 800$ K, D: $T_V = 500$ K, E: $T_V = T$. (b) Vibrational conductivity of oxygen in a mixture (air). (Notation of Fig. 16(a)).

The distribution function f_i^0, as discussed in Chapter 2 (Eqn. (2.24)), includes a vibrational population n_{i_v}, so that

$$\frac{n_{i_v}}{n} = \frac{\exp\left(-\frac{\varepsilon_{i_v}}{kT} + K i_v\right)}{Q'_V} \tag{6.34}$$

with

$$Q'_V = \sum_{i_v} \exp\left(-\frac{\varepsilon_{i_v}}{kT} + K i_v\right)$$

As also pointed out in Chapter 2, the macroscopic parameter K characterizes anharmonicity and non-equilibrium effects. These effects, however, may be separated if we write K in the following way:

$$K = \frac{\varepsilon_1}{k}\left(\frac{1}{T} - \frac{1}{\theta_1}\right) \tag{6.35}$$

where ε_1 represents the vibrational energy of the first level. So, we have

$$\frac{n_{i_v}}{n} = \frac{\exp\left(-\frac{\varepsilon_1 i_v}{k\theta_1}\right) \exp\left(\frac{\varepsilon_1 i_v - \varepsilon_{i_v}}{kT}\right)}{Q'_V} \qquad (6.36)$$

Thus, the non-equilibrium is represented by the first factor and the anharmonicity by the second. Here, θ_1 may be considered the 'temperature' of the first level.

The zero-order relaxation equation is Eqn. (2.25). If we linearize this equation,[46] we write $n_{i_v} = \bar{n}_{i_v}\left[1 + K\left(i_v - \bar{I}_V\right)\right]$, which takes the more classical following form:

$$\frac{dI_V}{dt} = \frac{\bar{I}_V - I_V}{\tau'_V} \qquad (6.37)$$

with

$$\tau'_V = \frac{\overline{I_V^2} - \bar{I}_V^2}{\sum_{i_v} a^{i_v}_{i_v+1} \bar{n}_{i_v+1}} \qquad (6.38)$$

At first order, we have

$$\phi_i = A_i \frac{1}{T}\frac{\partial T}{\partial r} \cdot \boldsymbol{u} + B_i \frac{\partial \boldsymbol{V}}{\partial r} : \overset{0}{\boldsymbol{u}\boldsymbol{u}} + D_i \frac{\partial \cdot \boldsymbol{V}}{\partial r} + F_i \frac{1}{\theta_1}\frac{\partial \theta_1}{\partial r} + G_i \qquad (6.39)$$

In addition to the Navier–Stokes equations, we have the following relaxation equation:

$$\frac{dI_V}{dt} + \frac{\partial \cdot \boldsymbol{q}_{I_V}}{\partial r} = \sum_{i_v} i_v \int_v \left(J^0_{TV} + J^1_{TV}\right) d\boldsymbol{v} \qquad (6.40)$$

where $\boldsymbol{q}_{I_V}$ represents a 'mean quantum number flux', such that

$$\boldsymbol{q}_{I_V} = -\lambda'\frac{\partial T}{\partial r} - \lambda''\frac{\partial \theta_1}{\partial r} \qquad (6.41)$$

In the right-hand side of this equation, the second term is dominant.

Appendix 6.4 A simplified technique: BGK method

In the Boltzmann equation, the collisional term J tends to bring the system back to an equilibrium state. It may then be modelled as the corresponding term of a classical relaxation equation (Landau–Teller, for example). Therefore, if we write

176 CHAPTER 6 GENERALIZED CHAPMAN–ENSKOG METHOD

that it is proportional to the deviation from the equilibrium distribution, for a gas with elastic collisions[47] we have

$$\frac{df}{dt} = J = \frac{f^0 - f}{\tau} \tag{6.42}$$

where τ is a relaxation time characteristic of these collisions, that is, of the order of τ_T. This non-linear equation is called the Bathnagar–Gross–Krook (BGK) equation.

Assuming a constant value for τ_T, we can obtain a solution, approximate but qualitatively correct, by expanding the distribution function as in the CE method, that is, $f = f^0(1 + \varphi)$. Thus, at zero order we have $f = f^0$, with Euler equations including the macroscopic quantities n, V, T defined using f^0. At first order, we have

$$\varphi = -\tau_T \frac{1}{f^0} \frac{df^0}{dt} \tag{6.43}$$

and the Navier–Stokes equations, with

$$\mathbf{P} = p\mathbf{I} - 2\mu \overline{\frac{\partial \mathbf{V}}{\partial \mathbf{r}}}^0$$

$$\mathbf{q} = -\lambda \frac{\partial T}{\partial \mathbf{r}}$$

where

$$\mu = \tau_T p \text{ and } \lambda = \frac{5}{2} \frac{k}{m} \tau_T p \tag{6.44}$$

For the Prandtl number, however, we find a value of 1 (instead of 2/3). This is not surprising, since we are using only one characteristic time. 'Refinements' are of course possible[13] (τ_T depending on T; the BGK equation modified to show a correct value for the Prandtl number, and so on).

For polyatomic gases, we can use two timescales, τ_{TR} and τ_V for example, and we may write the BGK equation in the following form:[48,14]

$$\frac{df_i}{dt} = \frac{f^0_{iTR} - f_i}{\tau_{TR}} + \frac{f^0_{iTRV} - f_i}{\tau_V} \tag{6.45}$$

where f^0_{iTR} and f^0_{iTRV} represent the translation–rotation equilibrium distribution function and the translation–rotation–vibration equilibrium distribution function respectively.

As we generally have $\tau_V \sim \theta \gg \tau_{TR}$, we can expand f_i as before. Thus, we have at zero order

$$f_i = f^0_{iTR} \tag{6.46}$$

APPENDIX 6.4 A SIMPLIFIED TECHNIQUE: BGK METHOD

This corresponds to the solution given by Eqn. (2.12), with n_{i_v} given by Eqn. (2.28) if the VV resonant collisions are included in the TR collisions. The corresponding Euler equations are closed with a relaxation equation of the following classical type:

$$\frac{dE_V}{dt} = \frac{\overline{E}_V - E_V}{\tau_V} \tag{6.47}$$

However, here we have $\overline{E}_V = \overline{E}_V(T_{TRV})$, and not $\overline{E}_V = \overline{E}_V(T_{TR})$ as in the Landau–Teller equation.

The first-order solution is

$$\varphi_i = -\tau_{TR} \frac{1}{f_i^0} \frac{df_i^0}{dt} + \frac{\tau_{TR}}{\tau_V} \left(f_{iTRV}^0 - f_{iTR}^0 \right) \tag{6.48}$$

The corresponding Navier–Stokes equations include the following transport terms:

$$\mathbf{P} = p\mathbf{I} - 2\mu \overline{\frac{\partial \mathbf{V}}{\partial \mathbf{r}}}^{0} - \eta \frac{\partial \cdot \mathbf{V}}{\partial \mathbf{r}} \mathbf{I} \tag{6.49}$$

with

$$\mu = \tau_{TR} p \quad \text{(see Eqn. (3.33))} \tag{6.50}$$

$$\eta = \frac{2}{3} \frac{C_R}{C_{IR}} \tau_{TR} p \quad \text{(see Eqn. (3.58))} \tag{6.51}$$

and

$$\mathbf{q} = \mathbf{q}_T + \mathbf{q}_R + \mathbf{q}_V$$

with

$$\mathbf{q}_T = -\tau_{TR} \frac{5}{2} \frac{k}{m} p \frac{\partial T}{\partial \mathbf{r}} = -\frac{5}{2} \mu \frac{k}{m} \frac{\partial T}{\partial \mathbf{r}}$$

$$\mathbf{q}_R = -\tau_{TR} \frac{C_R}{m} p \frac{\partial T}{\partial \mathbf{r}}$$

$$\mathbf{q}_V = -\tau_{TR} \frac{C_V}{m} p \frac{\partial T_V}{\partial \mathbf{r}} \tag{6.52}$$

These formulae should also be compared with those of Eqn. (3.69).

Here, we have

$$T = T_{TR} \quad \text{and} \quad p_r = 0. \tag{6.53}$$

The first-order relaxation equation has the usual form, i.e.

$$\frac{dE_V}{dt} = \frac{1}{n} \frac{\partial \cdot \mathbf{q}_V}{\partial \mathbf{r}} + \frac{\overline{E}_V - E_V}{\tau_V} \tag{6.54}$$

Here also, we have $\overline{E}_V = \overline{E}_V(T_{TRV})$.

Appendix 6.5 Boundary conditions for the Boltzmann equation

In Chapter 2, boundary conditions were used for the Boltzmann equation in order to define an isolated system (Appendix 2.1).

More generally, spatial boundary conditions can play an important role in solutions of the Boltzmann equation. Thus, as discussed in Chapter 2, far from the boundaries of the domain under study, the collisions between particles determine the structure of the distribution function (collisional regime). Close to the boundaries, the solutions of Chapman–Enskog (CE) type are influenced by the walls or interfaces that define the domain limits. It is therefore clear that, in the 'immediate' neighbourhood of the boundaries, the influence of the background on the distribution function is dominant. Thus, near the boundaries, we have to consider a region, called the Knudsen layer (Fig. 17), in which the CE solutions are not valid. However, this region is limited to a few mean free paths, so that in the continuum regime, it remains very close to the boundaries. The processes inside this region, however, may influence the boundary conditions for the conservation equations.

As the Navier–Stokes equations are widely used in gas dynamics, it seems natural to extend their validity up to the regions close to a wall or boundary. We take into account the existence of the Knudsen layer in the boundary conditions ($y = 0$) by considering that, while the state of the gas at $y = 0$ is different from the wall conditions (temperature, velocity, and so on), it is of course related to these same conditions.[4,5,49]

Thus, we must reconsider the complete Boltzmann equation in the Knudsen layer with a gas–wall interaction model at $y = 0$ and a CE distribution for $y \to +\infty$ as boundary conditions.

Here we give only an approximate solution to the problem by considering that the Knudsen layer is so thin that the incident molecules (going towards the

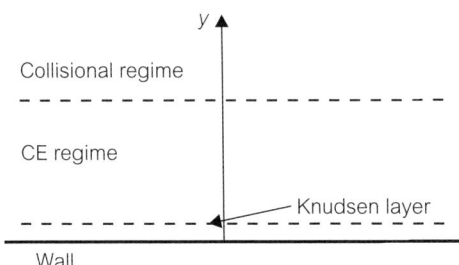

Figure 17. Various regimes near a wall.

APPENDIX 6.5 BOUNDARY CONDITIONS FOR THE BOLTZMANN EQUATION

wall) have a CE distribution. For a molecular gas with elastic collisions without macroscopic velocity, the problem is reduced to finding the gas temperature 'at the wall' (where $y = y_g$), which serves as a boundary condition for the system of Navier–Stokes equations and is characteristic of the thermal exchange between the gas and the wall at the temperature T_w.

The incident molecules at y_g have then a distribution derived from Eqn. (3.20), such that

$$f = f^0 \left[1 - \frac{15}{16} \frac{1}{nT\langle \gamma^4 \sin^2 \chi \rangle} \left(\frac{mu^2}{2kT} - \frac{5}{2} \right) \frac{\partial T}{\partial r} \cdot u \right]_g \tag{6.55}$$

corresponding to the heat flux:

$$q_g = -\frac{75}{32} \frac{k}{\langle \gamma^4 \sin^2 \chi \rangle} \left(\frac{kT}{m} \right) \frac{\partial T}{\partial r} \tag{6.56}$$

The total energy flux due to the incident particles F_i is

$$F_i = \int_{-\infty}^{+\infty} \int_{-\infty}^{0} \int_{-\infty}^{+\infty} \frac{1}{2} mu^2 u_y f \, du_x du_y du_z \tag{6.57}$$

where the coordinates x and z are in the plane of the wall. Therefore

$$F_i = nkT \left(\frac{2kT}{m} \right)^{1/2} + \frac{q_g}{2} \tag{6.58}$$

where the first term originates from the Maxwellian part of f, and the second from the non-Maxwellian part. In order to determine the state of the molecules reflected by the wall with an energy flux F_r, a statistical interaction model must be used. Thus, we define an 'accommodation coefficient' α as follows:

$$\alpha = \frac{F_i - F_r}{F_i - F_w} \tag{6.59}$$

where F_w would represent a reflected energy flux corresponding to molecules with an average temperature T_w equal to the wall temperature; these molecules are 'fully accommodated', that is, the wall (i.e. the background) has a dominant influence. In this case, $F_r = F_w$ and $\alpha = 1$; we say that the reflection is diffuse.

On the other hand, if $F_r = F_i$, we have $\alpha = 0$, and the wall plays only a geometrical role; there is no exchange, and we again find the 'specular reflection' discussed in Chapter 2. Generally, the practical situations are intermediate between these extreme cases, which may mask a complex physical reality.

Here, F_w may be easily calculated with f_w^0. In the same way, if we have no mass exchange, we have, for the incident molecule flux N_i and for the reflected fluxes

N_r and N_w:

$$N_i = N_r = N_w = n_g \sqrt{\frac{kT_g}{2\pi m}} = n_w \sqrt{\frac{kT_w}{2\pi m}}$$

Then we have $\left(n\sqrt{T}\right)_g = \left(n\sqrt{T}\right)_w$.

Finally, we find the following relation between T_g and T_w (Maxwell–Smoluchowski):

$$T_g = T_w + q_g \frac{2-\alpha}{\alpha} \left(\frac{\pi m}{2kT_w}\right)^{1/2} \frac{1}{n_w k} \tag{6.60}$$

This relation constitutes the boundary condition for the gas temperature 'at the wall' ($y = 0$). It is called a 'temperature jump'. When the heat flux is null ($q_g = 0$), the accommodation coefficient is also null ($\alpha = 0$). Of course, for high gaseous density, we again find $T_g = T_w$.

If the gas has a macroscopic velocity V, we similarly find a residual velocity 'at the wall', called 'slip velocity', equal to

$$V_g = \left(\frac{2}{mkT}\right)^{1/2} \frac{1}{n_w} \left(\mu \frac{\partial V}{\partial y}\right)_g \tag{6.61}$$

Thus, the slip velocity and the temperature jump are proportional to the normal velocity and temperature gradient, respectively.

More rigorous methods may of course be carried out; one involves solving the Boltzmann equation in the Knudsen layer by using the BGK collision operator approximation[50] (Appendix 6.4): however, the result is not very different, since Eqn. (6.60) remains valid, except that the term involving α is replaced with the term $\frac{2-a\alpha}{\alpha}$, where $a = 0.82$.

Other more sophisticated methods such as the direct simulation Monte Carlo method (DSMC; Appendix 6.7) give more accurate results, particularly in the case of low-density flow.[51] The main problem, however, remains how to model the gas–wall interaction while also taking into account possible physical processes[52,53] (adsorption, metastable states, and so on).

For polyatomic gases, the problem is more complicated because of a possibly different accommodation coefficient for each energy mode and because of possible exchanges between modes during the interaction process with the wall. This last point is related to the 'catalytic property' of the wall, favouring these exchanges more or less. Examples are presented in Chapter 10. If we assume an equilibrium distribution in the Knudsen layer and a single accommodation coefficient, we obtain for the temperature jump a formula identical to the previous case with a value for *a* varying from 0.82 for the case without internal energy to 0.86 for complex molecules.[54]

In the case of reactive flows, the phenomenon of catalycity becomes very important[55,56] and may have a strong influence on the gas–wall thermal exchanges. Examples are also given in Chapters 10 and 12.

Appendix 6.6 Free molecular regime

In contrast with the collisional regime ($Kn \to 0$), the free molecular regime corresponds to collisionless flow ($Kn \to \infty$), for which, in the case of identical monatomic particles, we have

$$\frac{df}{dt} = 0 \qquad (6.62)$$

In the absence of external forces, the flow preserves its initial distribution along the streamlines. The presence of obstacles does not modify the upstream flow, since there is no possible information in the absence of collisions.

Therefore the main problem is to calculate the possible exchanges between an obstacle and this flow. Assuming that the upstream distribution is Maxwellian (for example the case of a rapid expansion from a tank), the distribution function of the incident particles is the following:

$$f_\infty = n_\infty \left(\frac{m}{2\pi k T_\infty}\right)^{3/2} \exp\left(-\frac{mu^2}{2kT_\infty}\right)$$

The incident fluxes of the particles N_i, of the momentum $\boldsymbol{P}_i$, and of the energy F_i, may be written respectively in the following forms (Appendix 6.5):

$$N_i = \int_{-\infty}^{+\infty}\int_{-\infty}^{0}\int_{-\infty}^{+\infty} f_\infty v_{\infty y} \, dv_{\infty x} \, dv_{\infty y} \, dv_{\infty z}$$

$$\boldsymbol{P}_i = \int_{-\infty}^{+\infty}\int_{-\infty}^{0}\int_{-\infty}^{+\infty} f_\infty m \boldsymbol{v}_\infty v_y \, dv_{\infty x} \, dv_{\infty y} \, dv_{\infty z} \qquad (6.63)$$

$$F_i = \int_{-\infty}^{+\infty}\int_{-\infty}^{0}\int_{-\infty}^{+\infty} f_\infty \frac{1}{2} m v_\infty^2 v_y \, dv_{\infty x} \, dv_{\infty y} \, dv_{\infty z}$$

So we find the following expressions:[4,57]

$$N_i = n_\infty \left(\frac{RT_\infty}{2\pi}\right)^{1/2} \left[\exp(-S_\theta^2) + \pi^{1/2} S_\theta (1 + \text{erf } S_\theta)\right]$$

$$P_{ni} = \rho_\infty RT_\infty \left[\pi^{-1/2} S_\theta \exp(-S_\theta^2) + (1 + \text{erf } S_\theta)\left(\frac{1}{2} + S_\theta^2\right)\right]$$

$$P_{\tau i} = \rho_\infty RT_\infty S \cos\theta \left[\pi^{-1/2} \exp(-S_\theta^2) + S_\theta (1 + \text{erf } S_\theta)\right] \quad (6.64)$$

$$F_i = \rho_\infty (RT_\infty)^{3/2} (2\pi)^{-1/2} \left[(S^2 + 2) \exp(-S_\theta^2)\right.$$

$$\left. + \pi^{1/2} \left(S^2 + \frac{5}{2}\right) S_\theta (1 + \text{erf } S_\theta)\right]$$

with

$$S = \frac{V}{(2RT_\infty)^{1/2}} = \left(\frac{\gamma}{2}\right)^{1/2} M_\infty \text{ and } S_\theta = S \sin\theta$$

where θ represents the local angle between the wall and the flow.

As before, we can define accommodation coefficients for momentum and energy (Appendix 6.5). Thus for example, if we define a coefficient α for energy, the reflected flux may be written

$$F_r = F_i - \alpha (F_i - F_w)$$

The heat flux to the obstacle is then

$$q_w = \alpha (F_i - F_w)$$

If we also have $N_i = N_r = N_w$ (no mass transfer), we find for the heat flux:

$$q_w = 2W_i R\alpha (T_{wr} - T_w) \quad (6.65)$$

where $W_i = mN_i$ represents the incident mass flux and T_{wr} the adiabatic temperature of the wall (obtained for $q_w = 0$), that is:

$$T_{wr} = \frac{1}{4R}\left[V^2 + RT_\infty \left(4 + \frac{1}{1+\varphi}\right)\right] \quad (6.66)$$

with

$$\varphi = \frac{\exp(-S_\theta^2)}{\pi^{1/2} S_\theta (1 + \text{erf}(S_\theta))}$$

Of course, the minimum heat flux is obtained in the case of a plane wall ($\theta = 0$), for which we have

$$T_{wr} = T_\infty \left(1 + \frac{S^2}{2}\right) = T_\infty \left(1 + \frac{\gamma}{4} M_\infty^2\right) \quad (6.67)$$

Therefore, T_{wr} is always higher than the stagnation temperature of the continuum regime, T_0 (Chapter 7).

Accommodation coefficients are also defined for the normal and tangential components of the strain, i.e.

$$\alpha_n = \frac{P_{ni} - P_{nr}}{P_{ni} - P_{nw}} \quad \text{and} \quad \alpha_t = \frac{P_{ti} - P_{tr}}{P_{ti}}$$

which, in principle, enables us to know the pressure acting on the body, i.e.

$$p = P_{ni} + P_{nr} = (2 - \alpha_n) P_{ni} + \alpha_n P_{nw}$$

as well as the tangential force per surface unit:

$$\tau = P_{ti} - P_{tr} = \alpha_t P_{ti}$$

The complete formulae are easily deduced from the preceding expressions.

The intermediate regime between the collisional and the free molecular regimes (called the transitional regime) is more difficult to investigate. In fact, only DSMC methods are efficient. However, by considering the formulae obtained for the heat flux in the continuum regime (Chapter 8) and in free molecular flow, i.e. $q_w \sim (T_{wr} - T_w)$, we obtain a reasonable approximation for heat flux in the transitional regime, with the following formula written for the corresponding Stanton numbers (Eqn. (7.34)):

$$\frac{1}{St} = \frac{1}{St_c} + \frac{1}{St_{fm}} \tag{6.68}$$

where St_c and St_{fm} represent the Stanton numbers in the continuum and free molecular regimes, respectively.

Appendix 6.7 Direct simulation Monte Carlo methods

Rather than using 'conventional' equations to describe the state and evolution of a gaseous system, it is possible to simulate the behaviour of the system by directly considering an ensemble composed of several thousands or millions of molecules: the position, velocity, and internal state of each molecule are 'memorized' then modified with time by simultaneously following the molecules during their movement and taking into account their mutual interaction or

184 CHAPTER 6 GENERALIZED CHAPMAN–ENSKOG METHOD

their interaction with the domain boundaries. This method can be defined as an algorithm which, at each iteration, operates on a sample of N molecules with the distribution $f(\mathbf{r}, \mathbf{v}, t)$ and generates a new sample of molecules with the distribution $f(\mathbf{r}, \mathbf{v}, t + \Delta t)$.

The physical domain is decomposed in a net of cells $\Delta \mathbf{r}$ that include N molecules and have small dimensions compared with the distance along which the macroscopic quantities 'significantly' vary. Thus $\Delta r \leq \lambda$.

The choice of Δt is essential. The basic assumption lies in the uncoupling of the displacement phase from the collision phase, as in the Boltzmann equation, which takes into account these processes separately. Thus, the Monte Carlo simulation consists of the repetition, in each cell, of the following procedure:

- All particles are displaced by the distance they should cover during Δt with their initial velocity $\mathbf{v}$.

- The collisions which should have occurred during Δt are taken into account by a random process; the velocities (and the internal states) of the involved molecules are thus modified but not their location.

Here, Δt must not be too large, in order to respect the scheme of the displacement–collision separation. A choice criterion is to take it small as compared to the mean collision time, i.e. $Z \Delta t < 1$. The state of the molecules in the neighbourhood of the frontiers changes according to the chosen gas–wall interaction model. Macroscopic quantities such as state and transport parameters are calculated by averaging the corresponding quantities over the velocities and internal states of the N molecules of each cell.

Of course, the main problem is to correctly treat the collision process. Various methods are available,[58–60] one of which is briefly described below.[58]

The mean collision number during Δt is

$$N_{coll} = Z_0 \Delta t = \frac{n}{2} Z \Delta t \qquad (6.69)$$

where Z is given by Eqn. (1.56), which may also be written $Z = n\overline{Cg}$, with usual notations. Therefore, in the volume element $\Delta \mathbf{r}$, during Δt the mean collision number is the following:

$$N_c = \frac{N}{2} n \overline{Cg} \Delta t$$

The difficulties arising from the computation of $\overline{Cg}$ in $\Delta \mathbf{r}$ may be overcome by using a 'time counter', incremented with a time Δt_c at each collision between two molecules that have a relative velocity g such that

$$\Delta t_c = \frac{2}{nNCg}$$

Thus, the collision number N_c is determined by the following 'criterion':

$$\sum_{i=1}^{N_c} \Delta t_c(i) \geq \Delta t \qquad (6.70)$$

In particular, for elastic collisions with a spherical interaction potential $\varphi(r) = \frac{K}{r^{s-1}}$ (Eqn. (1.80)), we have

$$\Delta t_c = \frac{2}{N}\left[\pi \beta_0^2 \left(\frac{(s-1)K}{m_r}\right)^{2/(s-1)} ng^{s-5/s-1}\right]^{-1} \qquad (6.71)$$

where β_0 is the maximum value of the dimensionless impact parameter $\beta = \frac{b}{r}$ (cut-off). For the rigid elastic sphere model, we have

$$\Delta t_c = \frac{2}{N n\pi d^2 g} \qquad (6.72)$$

The couples of interacting particles must then be determined. The collision probability being proportional to the relative velocity, an 'accept–reject' method is used. Thus, the value of g_{max} is determined in each cell, a number X is drawn ($0 < X < 1$), and if g/g_{max} is smaller than X, the collision is rejected, otherwise it is accepted. In this last case, the counter time is increased in Δt_c, and

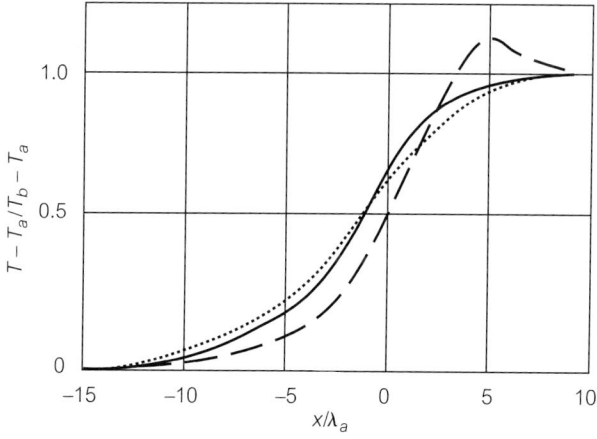

Figure 18. Temperature profiles across a shock wave [$M_s = 8$, $(n_A/n_{He})_a = 0.1$]. ———: T, ·····: T_{He}, — —: T_A ; a: upstream, b: downstream.

the post-collision velocities are computed. The procedure is repeated until the criterion given by Eqn. (6.70) is verified.

The particles interacting with the domain boundaries are treated in a similar way. Finally, the moments are computed. The case of inelastic collisions requires a particular treatment, and various models have been proposed.[60,61]

The DSMC methods have raised considerable interest and contribute to solving problems for which the Navier–Stokes equations are invalid (shock structure, Knudsen layer, transitional and rarefied regimes, and so on).

As a simple example, Fig. 18 shows the evolution of the temperatures of the components of a monatomic gas mixture (He/A) across a shock wave, resulting from a DSMC computation.[62,63] Thus, defining a specific temperature for every species, we may observe the different behaviour of the two species and, in particular, a temperature overshoot for the heavy component.

Appendix 6.8 Hypersonic flow regimes

It may be interesting to get an overview of the equations and methods used in the analysis of hypersonic flows as functions of velocity and density (altitude),

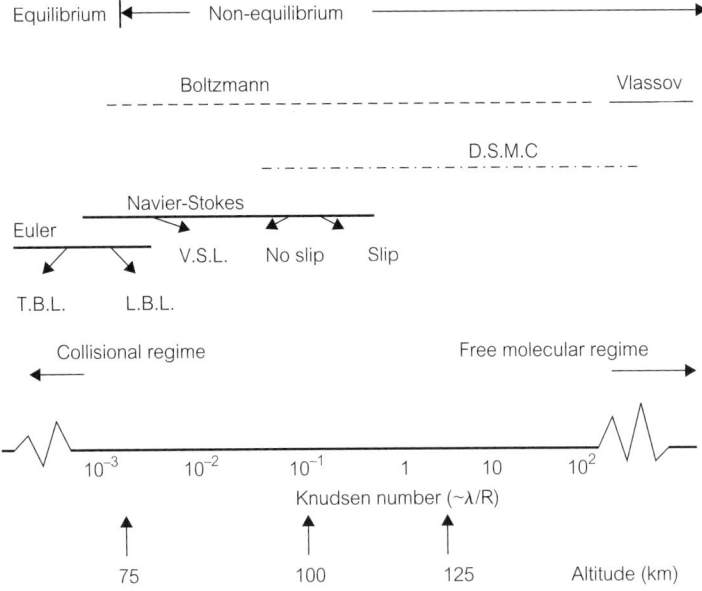

Figure 19. Mathematical and physical models for hypersonic flows. V.S.L.: Viscous shock layer, L.B.L.: Laminar boundary layer, T.B.L.: Turbulent boundary layer, D.S.M.C.: Direct simulation Monte Carlo method.

APPENDIX 6.8 HYPERSONIC FLOW REGIMES **187**

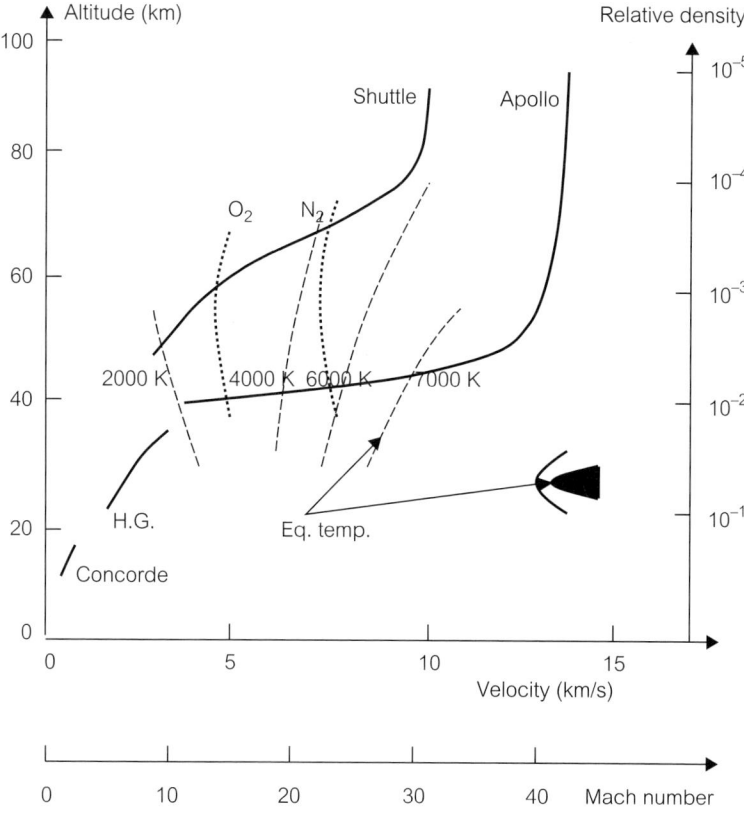

Figure 20. Examples of hypersonic trajectories. ——— Trajectories (H.G.: Hypersonic glider), – – – Equilibrium stagnation temperature (Eq. Temp.), ... 10% dissociated species.

and an overview of the physical processes taken into account.[64] Thus, for air, Fig. 19 gives qualitative information about the domain of validity of various equations and methods for a large range of Knudsen numbers. Similarly, Fig. 20 shows the domains where chemical processes for N_2 and O_2 are significant at the stagnation point of a re-entry body.

PART II

Macroscopic Aspects and Applications

Notations to Part II

A_n^m	Einstein coefficient (radiative transition $n \to m$)
A_p	symbolic notation for species p
B	magnetic induction
C	$\rho\mu/\rho_e\mu_e$
C_f	skin-friction coefficient
C'	specific heat (per unit mass)
C^p	specific heat at constant pressure (per unit mass)
Da	Damköhler number
E	Eckert number
Eu	Euler number
f'	u/u_e (dimensionless velocity)
F_p	vibrational number of the component p
g	h_0/h_{0e} (dimensionless stagnation enthalpy)
h_0	stagnation enthalpy (per unit mass)
I	spectral line intensity, ionization energy (per particle), MHD interaction parameter
J	current density
K_p	equilibrium reaction constant (partial pressures)
L	reference length, Lewis number
M	Mach number, mean molecular mass
M_p	molecular mass of the component p
n, N	unit vector normal to a surface
Nu	Nusselt number
N_p	mole fraction of the component p
P	Prandtl number
P, Q	Riemann parameters
r	body radius, recovery factor
R	element of vibration matrix
Re	Reynolds number
R_h	Hartmann number
S	Schmidt number, entropy (per unit mass), cross section
St	Stanton number
u	longitudinal component of flow velocity **V**
v	transverse component of flow velocity **V**
$\dot{w}_{Vp}$	mass production rate of vibrational energy (component p)

$\dot{x}$	chemical production rate of the quantity x
X	distance from the diaphragm (shock tube)
z_p	c_p/c_{pe} (dimensionless concentration)
α	dissociation rate, absorption rate
α, β	unit vectors
β	dimensionless quantity related to the pressure gradient
$\gamma, C^p/C'$	(specific-heat ratio)
δ	characteristic thickness of boundary layer
ϕ	shape parameter of spectral line
Φ	dissipation function
λ	wavelength
μ_p	chemical potential of the component p (per unit mass)
μ_0	magnetic permeability
ν_{nm}	frequency corresponding to the transition $n \to m$
ν_p	stoichiometric coefficient of the component p
σ	electrical conductivity
τ	stress tension
ξ, η	boundary layer coordinates

Subscripts

0	reference state, stagnation condition
1, 2	medium 1, medium 2
A, A_2	related to atoms A, to molecules A_2
c	charged species, critical value (sonic condition)
e	edge of boundary layer, electronic state
f	frozen state
id	ideal model
m	averaged quantity
r	related to a reflected shock wave
s	related to an incident shock wave, exit condition (nozzle)
t	turbulent
∞	free-stream conditions

Superscripts

j	0: plane flow, 1: axisymmetric flow
p	constant pressure
$*$	dimensionless quantity, excited state

Abbreviations

CFL	Courant–Friedrich–Levy
HP, LP	high pressure, low pressure
LLD	Levy–Lees–Dorodnitsyn
MHD	magnetohydrodynamics
RH	Rankine–Hugoniot
SSD	shock stand-off distance

SEVEN
General Aspects of Gas Flows

7.1 Introduction

The results presented in the preceding chapters give a detailed description of the flows of reactive gas mixtures, and in this chapter, we clarify the main points relating to this description. First, we revisit the general Navier–Stokes equations governing these gas systems, and provide a somewhat different point of view, related to the concept of a 'continuous medium'.

The fluid dynamic equations thus describe gas flows, possibly unsteady and multidimensional, generally dissipative, i.e. viscous, conductive, and diffusive, composed of reacting species and in thermodynamic and chemical non-equilibrium.

These governing equations, including the species conservation equations and the vibrational relaxation equations (Appendix 7.1), are then written in a dimensionless form: in these equations, 'dimensionless numbers' appear which represent ratios of particular forces or energies, thus characterizing the relative intensity of the phenomena or processes considered. Depending on the order of magnitude of these numbers, we can define specific flow regimes corresponding to the dominance of various processes.

In the last part of the present chapter, a number of basic flows are discussed; they are generally used as reference flows in the following chapters. However, the classical flows of traditional gas dynamics are described in Chapter 8.

7.2 General equations: macroscopic aspects and review

The three state quantities related to the concepts of mass, momentum, and energy, that is, ρ (or n), V, and e, are defined in Chapter 1, and in principle, they are determined from the conservation equations (1.26) and (1.29).

196 CHAPTER 7 GENERAL ASPECTS OF GAS FLOWS

In these equations, the transport quantities U_p, P, and q characterizing the local exchanges of mass, momentum, and energy are also present. These quantities may be expressed as functions of gradients of the corresponding state quantities dependent on the system under study. They can have different forms as explained in Chapters 3–6. A recapitulation of these equations and related equations are presented in Appendix 7.1.

7.2.1 Comments on the transport terms
Diffusion velocity

The mass flux of species p may be written in the following general form:

$$j_p = \rho_p U_p = \rho \sum_q \frac{M_p M_q}{M^2} D_{pq} \frac{\partial \xi_q}{\partial r} \tag{7.1}$$

As previously discussed, the 'multinary' diffusion coefficients D_{pq} may often be reduced to a binary coefficient D (Eqn. (4.18)).

Stress tensor

From a macroscopic point of view, the velocity gradients generate transformations of an elementary volume called the 'fluid particle'. These transformations consist of a deformation without any change in volume, and a dilatation without deformation. Thus, we define a tensor of deformation rate $\overline{\overline{\frac{\partial V}{\partial r}}}$ equal to

$$\overline{\overline{\frac{\partial V}{\partial r}}} = \frac{1}{2}\left(\frac{\partial V}{\partial r} + \overline{\frac{\partial V}{\partial r}}\right) \tag{7.2}$$

and a deformation rate tensor without volume change $\overset{0}{\overline{\overline{\frac{\partial V}{\partial r}}}}$, equal to

$$\overset{0}{\overline{\overline{\frac{\partial V}{\partial r}}}} = \overline{\overline{\frac{\partial V}{\partial r}}} - \frac{1}{3}\frac{\partial \cdot V}{\partial r}\mathbf{I} \tag{7.3}$$

and a dilatation rate tensor

$$\frac{\partial \cdot V}{\partial r}\mathbf{I} \tag{7.4}$$

The viscosity coefficients μ and η are then the proportionality factors between the viscous part of the stress tensor $\mathbf{P}' = \mathbf{P} - p\mathbf{I}$ and the deformation and dilatation rates (Eqns (3.42) and (4.33)). These linear relations correspond to the

7.2 GENERAL EQUATIONS: MACROSCOPIC ASPECTS AND REVIEW

first-order linear Boltzmann equation. In particular, they explain the simplifications resulting from the hypothesis of 'incompressible' flow, characterizing low velocity flows ($\frac{\partial \cdot V}{\partial r} = 0$).

Heat flux

As discussed in Chapters 3 and 4, there is a conduction (or thermal) flux q_c related to the various temperature gradients (T, T_V ...) and a diffusion flux q_d transporting the local 'internal energy'. This energy is the sum of translational, rotational, and vibrational energies plus stored potential energy (pressure): this sum constitutes the available energy or enthalpy. Thus:

$$q_d = \sum_p \rho_p \left(e_p + \frac{p_p}{\rho_p} \right) U_p = \sum_p \rho_p h_p U_p \qquad (7.5)$$

7.2.2 Particular forms of balance equations

Other forms of the usual conservation equations may shed light on diverse processes. Thus, multiplying the momentum conservation equation (1.26) by V, we obtain

$$\rho \frac{d}{dt}\left(\frac{V^2}{2}\right) = V \cdot \frac{\partial \cdot P}{\partial r} \qquad (7.6)$$

From Eqn. (7.6) we see that the rate of change of the kinetic energy of a fluid particle (along streamlines) is equal to the work done by the 'internal forces' (per unit time). This work may also be written

$$V \cdot \frac{\partial \cdot P}{\partial r} = \frac{\partial \cdot (pV)}{\partial r} + \frac{\partial \cdot (P'V)}{\partial r} - p\frac{\partial \cdot V}{\partial r} - P' : \frac{\partial V}{\partial r} \qquad (7.7)$$

The first two terms of the right-hand side of Eqn. (7.7) represent the part of this work that is exchanged with neighbouring particles and done by the pressure p and by the viscous strain P', respectively. The third term represents the dilatation work due to the pressure, and the fourth term is the part of the work 'lost' or dissipated (heat) because of the friction. This last term is called the dissipation function Φ, with

$$\Phi = -P' : \frac{\partial V}{\partial r} \qquad (7.8)$$

Another form of the energy conservation equation (1.26) may be obtained from the definition of enthalpy $h = \sum_p c_p h_p = e + \frac{p}{\rho}$, that is:

$$\rho \frac{dh}{dt} = \frac{dp}{dt} - \frac{\partial \cdot q}{\partial r} - P' : \frac{\partial V}{\partial r} \qquad (7.9)$$

with
$$dh = \sum_p c_p dh_p + \sum_p h_p dc_p$$

For a reactive mixture in thermodynamic equilibrium (TRV equilibrium), we have

$$dh_p = (C'_{TRVp} + R_p)dT = C^p_{TRVp}dT \qquad (7.10)$$

where

$$C'_{TRVp} = \frac{de_p}{dT} = \frac{C_{TRVp}}{m_p}$$

Here, $C^p_{TRVp} = C'_{TRVp} + R_p$ is the specific heat at constant pressure. Then

$$dh = \sum_p c_p C^p_{TRVp} dT + \sum_p h_p dc_p = C^p_f dT + \sum_p h_p dc_p \qquad (7.11)$$

Where $C^p_f = \sum_p c_p C^p_{TRVp}$ is the 'frozen' specific heat at constant pressure. In Eqn. (7.11), the first term arises from the temperature variations, and the second is from the concentration variations, so that in a non-reactive mixture of 'perfect' gases only the first term is present; this explains the use of the word 'frozen' for C^p_f.

In the same way, for a system in vibrational and chemical non-equilibrium, we have

$$dh_p = C'_{TRp}dT + C'_{Vp}dT_{Vp} + R_p dT = C^p_{TRp}dT + C'_{Vp}dT_{Vp}$$

or

$$dh = C^p_{TRf}dT + \sum_p c_p C'_{Vp}dT_{Vp} + \sum_p h_p dc_p \qquad (7.12)$$

where $C^p_{TRf} = \sum_p c_p C^p_{TRp}$, in this case, represents the frozen specific heat at constant pressure.

A stagnation enthalpy h_0 may also be defined, with

$$h_0 = h + \frac{V^2}{2} = e + \frac{p}{\rho} + \frac{V^2}{2} \qquad (7.13)$$

Thus, h_0 represents the total available energy. The corresponding balance equation is the following:

$$\rho \frac{dh_0}{dt} = \frac{\partial p}{\partial t} - \frac{\partial \cdot \mathbf{q}}{\partial \mathbf{r}} - \frac{\partial \cdot (\mathbf{P'V})}{\partial \mathbf{r}} \qquad (7.14)$$

We see that, in a non-dissipative steady flow, the stagnation enthalpy is preserved along the streamlines.

7.2.3 Entropy balance

For a gaseous reactive system, despite the previous conservation equations constituting a closed set, it is interesting to consider the entropy balance of the system.

Analogously with e and h, we have for the entropy per unit mass S:

$$S = \sum_p c_p S_p + S_0$$

therefore

$$dS = \sum_p c_p dS_p + \sum_p S_p dc_p \tag{7.15}$$

From Gibbs's theorem, we can define the entropy variation of the component p, dS_p, so that TdS_p represents the variation of the 'reversible' energies. Thus we have

$$TdS_p = de_p + pd\left(\frac{1}{\rho_p}\right) \tag{7.16}$$

where e_p corresponds to TRV equilibrium.

Then, the entropy balance equation may be written

$$T\frac{dS}{dt} = \frac{de}{dt} + p\frac{d}{dt}\left(\frac{1}{\rho}\right) + \sum_p (S_p T - h_p) \frac{dc_p}{dt} \tag{7.17}$$

where $h_p - S_p T = \mu_p$ represents the chemical potential of the species p.

By using the energy conservation equation, we may write Eqn (7.17) in the following form:

$$T\frac{dS}{dt} = -\frac{1}{\rho}\frac{\partial \cdot q}{\partial r} - \frac{1}{\rho}\mathbf{P'} : \frac{\partial V}{\partial r} - \sum_p \mu_p \frac{dc_p}{dt} \tag{7.18}$$

where $\frac{dc_p}{dt}$ is given by the species conservation equations (5.33).

Therefore, the entropy variation in a reactive flow is due to conduction, diffusion, dissipation, and chemical reactions. Thus, during all energy transformations taking place in such a medium (for example: kinetic energy ⇔ potential energy; internal energy ⇔ chemical energy), there are 'losses'. The losses are particularly important either when the medium is 'dissipative' (high values for μ, η, λ, D) or when the gradients of flow quantities are important. Similarly, the expression for $\frac{dS}{dt}$ is given in Appendix 7.1 for the case of vibrational and chemical non-equilibrium.

The medium is 'isentropic' if the entropy is preserved along the streamlines ($\frac{dS}{dt} = 0$): this is equivalent to the approximation of a 'perfect fluid' for which conduction, diffusion, dissipation, and chemical production are negligible. This is the case when, for example, velocity, temperature, and concentration gradients as well as the chemical production of species are small. In this case, the medium is chemically frozen or in equilibrium ($\dot{w}_p = 0$).

The flow is 'homoentropic' if the entropy remains constant in the whole medium.

The connection between the entropy S and the quantity $H = -\frac{S_D}{k}$ may be noted (Chapter 2); as we have $\frac{dH}{dt} \leq 0$, we have $\frac{dS}{dt} \geq 0$, and the stable solution corresponds to $\frac{dH}{dt} = \frac{dS}{dt} = 0$ (equilibrium).

7.2.4 Boundary conditions

Solving the system of Navier–Stokes equations as well as species evolution equations and/or vibrational relaxation equations requires coherent boundary conditions depending on the particular physical problem under study. Because of the dissipative terms, we observe that the equations have second-order derivative terms relatively to V, e (or T), and c_p.

The boundary conditions at interfaces in particular are of great importance, not only for knowledge of the flow but also for the determination of mass, momentum, and energy transfers across these interfaces. Rigorously, the conditions at an interface represent the coupling between the media located on each side and, therefore, include relations between the quantities governed by equation systems specific to each medium. Generally, we thus need two relations for each quantity governed by second-order differential equations. In fact, the influence of one medium (often the solid phase) is usually simply given by interface conditions or wall conditions considered as boundary conditions for the gaseous phase. This point of view is generally adopted later. However, examples of the simultaneous treatment of conservation equations in both media are given in Appendices 7.2 and 7.3; they correspond to important practical cases.

Thus, if the mass transfer at the level of an interface is null (general case), the normal velocity of the gas is also null. If the normal component is finite, this arises from an injection, suction, or physical process such as melting, evaporation, sublimation, or chemical reaction between both media.

For momentum transfer, it is clear that such a transfer occurs when there is a mass transfer due to a normal velocity. Otherwise, a specific condition for longitudinal velocity in the plane of the interface is required: this condition may be either the equality for the velocity of each medium at the interface or a value imposed by one of them (flow velocity null on a solid wall, for example). The wall skin friction is then deduced from this condition. In a few cases (transitional flows, for example), a 'slip velocity' should be taken into account (Appendix 6.5).

The energy transfer, knowledge of which is generally important, depends on many parameters, such as the conductivity of the medium adjacent to the gas flow and its specific energetic level. Moreover, in a reactive medium, the interfacial heat flux depends on the catalytic properties of the interface in relation to the chemical reactions.

The various boundary conditions briefly discussed above have applications in all the cases or examples treated later.

7.3 Physical aspects of the general equations

As discussed above, the terms of the general equations represent the quantitative influence of the various phenomena involved in the different balances. Thus, it is important, before solving the Navier–Stokes system, to know a priori the order of magnitude of their relative importance in order to at least simplify the system.

The general method used involves writing each physical quantity as the product of a dimensionless quantity and of an estimated average value (called the 'characteristic' or 'reference' value) taken by this quantity in the physical problem under study. This characteristic value gives the order of magnitude of the corresponding quantity, whereas the dimensionless values remain of the order of 1. Thus, in the balance equations, including n terms and written with these dimensionless quantities, there are dimensionless groups of characteristic quantities in $n-1$ terms. The value of these dimensionless numbers, in principle well-known, gives the relative order of magnitude of the different terms and therefore the relative importance of the corresponding phenomena.

However, finding the value of the characteristic quantities a priori is not always an easy task, because sometimes the corresponding quantities may vary strongly in space and time, so that it is necessary in each case to define the problem clearly.

7.3.1 Characteristic quantities

The independent variables r and t are made dimensionless with the characteristic length and time L_0 and t_0 respectively, so that

$$r^* = \frac{r}{L_0} \quad \text{and} \quad t^* = \frac{t}{t_0}$$

Independently of the particular problem under consideration, we choose L_0 and t_0 so that $L_0 = V_0 t_0$, where V_0 represents the characteristic velocity, with $V^* = \frac{V}{V_0}$.

Thus, L_0 represents the distance covered during the time t_0 by a fluid particle that has velocity V_0.

It is probably not necessary to enumerate the other characteristic quantities, except maybe for the temperature, since its definition must take into account eventual important thermal exchanges with the background. Thus, if the estimated temperature of the interface is T_w, and if T_0 is a characteristic temperature of the gaseous medium under study, we may choose a dimensionless temperature as

$$T^* = \frac{T - T_0}{T_w - T_0} = \frac{\Delta T}{\Delta T_0}$$

Of course, in this case, the transport terms are chosen in the same way. Below, we also examine the choice of the characteristic pressure p_0.

7.3.2 Dimensionless conservation equations

As discussed above, those regions that include strong gradients such as shock waves or Knudsen layers are excluded from the domain of validity of the Navier–Stokes equations. Thus, we may assume that the differential operators of these equations do not change the order of magnitude of the corresponding quantities. Then, in order to obtain dimensionless equations, we simply replace each quantity by the product of their characteristic value by the dimensionless corresponding quantity.

The mass conservation equation remains formally unchanged, i.e.

$$\frac{\partial \rho^*}{\partial t^*} + \frac{\partial \cdot (\rho^* \mathbf{V}^*)}{\partial \mathbf{r}^*} = 0 \tag{7.19}$$

If we do not make any particular assumption (incompressible flow $\rho = $ const., for example), both terms of Eqn. (7.19) have the same order of magnitude.

The momentum conservation equation may be written in the following way, without external forces:

$$\rho^* \frac{d\mathbf{V}^*}{dt^*} = -\frac{1}{Eu} \frac{\partial p^*}{\partial \mathbf{r}^*} - \frac{1}{Re} \frac{\partial \cdot \mathbf{P}'^*}{\partial \mathbf{r}^*} \tag{7.20}$$

In this equation, μ and η are referenced to μ_0, and the relaxation pressure is neglected.

Two groups of characteristic quantities appear in Eqn. (7.20), the Euler number Eu and the Reynolds number Re, with

$$Eu = \frac{\rho_0 V_0^2}{p_0} \quad \text{and} \quad Re = \frac{\rho_0 V_0 L_0}{\mu_0} = \frac{\rho_0 V_0^2 / L_0}{\mu_0 V_0 / L_0^2} \tag{7.21}$$

The balance of forces acting on a fluid particle represented in Eqn. (7.20) includes the inertial force (left-hand side), generating kinetic energy and pressure, and viscosity forces (right-hand side), so that the inverse of the Euler

7.3 PHYSICAL ASPECTS OF THE GENERAL EQUATIONS

and Reynolds numbers represent the ratio of pressure and viscosity forces, respectively, to the inertial force.

In the expression of *Eu*, the choice of p_0 must take into account the change of pressure between its stagnation value (or its 'reservoir' value) and its average local value. The difference is essentially due to the transformation of this energy into kinetic energy. Thus, these energies have generally the same order of magnitude, so that the Euler number remains of the order of 1.

In contrast, the Reynolds number may take very different values depending on whether the viscosity forces (stress, friction) are more or less important in comparison with the inertial force or whether the viscous energy exchanged is more or less important in comparison with the kinetic energy.

The energy conservation equation in the form of Eqn. (7.9) may be easily written in a dimensionless form, and assuming thermodynamic equilibrium, we have

$$\rho^* \frac{dh^*}{dt^*} = \rho^* C_f^{p^*} \frac{dT^*}{dt^*} + \rho^* \sum_p h_p^* \frac{dc_p}{dt^*}$$

$$= E \frac{dp^*}{dt^*} + \frac{1}{ReP} \frac{\partial \cdot}{\partial r^*} \left(\lambda^* \frac{\partial T^*}{\partial r^*} \right)$$

$$+ \frac{1}{ReS} \frac{\partial \cdot}{\partial r^*} \left(\rho^* D^* \sum_p h_p^* \frac{\partial c_p}{\partial r^*} \right) + \frac{E}{Re} \Phi^* \quad (7.22)$$

Here, a reference enthalpy h_0 has of course been chosen, with $h_0 = C_{f_0}^p T_0$.

The case with simultaneous vibrational non-equilibrium is treated in Appendix 7.1.

In Eqn. (7.22), we find the Reynolds number and the other following dimensionless numbers:

$$\text{Eckert number:} \quad E = \frac{V_0^2}{C_{f_0}^p T_0} \simeq \frac{p_0}{\rho_0 C_{f_0}^p T_0} \quad (7.23)$$

$$\text{Prandtl number:} \quad P = \frac{C_{f_0}^p \mu_0}{\lambda_0} \quad (7.24)$$

In a reactive flow, this is a 'frozen' Prandtl number.

$$\text{Schmidt number:} \quad S = \frac{\mu_0}{\rho_0 D_0} \quad (7.25)$$

Thus, we see that the groups of terms E, $\frac{1}{ReP}$, $\frac{1}{ReS}$, $\frac{E}{Re}$ represent ratios of pressure energy (or kinetic energy), conduction energy, diffusion energy, and viscous dissipation energy respectively to the enthalpy transported (convected) in the flow.

Of course, the order of magnitude of the ratio of two peculiar energies appearing in Eqn. (7.22) is obtained by writing the ratio of the corresponding dimensionless groups. Thus, for example, the ratio of the kinetic energy to the energy dissipated by viscosity is equal to $E/\frac{E}{Re} = Re$. We again find this ratio in the momentum conservation equation.

The species conservation equation may be written in the following dimensionless form:

$$\rho^* \frac{dc_p}{dt^*} = \frac{1}{ReS}\left[Da\, \dot{w}_p^* + \frac{\partial \cdot}{\partial r^*}\left(\rho^* D^* \frac{\partial c_p}{\partial r^*}\right)\right] \qquad (7.26)$$

In addition to the preceding numbers Re, S, a new number appears, i.e.

$$Da = \frac{\dot{w}_{p0} L_0^2}{\rho_0 D_0} \qquad (7.27)$$

Here, Da is the Damköhler number.

Thus, in Eqn. (7.26), $\frac{Da}{ReS}$ gives the order of magnitude of the ratio of the mass rate of species p created by chemical reactions to the mass rate transported in the flow, that is $\dot{w}_{p0}/\frac{\rho_0 V_0}{L_0}$. Similarly, $1/ReS$ represents the ratio of the transported mass rate to the diffused mass rate. This ratio, of course, has also been found in the energy balance equation.

7.3.3 Dimensionless numbers: flow classification

In the following, we examine the physical meaning of the dimensionless numbers defined above, and we analyse the consequences of their order of magnitude on the structure of conservation equations.

Reynolds number

The Reynolds number is related to a characteristic length which may vary considerably according to the specific problem under consideration. Thus, practically, it may vary from 10^{-1} to 10^7.

Independently of those (rare) cases in which the viscous terms are predominant (Stokes–Oseen flow, for example), there are many flow regimes where the convection and pressure terms have the same order of magnitude as the viscous terms. This is particularly the case for high-speed flows, more or less rarefied, in wind tunnels or in real flight at high altitude.

For denser flows, the Reynolds number referring to a characteristic dimension of the flow may be high enough to neglect the viscous terms. Independently of the stability problems that may then arise (see Appendix 8.3, for example), we again find the approximation of 'perfect fluid'. This approximation of course may also be applied to the energy conservation equation in which conduction, diffusion, and dissipation terms may be neglected when Prandtl and Schmidt numbers (as well as the Eckert number) are of the order of 1, which is the general case for gases: the conservation equation system is then the Euler system, closed using the species conservation equations in which the diffusion term is also neglected. The influence of chemical reactions disappears only if $Da \ll ReS$, that is if $\dot{w}_{p0} \ll \frac{\rho_0 V_0}{L_0}$. In this case, the species mass rate created by chemical reaction is very low compared with the mass rate transported in the flow.

Eckert number

When the value of this term is low, the average kinetic energy of the fluid is not very important compared with the internal energy transported by convection. This is for example the case for cooling flows at moderate velocities in exchangers or for flows behind strong shock waves (Chapter 9). Then, for a flow without chemical reaction, the energy conservation equation is simply written

$$\rho C_f^p \frac{dT}{dt} = \frac{\partial \cdot}{\partial r}\left(\lambda \frac{\partial T}{\partial r}\right) \tag{7.28}$$

This equation is all the more 'exact', as the Prandtl number is small.

The case of high Eckert numbers corresponds generally to high-velocity flows where compressibility effects are important and the viscous dissipation non-negligible. These effects are of course increased if the Reynolds number is small, that is, for example, for low-density flows. Generally, however, the Eckert number remains of the order of 1.

Prandtl and Schmidt numbers

These two numbers are essentially related to the state of the fluid and are, at least directly, independent of the velocity.

The product $Br = EP$ (Brinkmann number) represents the ratio of the energy dissipated by viscosity to the conduction energy. Equation (7.28) corresponds to the case $Br \ll 1$, whereas the case $Br \gg 1$ corresponds to viscous and non-conductive flows. Similarly, the product ES represents the ratio of the energy dissipated to the diffusion energy. The product $Pe = ReP$ is the Peclet number.

The ratio of the Prandtl number to the Schmidt number is the Lewis number L,

$$L = \frac{P}{S} = \frac{\rho_0 D_0 C_{f0}^p}{\lambda_0} = \frac{\rho_0 D_0 C_{f0}^p T_0/L_0^2}{\lambda_0 T_0/L_0^2} \qquad (7.29)$$

which represents the ratio of diffusion energy to conduction energy.

Damköhler number

In reactive gas flows, the order of magnitude of the Damköhler number determines different flow categories. As pointed out above, it represents the ratio of chemical reaction rates to the diffusion velocities, and $\frac{Da}{ReS}$ represents the ratio of chemical reaction rates to the macroscopic velocities (convection).

Thus, if $\frac{Da}{ReS} \gg 1$, the convection term may be neglected. Moreover, if the value of Da is much higher than unity, the species conservation equation reduces to

$$\dot{w}_p = 0 \qquad (7.30)$$

In this case, we have a flow in chemical equilibrium. The reaction rates are much faster than the diffusion and convection velocities. The local concentrations are then determined only by the chemical reactions, which, however, depend on local parameters given by the Euler or Navier–Stokes equations.

The opposite case corresponds to frozen flows: the reaction rates are too slow to have an influence on the local concentrations which preserve their value along the streamlines if the Reynolds and Schmidt numbers are not too low.

Non-independent dimensionless numbers

There are numbers widely used but in fact non-independent; they are combinations of the previously defined numbers but may also represent ratios of forces or energies.

We have seen examples with Lewis and Peclet numbers. Another well-known dimensionless number is the Mach number M, important in gas dynamics because its value delimits flow regions with different characteristics (Chapter 8). Thus, for a perfect gas in TRV equilibrium, we have

$$M = \frac{V_0}{a_0} = \left(\frac{V_0^2}{\Gamma RT_0}\right)^{1/2} = \left(\frac{E}{\Gamma - 1}\right)^{1/2} \qquad (7.31)$$

with

$$\Gamma = \frac{C_{TRV}^p}{C_{TRV}'}$$

Other dimensionless numbers

Independently of the dimensionless numbers related to external forces (gravity, magnetic field, and so on), there are numbers simply used for the representation of values of a particular quantity (in a complex situation) by comparison with values we would obtain in a simpler well-known (but virtual) situation.

Thus, at an interface, we define a skin-friction coefficient representing the transfer of longitudinal momentum in a dimensionless form: for example, in the case of a solid wall limiting the flow, this coefficient is chosen as the ratio of the modulus of the stress tension in the plane of the wall τ_w to a reference 'dynamic pressure' $\frac{1}{2}\rho_0 V_0^2$ (Chapter 8), i.e.

$$C_f = \frac{\tau_w}{\frac{1}{2}\rho_0 V_0^2} \tag{7.32}$$

Similarly, the normal component of the heat flux q_w referenced to a purely conductive reference flux $\frac{\lambda_0 \Delta T_0}{L_0}$ is written in the form of the Nusselt number Nu, defined as

$$Nu = \frac{-q_w}{\lambda_0 \Delta T_0 / L_0} \tag{7.33}$$

Here, q_w may also be referenced to a transported (convected) reference heat flux $\rho_0 C_{f0}^p V_0 \Delta T_0$: in this way, we define the Stanton numbers St as

$$St = \frac{-q_w}{\rho_0 C_{f0}^p V_0 \Delta T_0} \tag{7.34}$$

This way of representing the characteristic quantities of a process (heat flux, skin friction, and so on) may be simply deduced from a dimensional analysis independent of the conservation equations. An example is given in Appendix 7.4.

7.4 Characteristic general flows

A few simple typical flows are often used as reference flows, and their essential aspects are described below.

7.4.1 Steady flows

These are also called permanent flows ($\frac{\partial}{\partial t} = 0$): the flow quantities depend on spatial coordinates only.

Global balances

It is sometimes unnecessary to know the details of the flows or the 'profiles' of various quantities. Global balances are then sufficient and are applied to a finite volume of fluid D limited by a closed surface S by integrating the general equations in D. Thus, the mass conservation gives

$$\int_D \frac{\partial \cdot \rho V}{\partial r} dr = 0 \text{ or } \int_S \rho V \cdot N\, dS = 0 \quad \text{(Ostrogradski theorem)} \tag{7.35}$$

The global mass flux is null if there is no mass source in D.

For the momentum, we have

$$\int_D \rho V \cdot \frac{\partial V}{\partial r} dr = -\int_D \frac{\partial \cdot P}{\partial r} dr \tag{7.36}$$

With the continuity equation, we deduce the following equation:

$$\int_S \rho V V \cdot N\, dS = -\int_S P \cdot N\, dS \tag{7.37}$$

This is the 'momentum theorem', the basis of the theory of propellers, wind engines, jet engines, and so on: the momentum flux is equal to the sum of forces (here, internal forces only) applied to the system.

Finally, for energy, we have

$$\int_D \frac{\partial \cdot \rho V h_0}{\partial r} dr = -\int_D \frac{\partial \cdot}{\partial r} \left(q + P' \cdot V\right) dr \tag{7.38}$$

or

$$\int_S \rho V h_0 \cdot N\, dS = -\int_S \left(q + P' \cdot V\right) \cdot N\, dS \tag{7.39}$$

For a perfect fluid, the right-hand side of Eqn. (7.39) is null.

Perfect fluid flows

$$\frac{dh_0}{dt} = 0 \tag{7.40}$$

That is:

$$h_0 = e + \frac{p}{\rho} + \frac{V^2}{2} = \text{const.} \tag{7.41}$$

Total enthalpy is preserved along trajectories. Temperature T_0 and pressure p_0 are the 'reservoir' temperature and pressure respectively ($V = 0$). The flow is 'isoenergetic', and Eqn. (7.40) shows only the possible transformations undergone by a perfect fluid without losses.

A similar equation arises from the momentum conservation equation, that is, along trajectories:

$$\int \frac{dp}{\rho} + \frac{V^2}{2} = \text{const.} \qquad (7.42)$$

with

$$dh = \frac{dp}{\rho}$$

For an isentropic flow (perfect fluid in equilibrium), h is a function only of T, and $p = \rho RT$. Therefore $p = f(\rho)$, and we have

$$\frac{dp}{\rho} = \left(\frac{dp}{d\rho}\right)_S \frac{d\rho}{\rho} \qquad (7.43)$$

with

$$\left(\frac{dp}{d\rho}\right)_S = g(T)$$

This term, of course, depends on the gas species.

7.4.2 Unsteady flows

One basic problem concerns the propagation of waves in gaseous media. If these waves are assumed to be of weak intensity, their behaviour may be deduced from the continuity and momentum equations of the Euler system, since in that case, the regime remains isentropic. Then, we can linearize the quantities p, ρ, T from the values of p_0, ρ_0, T_0 corresponding to the medium at rest ($V_0 = 0$), so that $p = p_0 + \tilde{p}$ and so on. Then we find the following equation:

$$\frac{\partial^2 \tilde{\rho}}{\partial t^2} = \left(\frac{dp}{d\rho}\right)_S \frac{\partial^2 \tilde{\rho}}{\partial r^2} \qquad (7.44)$$

where

$$\frac{\partial^2 X}{\partial r^2} = \frac{\partial \cdot}{\partial r}\left(\frac{\partial X}{\partial r}\right)$$

We find similar equations for $\tilde{p}$, $\tilde{V}$, $\tilde{T}$, and so on.

We set

$$\left(\frac{dp}{d\rho}\right)_S = a^2 \qquad (7.45)$$

where a depends on temperature and on gas species.

The solutions of Eqn. (7.44) have the following form:

$$\tilde{\rho} = F(r - at) + G(r + at) \qquad (7.46)$$

and a represents the speed of propagation of these small perturbations (sound velocity). Each term of the right-hand side of Eqn. (7.46) represents a 'simple wave' propagating without deformation towards $r > 0$ and $r < 0$ respectively (spherical waves). Developments concerning this type of wave in a moving medium are given in Chapter 8.

7.4.3 Simplified flow models

One-dimensional approximation

In many cases, in particular for 'internal' flows (limited by walls), there is a preferential direction, or principal direction, of the flow.

For an isentropic flow, we may then assume that the flow quantities depend only on this direction (x for example). In fact, they represent an average value in each cross section $S(x)$ of the flow, that is $p_m(x)$, $T_m(x)$..., and so on. Then, the Euler equations remain valid except for the continuity equation, which must be modified by writing the mass flow rate conservation, i.e.

$$\int_S \rho V \cdot N \, ds = \rho_m V_m S = \text{const.} \tag{7.47}$$

The momentum conservation equation is written as follows:

$$V_m \frac{dV_m}{dx} = -\frac{1}{\rho_m} \frac{dp_m}{dx} = -a_m^2 \frac{1}{\rho_m} \frac{d\rho_m}{dx} \tag{7.48}$$

In dissipative and fully developed flows, a one-dimensional approximation is also usable by defining average transport terms characterizing the average energy dissipation or momentum and energy exchange with the background. Typical and simple examples are represented by Fanno and Rayleigh flows.

Two-dimensional approximation

Two-dimensional flows may be described using a coordinate system depending on the flow geometry. The most usual examples are the 'plane' flows (Cartesian coordinates x, y or polar coordinates r, θ) and the axisymmetric flows (semi-polar coordinates r, z). Examples are given in the next chapters.

Another type of description of some flows uses 'intrinsic' (or 'natural') coordinates consisting of trajectories s and straight lines normal to these trajectories, n, with

$$s = s\boldsymbol{\alpha} \quad \text{and} \quad n = n\boldsymbol{\beta} \tag{7.49}$$

where $\boldsymbol{\alpha}$ and $\boldsymbol{\beta}$ are unit vectors.

Then we have

$$V = \frac{d\mathbf{r}}{dt} = \frac{ds}{dt}\boldsymbol{\alpha} = V\boldsymbol{\alpha} \qquad (7.50)$$

and

$$\frac{d\mathbf{V}}{dt} = \frac{d^2\mathbf{r}}{dt^2} = \frac{dV}{dt}\boldsymbol{\alpha} + V\frac{d\boldsymbol{\alpha}}{dt} = \frac{dV}{dt}\boldsymbol{\alpha} + \frac{V^2}{R}\boldsymbol{\beta} \qquad (7.51)$$

where R is the local curvature radius of the trajectory.

For a permanent isentropic flow, the components of the momentum equation on s and n are, respectively:

$$\rho V \frac{\partial V}{\partial s} = -\frac{\partial p}{\partial s} \quad \text{and} \quad \rho \frac{V^2}{R} = -\frac{\partial p}{\partial n} \qquad (7.52)$$

Thus, when the curvature of the flow is not very important, the pressure is constant along perpendicular lines to the trajectories. Furthermore, as we have $\frac{\partial S}{\partial s} = \frac{\partial h_0}{\partial s} = 0$, we also have

$$T\frac{dS}{dn} = \frac{dh_0}{dn} + \left(\frac{V}{R} - \frac{\partial V}{\partial n}\right)V \qquad (7.53)$$

Here, $\zeta = \frac{V}{R} - \frac{\partial V}{\partial n}$ is the flow 'vorticity', and if the total enthalpy is constant (general case), the entropy change is due to this vorticity, representing the dissipative exchange factor between the fluid trajectories.

7.4.4 Stability of the flows: turbulent flows

When, among other factors, the Reynolds number of a dissipative flow increases beyond a 'critical' value, the flow becomes unstable and typically unsteady. The resulting perturbations present varying frequencies but generally small amplitudes; the structure of the flow, however, is completely modified. The deterministic preceding equations remain locally valid but not at all suitable. A statistical analysis may be generally applied, and the flow is then considered as the superposition of an 'average' (but fictitious) flow and of a fluctuating flow.

The description of the average flow is of course essential for knowledge of the global properties. However, between the fictitious streamlines of this flow, mass, momentum, and energy exchanges take place because of these fluctuations. Turbulent transport processes are then induced by these exchanges and are added to molecular transport phenomena. They become generally dominant, and their modelling is necessary. Complete description of these turbulent flows is beyond the scope of this book, but an overview of a few of them, sometimes encountered in gas dynamics, is given in Appendix 8.3. We should be aware that such a statistical treatment is not always possible, because of the non-random character of certain types of fluctuations.

Appendix 7.1 General equations: review

Navier–Stokes equations

$$\text{Mass balance:} \quad \frac{\partial \rho}{\partial t} + \frac{\partial \cdot (\rho V)}{\partial r} = 0 \tag{7.54}$$

$$\text{Momentum balance:} \quad \rho \frac{dV}{dt} = -\frac{\partial p}{\partial r} + \frac{\partial \cdot}{\partial r}\left(2\mu \overline{\frac{\partial V}{\partial r}}^{0}\right) + \frac{\partial}{\partial r}\left(\eta \frac{\partial \cdot V}{\partial r}\right) \tag{7.55}$$

$$\text{Energy balance:} \quad \rho \frac{dh}{dt} = \frac{dp}{dt} - \frac{\partial \cdot q}{\partial r} + \left(2\mu \overline{\frac{\partial V}{\partial r}}^{0} + \eta \frac{\partial \cdot V}{\partial r}\mathbf{I}\right) : \frac{\partial V}{\partial r} \tag{7.56}$$

with

$$p = \rho RT$$

$$h = e + \frac{p}{\rho}$$

$$p = \sum_p p_p, \quad \rho = \sum_p c_p \rho_p, \quad R = \sum_p c_p R_p, \quad h = \sum_p c_p h_p$$

$$h_p = e_p + \frac{p_p}{\rho_p} = e_{TRp} + e_{Vp} + R_p T = \frac{7}{2} R_p T + e_{Vp}$$

$$q = q_{TR} + \sum_p q_{Vp} = -\lambda_{TR} \frac{\partial T}{\partial r} - \sum_p \lambda_{Vp} \frac{\partial T_{Vp}}{\partial r} - \rho D \sum_p h_p \frac{\partial c_p}{\partial r}$$

In the case of thermodynamic (TRV) and chemical equilibrium, we have

$$e_{Vp} = \bar{e}_{Vp} \text{ and } (= f(T)) \tag{7.57}$$

$$q = -\lambda_{TRV} \frac{\partial T}{\partial r} - \rho D \sum_p h_p \frac{\partial c_p}{\partial r} \tag{7.58}$$

since

$$T_{Vp} = T$$

and

$$\dot{w}_p = 0 \tag{7.59}$$

In thermodynamic equilibrium but in chemical non-equilibrium, Eqns (7.57) and (7.58) remain valid, but Eqn. (7.59) is replaced by

$$\rho \frac{dc_p}{dt} = \dot{w}_p + \frac{\partial \cdot}{\partial r}\left(\rho D \frac{\partial c_p}{\partial r}\right) \tag{7.60}$$

In a general way, $\dot{w}_p$ is given by Eqn. (9.36) when several reactions are involved. The rate constants of these reactions must eventually take into account vibrational non-equilibrium (Chapters 5 and 9).

In thermodynamic and chemical non-equilibrium, Eqn. (7.59) must be replaced by Eqn. (7.60), and Eqn. (7.57) by the following:

$$\rho_p \frac{de_{Vp}}{dt} = \dot{w}_{Vp} - \frac{\partial \cdot q_{Vp}}{\partial r} \tag{7.61}$$

with

$$q_{Vp} = -\lambda_{Vp}\frac{\partial T_{Vp}}{\partial r} - \rho D e_{Vp}\frac{\partial c_p}{\partial r}$$

The source term $\dot{w}_{Vp}$ takes into account the vibrational balance of the TV and VV collisions and that of the reactive collisions (Chapters 5 and 9).

The preceding equations take into account the simplifications pointed out in the first part of this book. In particular, the heat fluxes due to the different energy modes depend only on the corresponding temperature gradients, and the relaxation pressure is neglected. However, the neglected terms may be considered as perturbations and eventually calculated (Chapter 9). Physical models are also necessary for calculating the terms $\dot{w}_p$ and $\dot{w}_{Vp}$; thus, a few of these were presented above (Chapter 5).

Dimensionless vibrational relaxation equation

In dimensionless form, Eqn. (7.61) may be written

$$\rho_p^* \frac{de_{Vp}^*}{dt^*} = \frac{\dot{w}_{Vp0}L_0}{\rho_0 e_{Vp0} V_0}\dot{w}_{Vp}^* + \frac{\lambda_{Vp0}}{\rho_0 C_{Vp0} V_0 L_0}\frac{\partial \cdot}{\partial r^*}\left(\lambda_{Vp}^* \frac{\partial T_{Vp}^*}{\partial r^*}\right)$$
$$+ \frac{D_0}{V_0 L_0}\frac{\partial \cdot}{\partial r^*}\left(\rho^* D^* e_{Vp}^* \frac{\partial c_p}{\partial r^*}\right) \tag{7.62}$$

The group of terms $X_p = \frac{\dot{w}_{Vp0}L_0}{\rho_0 e_{Vp0} V_0}$ represents the ratio of the vibrational energy rate created by collisions to the vibrational energy rate transported in the flow. The ratio of this created energy rate to the diffused vibrational energy rate defines, as in the case of the species conservation equation, a 'vibrational Damköhler number' Da_V, such that

$$Da_V = X_p ReS$$

The dimensionless relaxation equation may then be written

$$\rho_p^* \frac{de_{Vp}^*}{dt^*} = \frac{1}{ReS}\left[Da_V \dot{w}_{Vp}^* + \frac{\partial \cdot}{\partial r^*}\left(\frac{1}{L_{Vp}}\lambda_{Vp}^*\frac{\partial T_{Vp}^*}{\partial r^*} + \rho^* D^* e_{Vp}^* \frac{\partial c_p}{\partial r^*}\right)\right] \quad (7.63)$$

where $L_{Vp} = \frac{\rho_0 D_0 C_{Vp0}}{\lambda_{Vp0}}$ is the 'vibrational Lewis number' of the species p, representing the ratio of the diffused vibrational heat flux to the conducted vibrational heat flux. This number is generally of the order of 1, as are most of the other dimensionless numbers.

In the energy conservation equation governing a flow in vibrational and chemical non-equilibrium (not written here), we find a (non-independent) dimensionless number, that is the ratio of the frozen Lewis number to the vibrational Lewis number:

$$F_p = \frac{C_{TR0}^p \lambda_{Vp0}}{C_{Vp0} \lambda_{TR0}} = \frac{L_f}{L_{Vp}} \quad (7.64)$$

The number F_p is called the 'vibrational number' of the species p.

Entropy balance in thermodynamic and chemical non-equilibrium media

In such media, the Gibbs relation may be written

$$TdS_p = de_{TRp} + p_p d\left(\frac{1}{\rho_p}\right) \quad (7.65)$$

We also have the following obvious relations:

$$T_{Vp} dS_{Vp} = de_{Vp}$$

$$\mu_{TRp} = e_{TRp} + \frac{p_p}{\rho_p} - TS_{TRp}$$

$$\mu_{Vp} = e_{Vp} - T_{Vp} S_{Vp}$$

$$\mu_p = \mu_{TRp} + \mu_{Vp}$$

Then, we find the following balance:

$$T\frac{dS}{dt} = \frac{de}{dt} + p\frac{d}{dt}\left(\frac{1}{\rho}\right) + \sum_p\left[\mu_p \frac{dc_p}{dt} + (T - T_{Vp})\frac{d}{dt}(c_p S_{Vp})\right] \quad (7.66)$$

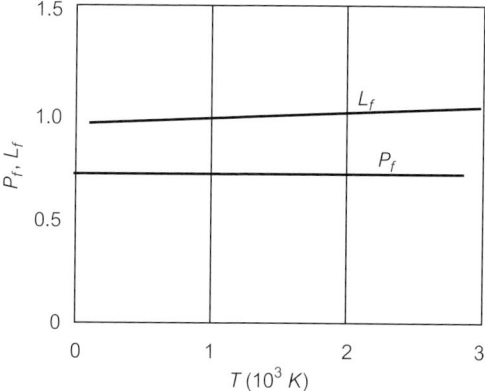

Figure 21. Frozen Prandtl and Lewis numbers (air).

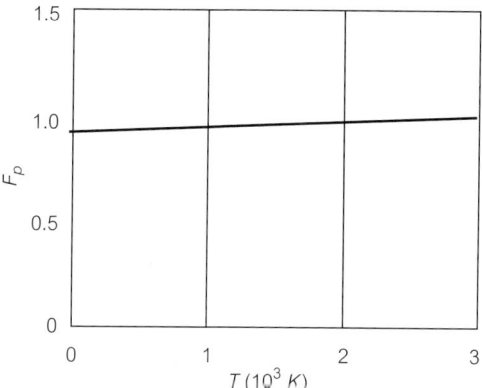

Figure 22. Vibrational number (nitrogen in air).

In thermodynamic equilibrium, we again find Eqn. (7.17).

We can also develop Eqn. (7.66) by using the energy and species conservation equations and the vibrational relaxation equations.

Representation of dimensionless numbers

As an example, the frozen Prandtl and Lewis numbers for air, i.e.

$$P_f = \frac{\mu C^p_{TRf}}{\lambda_{TR}} \quad \text{and} \quad L_f = \frac{\rho D C^p_{TRf}}{\lambda_{TR}} \tag{7.67}$$

are represented in Fig. 21.

Similarly, the vibrational number for nitrogen in air is shown in Fig. 22. The computations for moderate temperatures are made from results obtained in preceding chapters.[28]

In the same temperature range, we observe that these numbers remain practically constant (no chemical change).

Appendix 7.2 Unsteady heat flux at a gas–solid interface

It is often important to determine the unsteady wall heat flux $q_w(t)$ arising from a gaseous medium that has a characteristic temperature higher than that of the wall material (generally insulation material). This may be done by measuring the temperature increase $\Delta T_w(t)$ of a thin metallic film deposited on the wall. This temperature increase induces a resistance variation and therefore a voltage variation if the film is crossed by a constant electric current. The problem is therefore to find a relation between $q_w(t)$ and $\Delta T_w(t)$.[65,66]

To do this, we assume a one-dimensional heat conduction in the wall material normal to the gas flow (coordinate y; the wall surface is at $y = 0$). If ρ_s, c_s, and λ_s are the density, specific heat, and thermal conductivity of the material, assumed to have constant values, we can write the energy equation in the solid in the form

$$\rho_s c_s \frac{\partial T_s}{\partial t} = \lambda_s \frac{\partial^2 T_s}{\partial y^2} \tag{7.68}$$

with initial and boundary conditions:

$$t = 0 \quad T_s = T_a$$
$$y \to \infty \quad T_s \to T_a$$
$$y = 0 \quad q = -\left(\lambda_s \frac{\partial T_s}{\partial y}\right)_0 = -q_w$$

Equation (7.68) may be easily solved, for example with the Laplace–Carson transformation, so that the relation between q_w and ΔT_w may be written

$$\frac{q_w}{K_s} = \frac{\Delta T_{w0}}{\sqrt{t}} + \int_0^t \left(\frac{d(\Delta T_w)}{du}\right)(t-u)^{-1/2} du \tag{7.69}$$

with ΔT_{w0} being the temperature increase at $t = 0$, eventually non-zero (passage of a shock wave, for example), and with $K_s = \sqrt{\frac{\rho_s c_s \lambda_s}{\pi}}$.

We may also transform Eqn. (7.69) in order to avoid the use of the derivative of an experimental quantity. Thus, we find

$$\frac{q_w}{K_s} = \frac{\Delta T_w(t)}{\sqrt{t}} + \frac{1}{2} \int_0^t \frac{\Delta T_w(t) - \Delta T_w(u)}{(t-u)^{3/2}} du \qquad (7.70)$$

Two important practical cases should be pointed out:

- If the heat flux is constant $(q_w = q_0)$ from $t = 0$, we simply have:

$$\Delta T_w = \frac{2q_0}{\pi} \frac{\sqrt{t}}{K_s} \qquad (7.71)$$

and the temperature increase is parabolic.

- If the heat flux is proportional to $t^{-1/2}$ (Chapter 8), we have:

$$\Delta T_w = \text{const.} \qquad (7.72)$$

The finite heat capacity of the film may be the cause of a time lag in the temperature signal (Chapter 11).

Remark

The theoretical determination of an unsteady heat flux in a non-reactive gas flow is generally carried out by assuming either a constant wall temperature (conductive material) or a zero heat flux (insulating material, adiabatic wall). In the first case, the wall temperature increase is supposedly null so that the calculated temperature increase, deduced from the computed heat flux ΔT_w, is compared to the measured value of ΔT_w. This increase of course must be low enough to validate the result of the computation. Iterations may be necessary.

Appendix 7.3 Gas–liquid interfaces

Compared with the gas–solid interfaces, the gas–liquid interfaces pose more problems, essentially because the momentum transfer may give rise to instabilities of these interfaces and modifications of their structure. A typical example is the gravity flow of a liquid film along the wall of the inner cylinder in the annular space separating two vertical concentric cylinders, while a gaseous counterflow goes upwards along the outer cylinder[67] (Fig. 23). For high gas velocities, a part of the film may be swept along upwards, and the interface undergoing important stresses may warp.

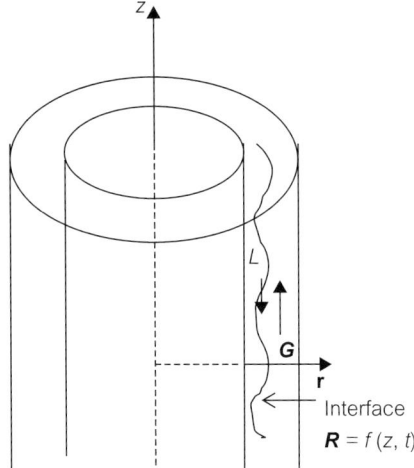

Figure 23. Outline of a gas–liquid interface.

The usual Navier–Stokes equations written in cylindrical coordinates r, z are intended to govern the flow of each phase k (l: liquid; g: gas). The matching conditions at the interface essentially concern the momentum transfer. Thus, if u_k and v_k are the axial and transverse components of the velocity V_k in each phase, we have at the interface:

$$u = u_k$$

$$v_k = \frac{\partial f}{\partial t} + u_k \frac{\partial f}{\partial z}$$

where $R = f(z, t)$ represents the equation of the interface, and $k = (l, g)$.

If we neglect mass exchanges and surface-tension effects, the momentum balance at the interface is simply

$$\boldsymbol{n}_l \cdot \mathbf{P}_l + \boldsymbol{n}_g \cdot \mathbf{P}_g = 0 \tag{7.73}$$

where $\boldsymbol{n}_l$ and $\boldsymbol{n}_g$ represent the vector units normal to the interface. The components of this equation on the tangent plane and on the straight line normal to the surface respectively give two conditions for the components of the velocity at the interface, i.e. for a monatomic gas and an incompressible liquid:

$$\mu_l \left[2 \left(v_{lr} - u_{lz} \right) f_z + \left(v_{lz} + u_{lr} \right) \left(1 - f_z^2 \right) \right]$$
$$= \mu_g \left[2 \left(v_{gr} - u_{gz} \right) f_z + \left(v_{gz} + u_{gr} \right) \left(1 - f_z^2 \right) \right]$$

and

$$p_l - 2\mu_l \left[v_{lr} - \left(v_{lz} + u_{lr} \right) f_z + u_{lz} f_z^2 \right] \left(1 + f_z^2 \right)^{-1}$$
$$= p_g - 2\mu_g \left[v_{gr} - \left(v_{gz} + u_{gr} \right) f_z + u_{gz} f_z^2 \right] \left(1 + f_z^2 \right)^{-1}$$

where the indices r and z correspond to derivatives with regard to these variables respectively.

If the two phases have very different average temperatures, the energy balance at the interface may be written

$$q_l \cdot n_l + q_g \cdot n_g = 0 \qquad (7.74)$$

These equations are difficult to solve, and generally only approximate solutions are available, including a periodic structure of the interface, which constitutes an ideal case. Stability analyses are also necessary in order to understand the influence of various parameters.[68]

The importance of this problem is particularly clear in the following two examples. The first example concerns the security problems of nuclear reactors where the metallic coating of the fissile nuclear material contained in vertical cylindrical 'needles' begins to melt and flows downwards. This flow interacts with an upwards gaseous flow resulting from evaporation of the cooling liquid.[69]

The second example is somewhat different and concerns the entry of a hypersonic vehicle into a dense atmosphere.[70] If the entry is too 'steep', the thermal coating of the vehicle may melt from the stagnation point where thermal transfer is most intense. Then, a molten film flows along the vehicle and interacts with the gaseous flow.

In both cases, if there is no mass transfer between phases (no wrench), energy and momentum transfers are important, the structure of the interface changes, surface waves appear, and the flowing liquid may solidify again in colder zones.

Appendix 7.4 Dimensional analysis

In situations where the use of the general conservation equations is difficult (complex geometry, for example) but where experiments may be carried out, the problem is how to relate the measured quantity to the characteristic quantities of the flow. A dimensional analysis may then be used, and the non-dimensional quantity is found to depend on characteristic dimensionless numbers such as those defined above.

As an example, if the measured quantity is the wall heat flux q_w, we first list the general parameters that have an influence on this flux. Thus, a priori, these parameters are V, characteristic velocity of the fluid, L, characteristic length (duct or body), ρ, C^p, μ, λ, state and transport properties of the fluid, and $\Delta T = T_0 - T_w$, characteristic temperature difference between the fluid and the wall. We may then write

$$q_w = f\left(V, \rho, L, C^p, \mu, \lambda, \Delta T\right)$$

Assuming a dependence of q_w in the form of product of terms, we have

$$q_w = K V^x \rho^y L^z C^{pt} \mu^u \lambda^v \Delta T^w \tag{7.75}$$

where K is an undetermined constant.

If each physical quantity is written as a function of the fundamental quantities: mass M, length L, time ϑ, temperature T, we write

$$q_w \sim M\vartheta^{-3}, \quad V \sim L\vartheta^{-1}, \quad \rho \sim ML^{-3}, \quad C^p \sim L^2\vartheta^{-2}T^{-1},$$

$$\mu \sim ML^{-1}\vartheta^{-1}, \text{ and } \lambda \sim ML\vartheta^{-3}T^{-1}$$

Putting these expressions into Eqn. (7.75) and equating the homologous exponents, we obtain the following system:

$$y + u + v = 1$$
$$x - 3y + z - u + 2t + v = 0$$
$$x + u + 2t + 3v = 3$$
$$t + v - w = 0$$

We can eliminate four exponents from this system, for example x, y, z, v, so that we obtain

$$q_w = K\rho V C^p \Delta T \, (Re)^z \, (P)^{u-w} \, (E)^{1-w}$$

where Re, P, and E are the dimensionless numbers previously defined.

More generally, we can therefore write

$$St = f_1 \, (Re, P, E)$$

We find a similar equation for the Nusselt number, i.e.

$$Nu = f_2 \, (Re, P, E)$$

Thus, a priori, we know how to express and to represent the experimental values of the heat flux. Any other experimental quantity, of course, may be similarly analysed.

Appendix 7.5 Generalities on total balances

Control volume

In a general unsteady regime, if we consider a 'control volume' D, limited by a closed surface S, the total rate of change $\tilde{A}$ of a local extensive property a is equal to the sum of the rate of change of the property inside D and of the net flux of this property across S, i.e.

$$\tilde{A} = \frac{\partial}{\partial t} \int_D a \, d\mathbf{r} + \int_S a\mathbf{V} \cdot \mathbf{N} \, ds \qquad (7.76)$$

Thus, if $a = \rho, \rho \mathbf{V}, \rho h_0$, without external forces, we obtain the following global conservation equations:

$$\frac{\partial}{\partial t} \int_D \rho \, d\mathbf{r} + \int_S \rho \mathbf{V} \cdot \mathbf{N} \, dS = 0$$

$$\frac{\partial}{\partial t} \int_D \rho \mathbf{V} \, d\mathbf{r} + \int_S \rho \mathbf{V} \mathbf{V} \cdot \mathbf{N} \, dS + \int_S \mathbf{P} \cdot \mathbf{N} \, dS = 0$$

$$\frac{\partial}{\partial t} \int_D \rho h_0 \, d\mathbf{r} + \int_S \rho \mathbf{V} h_0 \cdot \mathbf{N} \, dS + \int_S (\mathbf{q} + \mathbf{P}' \cdot \mathbf{V}) \cdot \mathbf{N} \, dS = 0 \qquad (7.77)$$

In the steady regime, we again find Eqns (7.35)–(7.37).

Control mass

Instead of a given volume D, let us consider a given fluid mass[71] ('control mass') contained at an instant t in a volume $D(t)$. The total variation $\frac{dA}{dt}$ of a corresponding local property a is equal to

$$\frac{dA}{dt} = \frac{d}{dt} \int_{D(t)} a \, d\mathbf{r}$$

After a few transformations, we obtain

$$\frac{dA}{dt} = \int_{D(t)} \left(\frac{da}{dt} + a \frac{\partial \cdot \mathbf{V}}{\partial \mathbf{r}} \right) d\mathbf{r} = \int_{D(t)} \frac{\partial a}{\partial t} d\mathbf{r} + \int_{S(t)} a\mathbf{V} \cdot \mathbf{N} \, dS \qquad (7.78)$$

The principles of mass, momentum, and energy conservation may of course be applied to this mass of fluid, and from Eqn. (7.78), we can easily find the corresponding conservation equations. Moreover, as Eqn. (7.78) is valid for any volume, we can again find the usual 'local' conservation equations.

Equation (7.78) constitutes the 'transport theorem'.

Appendix 7.6 Elements of magnetohydrodynamics

The general conservation equations (Eqns (7.54)–(7.61)) do not take into account the influence of external forces capable of modifying the structure of the flows. It is clear that the gravitational forces are generally negligible in high-speed flows. Then, only electric and magnetic fields can have a significant influence on gaseous systems that include charged species, either in the case of 'natural' ionization, arising from high temperatures encountered in high-enthalpy flows, or in the case of 'artificial' ionization produced by external sources (electric discharge, seeding, and so on).

It is beyond the scope of this book to propose a detailed description of ionized media, which may be found in many books dealing with 'plasma physics',[72–75] but in the present framework of reactive flows, the general aspects of 'magnetohydrodynamic flows' (MHD flows) cannot be ignored.

The behaviour of ionized gas flows including molecules, atoms, ions, and electrons may be derived from the Boltzmann equation, as may that of other reactive gas mixtures, without or with the presence of external fields. In this case, the usual set of moment equations may be closed by other equations governing the electromagnetic parameters appearing in the moment equations.[76–78]

As it is practically impossible to account for all phenomena related to the relative movement and interactions of charged particles, to a first approximation, we consider that the net charge density is null, that is, the plasma is electrically neutral. Similarly, the displacement current is neglected compared to the conduction current. Moreover, a linear Ohm's law is used to connect the current to the electric field, thus neglecting the influence of the electronic pressure gradient and the Hall effect.[79]

Thus, we have the following macroscopic conservation equations:

$$\frac{\partial \rho}{\partial t} + \frac{\partial \cdot (\rho V)}{\partial r} = 0 \quad \text{(mass conservation)} \tag{7.79}$$

$$\rho \frac{dV}{dt} = -\frac{\partial \cdot P}{\partial r} + J \wedge B \quad \text{(momentum conservation)} \tag{7.80}$$

where B is the total magnetic field, and $J \wedge B$ is the Laplace force.

With a simplified Ohm's law, we have

$$J = \sigma (E + V \wedge B) \tag{7.81}$$

The energy conservation equation is written as follows:

$$\rho \frac{dh}{dt} = \frac{dp}{dt} + \sum_p \frac{\partial \cdot}{\partial r}\left(\lambda_p \frac{\partial T_p}{\partial r}\right) - \sum_p \frac{\partial \cdot}{\partial r}(\rho_p U_p h_p) + \Phi + \frac{J^2}{\sigma} \tag{7.82}$$

This equation differs from its usual form by the term depending on the electric current (Joule heating) and by the conduction terms where a possible non-equilibrium between electrons (temperature T_e) and heavy species (temperature T_i) may occur.

Of course, particular momentum and energy equations for the various species may be derived from the Boltzmann equation.[79]

Equations (7.79)–(7.82) are closed with the Maxwell equations, i.e.

$$\frac{\partial}{\partial r} \wedge B = \mu_0 J \quad \text{with} \quad \frac{\partial \cdot B}{\partial r} = 0 \qquad (7.83)$$

$$\frac{\partial}{\partial r} \wedge E = -\frac{\partial B}{\partial t} \quad \text{with} \quad \frac{\partial \cdot E}{\partial r} = 0 \qquad (7.84)$$

Some other possible forms of the previous equations are often used. Thus, using Eqn. (7.83), the $J \wedge B$ term in the momentum equation (Eqn. (7.80)) may be written

$$J \wedge B = \frac{1}{\mu_0} B \cdot \frac{\partial B}{\partial r} - \frac{\partial}{\partial r}\left(\frac{B^2}{2\mu_0}\right) \qquad (7.85)$$

so that, defining the tensor $T = -\frac{BB}{\mu_0} + \frac{B^2}{2\mu_0} I$, Eqn. (7.80) becomes

$$\rho \frac{dV}{dt} = -\frac{\partial \cdot}{\partial r}(P + T) \qquad (7.86)$$

where $\frac{B^2}{2\mu_0}$ is the 'magnetic pressure', which is added to the static pressure p.

Eliminating E from previous equations, we obtain a single equation for the evolution of B:

$$\frac{\partial B}{\partial t} = \frac{\partial}{\partial r} \wedge (V \wedge B) + \frac{1}{\mu_0 \sigma}\frac{\partial^2 B}{\partial r^2} \qquad (7.87)$$

where the right-hand terms comprise a convective term and a diffusive term.

The above closed equation system necessitates explicit expressions for the state and transport terms requiring 'appropriate' physical models.

Writing Eqns (7.80) and (7.87) in a dimensionless form, we find new dimensionless numbers, such as

$$R_h = B_0 L_0 \sqrt{\frac{\sigma}{\mu_{o0}}} \quad \text{(Hartmann number)}$$

which is associated with the $J \wedge B$ term in the momentum equation (Eqn. (7.80)), and

$$R_\sigma = \mu_{o0} \sigma_0 V_0 L_0 \quad \text{(magnetic Reynolds number)}$$

which is associated with the diffusion term of Eqn. (7.87).

Numerous applications of the MHD equations are encountered in many fields of physics for practical purposes: plasma confinement by magnetic fields in controlled thermonuclear research (pinch effect), electric energy extraction from seeded plasmas (MHD generators), acceleration of ionized flows by cross electric and magnetic fields, and so on.

A simple example of the action of external fields on the boundary layer of an ionized gas flow is presented in Appendix 8.7.

EIGHT
Elements of Gas Dynamics

8.1 Introduction

'Traditional' gas dynamics is based on a very simple physical gas model that, nevertheless, can provide a qualitative description of gaseous high-speed flows. In the past, this description has been made without taking into account high-temperature effects encountered, for example, in flows related to the flight of hypersonic vehicles in dense atmospheres.

Thus, before examining 'real gas effects' resulting from these high temperatures, it is necessary to give a brief presentation of the general features of high-speed flows based on this simple physical 'ideal model'.

For this model, there is no chemical reaction, equilibrium is assumed, and moreover, specific heat is assumed to be constant. Vibrational energy is therefore neglected or assumed to be fully excited. This model simplifies the conservation equations, and despite its roughness, it may give a qualitatively correct description of high-speed flows. Thus, with this model, we can analyse the steady or unsteady Eulerian flows, associated wave systems, discontinuities such as shock waves or contact surfaces, and strong gradient dissipative flows.[80–82] The essential results about these flows are presented in this chapter, while the limitations of this ideal gas model are also pointed out.

8.2 Ideal gas model: consequences

In the definition of internal energy $dE = CdT$, $(de = C'dT)$, we assume with this model that specific heat C is constant. Thus, we can have the following values: (1) $C = C_T = \frac{3}{2}k$ for a monatomic gas; (2) $C = C_{TR} = \frac{5}{2}k$ for a diatomic gas at moderate temperatures; (3) $C = C_{TRV} = \frac{7}{2}k$ for a non-dissociated diatomic gas at high temperature; and (4) any constant value related to a particular case.

Then, $dh = C^p dT$, where for the first three preceding cases, we have respectively $C^p = C_T^p$, $C^p = C_{TR}^p$, $C^p = C_{TRV}^p$.

8.2 IDEAL GAS MODEL: CONSEQUENCES

The essential simplification resulting from this assumption concerns the energy conservation equation, which for a perfect gas may be written (Eqn. (7.42)):

$$C^P dT = \frac{dp}{\rho} \tag{8.1}$$

From Eqn. (8.1), we can deduce the equivalent following relations:

$$\frac{p}{\rho^\gamma} = \text{const.}, \quad \frac{T}{\rho^{\gamma-1}} = \text{const.}, \quad \text{and} \quad \frac{p}{T^{\gamma/\gamma-1}} = \text{const.} \tag{8.2}$$

where $\gamma = C^P/C$.

A simple expression for sound velocity a may also be found. Thus, from Eqn. (7.45), we have

$$dp = a^2 d\rho$$

so that

$$a^2 = \gamma RT = \gamma \frac{p}{\rho} \tag{8.3}$$

where $\gamma = \frac{5}{3}, \gamma = \frac{7}{5}, \gamma = \frac{9}{7}$ for the three preceding cases, respectively.

Thus, for the steady flow of an ideal perfect gas, taking into account Eqn. (7.41), we have

$$T_0 = T + \frac{mV^2}{2C^P} = T\left(1 + \frac{\gamma-1}{2}M^2\right) \tag{8.4}$$

where T_0 is the 'stagnation temperature' ($V = 0$).

We have also the following relations:

$$p_0 = p\left(1 + \frac{\gamma-1}{2}M^2\right)^{\gamma/\gamma-1}$$

$$\rho_0 = \rho\left(1 + \frac{\gamma-1}{2}M^2\right)^{\gamma/\gamma-1} \tag{8.5}$$

The local Mach number M and the ratio of specific heats γ are the dimensionless variables characterizing these types of flow (steady, isentropic, ideal). The complete determination of the flow also requires the solution of the continuity equation depending on the specific problem under study.

8.3 Isentropic flows

8.3.1 One-dimensional steady flows

The relations of Eqns (7.4) and (7.5) are valid along the abscissa x in a cross section $S(x)$ of the quasi-one-dimensional flow described in Chapter 7. The continuity equation (7.47) simply becomes

$$\frac{d(\rho V S)}{dx} = 0 \tag{8.6}$$

The Hugoniot relation is then deduced, i.e.

$$\frac{dV}{V} = \frac{1}{M^2 - 1}\frac{dS}{S} \tag{8.7}$$

Knowing $S(x)$, we can calculate the local average velocity $V(x)$ and the other quantities using the relations of Eqns (8.4) and (8.5).

Thus, two different regimes are defined: a subsonic regime ($M < 1$) and a supersonic regime ($M > 1$). The sonic (or critical) value for the Mach number, $M = 1$, is obtained for the minimum of the cross section S_C. From Eqn. (8.7), we see that a subsonic flow upstream from the minimum cross section accelerates in the convergent part and slows down in the divergent part. The inverse is of course true for an upstream supersonic flow. The passage through a minimum cross section may correspond to the sonic regime if the pressure at the exit of the divergent part is 'sufficiently' low.

An application of these properties may be found in nozzles that include a convergent part and a divergent part connected by a throat (minimum cross section). At the exit of the nozzle, we may thus obtain a supersonic flow with a Mach number depending only on the ratios S/S_C and γ (Appendix 8.2).

Well-known approximate expressions for pressure (Bernoulli) may easily be deduced from Eqn. (8.5) for $M \ll 1$.

8.3.2 Multidimensional steady flows

In the case of internal flows, according to the type of nozzle (plane or axisymmetric), we use of course the most convenient coordinate system: either a Cartesian coordinate system x, y or a polar system r, θ for plane flows, and a cylindrical system r, z for axisymmetric flows. However, the expected improvement with respect to the one-dimensional case is not very important with the ideal gas model, because this type of flow requires a realistic model that takes into account the influence of the walls (Chapter 10). The same is true for external flows (around bodies), which must be computed with realistic boundary conditions. The ideal gas model is generally used for relatively simple flows in order to point out the essential features of these flows.

8.3.3 One-dimensional unsteady flows

This classical case enables us to understand the propagation of waves in gaseous flows, especially in the supersonic regime, and to develop a computational method for the corresponding flows.[83,84]

Starting from the continuity and momentum equations written for one-dimensional unsteady flow (one single velocity component u along x), we have

$$\frac{\partial \rho}{\partial t} + \frac{\partial \rho u}{\partial x} = 0$$
$$\frac{\partial u}{\partial t} + u\frac{\partial u}{\partial x} = -\frac{1}{\rho}\frac{\partial p}{\partial x} \quad (8.8)$$

Using the relations

$$p = \rho RT \quad \text{and} \quad dp = a^2 d\rho$$

we obtain the following equations:

$$\frac{\partial}{\partial t}\left(u \pm \frac{2}{\gamma - 1}a\right) + (u \pm a)\frac{\partial}{\partial x}\left(u \pm \frac{2}{\gamma - 1}a\right) = 0 \quad (8.9)$$

Thus, in an x, t plane along the directions $u + a$ and $u - a$, called 'characteristics', that is, $\frac{dx}{dt} = u \pm a$, the quantities $P = u + \frac{2}{\gamma - 1}a$ and $Q = u - \frac{2}{\gamma - 1}a$ (Riemann variables) remain constant.

Setting $\frac{d_+}{dt} = \frac{\partial}{\partial t} + (u + a)\frac{\partial}{\partial x}$ and $\frac{d_-}{dt} = \frac{\partial}{\partial t} + (u - a)\frac{\partial}{\partial x}$ (variations along $u + a$ and $u - a$ respectively), we have:

$$\frac{d_+ P}{dt} = 0 \quad \text{and} \quad \frac{d_- Q}{dt} = 0 \quad (8.10)$$

For an isentropic flow, we also have $\frac{dS}{dt} = 0$ along the trajectories $\frac{dx}{dt} = u$. These trajectories are also characteristic directions.

Thus, P, Q, and S remain constant along the corresponding characteristics. Of course, in the homoentropic case, S is constant everywhere.

A numerical computational method may be deduced from these properties in the case of simple flow configurations, and constitutes the basis for actual numerical solvers. Examples are given in Appendix 8.1.

Physical aspects

In a flow-fixed coordinate system, it is clear that the characteristic directions represent the directions along which the perturbations propagate provided that the corresponding waves have a small amplitude, analogous to sound waves. For example, these perturbations may arise from boundary conditions. As seen above, we can distinguish the P waves and the Q waves.

The most classical example concerns the one-dimensional propagation of these waves in a tube in which perturbations may take place, such as cross section discontinuities, reflections on the end wall, and the variation of entry conditions. An important application is found in shock tubes (Chapter 11).

Thus, we consider a tube where u and a are constant everywhere (uniform state). If, for example, a pressure perturbation occurs upstream, a wave $u + a$ propagates downstream and modifies the local value of the pressure p, so that

$$P = u + \frac{2}{\gamma - 1} a \text{ is constant along } u + a$$

$$Q = u - \frac{2}{\gamma - 1} a \text{ is constant everywhere}$$

Therefore, u and a are constant along P, so that the waves propagating in an unperturbed region have characteristics which are straight lines in an x, t diagram. The same is true for Q waves if the perturbation originates from downstream.

We can distinguish the expansion waves (or rarefaction waves) and the compression waves, depending on whether they bring a pressure decrease or an increase. In the preceding example, if the perturbation is an expansion wave, the pressure, and therefore the temperature (Eqn. (7.45)), decrease locally; the sound speed a is then also decreasing, and as Q is constant, the velocity u decreases also, as well as $u + a$. The successive P waves propagate more and more slowly, and in an x, t diagram, the expansion wave as a whole is composed of a divergent bundle of characteristics.

Of course, the opposite is true for a compression wave: in this case, each wave element propagates in a medium where the temperature increases, so that its velocity increases. The characteristics may then merge and form a wavefront, through which macroscopic quantities undergo a discontinuity. This constitutes the process of the formation of a shock wave, also provoking an entropy change, whereas the flow remains isentropic in the case of an expansion wave.

Simple examples, including various types of interaction, are presented in Appendix 8.3. The computation of more complex flows is of course carried out using numerical techniques, but the general features of such flows are consistent with the present scheme.

8.4 Shock waves and flow discontinuities

As discussed above, in an unsteady flow, waves of small amplitude (acoustic waves) may coalesce and create a discontinuous front or shock wave. On each side of the front, the flow may remain Eulerian, but the properties discontinuously change across the shock wave, which propagates with a supersonic velocity with respect to the upstream medium. The 'thickness' of the shock wave is limited to a few mean free paths, so that a strong non-equilibrium prevails within the shock wave ($\varepsilon \geq 1$), which therefore cannot be described with the Navier–Stokes equations when the Mach number of the shock wave (defined with respect to the sound speed in the upstream medium) is greater than 1.1–1.2. For a realistic description, direct Monte Carlo simulations,[58] or simplifying assumptions such as BGK[29] or Mott-Smith[85] approximations, are necessary.

Shock waves are also present in supersonic steady flows, particularly in front of various objects placed in these flows, or similarly in front of vehicles moving at supersonic speed. This is because the perturbations arising from the presence of the obstacle cannot propagate upstream and the corresponding waves coalesce in front of the obstacle. They give rise to a steady shock wave depending on the shape of the obstacle and on the flow conditions. However, in the subsonic regime, the perturbations can travel upstream and progressively reduce the velocity of the flow.

8.4.1 Straight shock wave: Rankine–Hugoniot relations

In order to conveniently examine the essential macroscopic properties of shock waves, we consider the typical case of a wave normal to the direction of the flow (called a straight shock wave) and appearing for example in a one-dimensional flow in a tube. If, after the coalescence of compression waves, the shock wave moves in a medium that has constant properties, its velocity U_s remains constant, and the coordinate system may be fixed to the shock wave, which is equivalent to a steady regime in this system (Fig. 24).

Assuming an Eulerian flow on each side of the shock wave (regions 1 and 2), we have a Maxwellian distribution for the velocities. Without any hypothesis

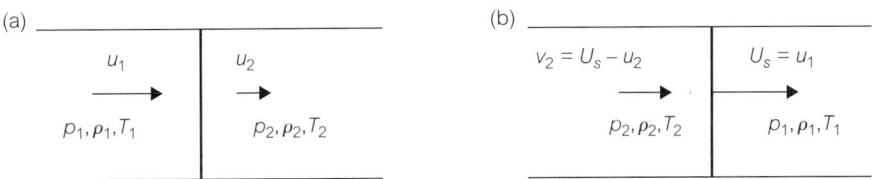

Figure 24. Flow with straight shock wave. (a) Steady shock wave; (b) Unsteady shock wave.

concerning the energy, equilibrium or non-equilibrium, we obtain equations relating the quantities of regions 1 and 2 by writing the conservation of mass, momentum, and energy flow rates through the shock wave, i.e.

$$\sum_{i,p} \int_{v_p} f^0_{ip(1)} v_{px} \begin{vmatrix} m_p \\ m_p v_{px} \\ \frac{1}{2} m_p v_p^2 + \varepsilon_{ip} \end{vmatrix} d\boldsymbol{v}_p = \sum_{i,p} \int_{v_p} f^0_{ip(2)} v_{px} \begin{vmatrix} m_p \\ m_p v_{px} \\ \frac{1}{2} m_p v_p^2 + \varepsilon_{ip} \end{vmatrix} d\boldsymbol{v}_p$$

(8.11)

Thus, we find

$$\rho_1 u_1 = \rho_2 u_2$$
$$p_1 + \rho_1 u_1^2 = p_2 + \rho_2 u_2^2$$
$$h_1 + \frac{u_1^2}{2} = h_2 + \frac{u_2^2}{2} \qquad (8.12)$$

Equation (8.12) constitutes the Rankine–Hugoniot relations and can also be obtained from the conservation equations (Eqns (7.36)–(7.38)). Thus, the properties of medium 2 may be determined from those of medium 1 and from the shock-wave velocity, a priori known. The enthalpy h_2 should be expressed as a function of the expected characteristics of medium 2, whether reactive or not, and whether in or out of equilibrium.

8.4.2 Ideal gas model

In the case of the ideal gas model considered in this chapter, we find simply:

$$\frac{\rho_2}{\rho_1} = \frac{u_1}{u_2} = \frac{\gamma + 1}{\gamma - 1 + 2/M_s^2}$$

$$\frac{p_2}{p_1} = 1 + \frac{2\gamma}{\gamma + 1}(M_s^2 - 1) \qquad (8.13)$$

$$\frac{T_2}{T_1} = 1 + \frac{2(\gamma - 1)}{(\gamma + 1)^2}(M_s^2 - 1)\left(\gamma + \frac{1}{M_s^2}\right) \qquad (8.14)$$

Moreover, the stagnation temperature remains constant across the shock wave (conservation of energy), i.e.

$$T_{01} = T_{02} \qquad (8.15)$$

For the flow Mach number behind the shock wave, we have

$$M_2 = \frac{(\gamma - 1) M_s^2 + 2}{2\gamma M_s^2 - (\gamma - 1)} < 1 \qquad (8.16)$$

Thus, the flow is subsonic in region 2 (in the coordinate system attached to the shock wave).

According to the chosen example, we have $\gamma = \frac{5}{3}, \frac{7}{5}, \text{or } \frac{9}{7}$.

In the hypersonic regime ($M_s \to \infty$), we have

$$\frac{p_2}{p_1} \to \infty \quad \text{and} \quad \frac{T_2}{T_1} \to \infty \tag{8.17}$$

but

$$\frac{\rho_2}{\rho_1} \to \frac{\gamma+1}{\gamma-1} \quad \text{and} \quad M_2 \to \left(\frac{\gamma-1}{\gamma+1}\right)^{1/2} \tag{8.18}$$

For high temperatures, these results are obviously academic (Chapter 9).

When $M_s \to 1$, we again find the relations for the isentropic flow.

If we calculate the entropy of each medium from Eqn. (7.17), for the ideal gas model, we find

$$S_2 - S_1 = RLog\frac{p_{01}}{p_{02}} = RLog\left[\frac{p_1}{p_2}\left(\frac{T_2}{T_1}\right)^{\gamma/\gamma-1}\right] \tag{8.19}$$

where p_0 is the stagnation pressure of each medium (Eqn. (8.5)).

We also have

$$S_2 > S_1 \quad \text{and} \quad p_{02} < p_{01}$$

Other properties of shock waves are given in Appendix 8.3, along with various interaction processes.

8.5 Dissipative flows

The computation of dissipative flows, internal or external, is of course made with the Navier–Stokes equations, and various results, essentially obtained by numerical methods and taking into account realistic physical models, are given in Chapter 10.

However, the general features of such flows are presented below, owing to the concept of the 'boundary layer' suggested by Prandtl.

8.5.1 Domain of influence: boundary layer

As previously discussed, dissipative phenomena are particularly important in the neighbourhood of interfaces separating the flow under study from the background, and they are connected in particular to the transverse gradients of various quantities. It is therefore often convenient to consider two distinct regions in a flow: one relatively far from the interface (wall, body, and so on) where the gradients are weak enough to neglect the corresponding dissipative phenomena (perfect fluid), and another region close to this interface where these gradients are important (Prandtl). Of course, this separation is not always possible, in particular in the case of fully developed regimes (narrow tubes) and in the case of flows where the dimensionless quantities Re, P, S are small.

We therefore consider non-developed regimes, such as the flow around a body (Fig. 25). The first challenge is to define the domain of influence of the dissipative effects along the body.[83] The solution may be found from the conservation equations themselves. Thus, in these equations, we neglect effects other than strictly dissipative effects such as compressibility and chemical production ($E \sim 0, Da \to \infty$, and so on), and we assume a constant value for all coefficients $\mu, \lambda, C_p \ldots$. Then, keeping only those terms that depend on transverse gradients, we find the following equations for the conservation of momentum, energy, and species respectively:

$$\frac{du}{dt} = \frac{\mu}{\rho}\frac{\partial^2 u}{\partial y^2}$$

$$\frac{dT}{dt} = \frac{\lambda}{\rho C_p}\frac{\partial^2 T}{\partial y^2}$$

$$\frac{dc_p}{dt} = D\frac{\partial^2 c_p}{\partial y^2} \qquad (8.20)$$

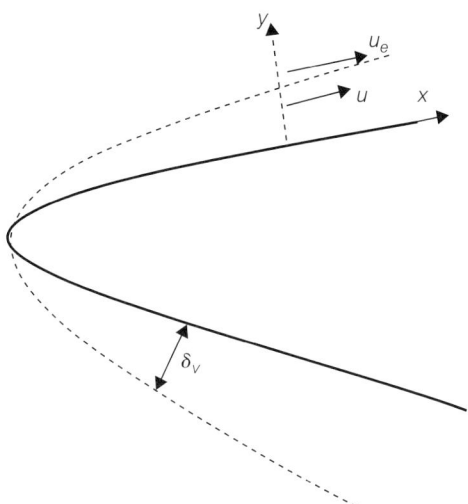

Figure 25. Development of the boundary layer along a body.

These equations are diffusion equations and illustrate the analogy between the transfer of mass, momentum, and energy.

Now we write these equations in a dimensionless form as in Chapter 7. However, the dimensionless transverse coordinate y^* is defined as $y^* = \frac{y}{\delta_v}$, where δ_v is a boundary layer 'characteristic thickness'. Thus, we obtain for the momentum equation along the body (x coordinate and L characteristic length):

$$\frac{du^*}{dt^*} = \frac{1}{Re}\left(\frac{L^2}{\delta_v^2}\right)\frac{\partial^2 u}{\partial y^2} \tag{8.21}$$

with

$$Re = \frac{\rho_0 V_0 L}{\mu_0}$$

We may deduce from Eqn. (8.21) that the dissipative effects related to viscosity are important in a region that is close to the body and that has a thickness δ_v, such that

$$\frac{\delta_v}{L} \sim \frac{1}{\sqrt{Re}} \tag{8.22}$$

This relation defines the boundary-layer domain.

The other two equations (energy, species) are also written in a dimensionless form (Chapter 7) by setting $y^* = \frac{y}{\delta_t}$ and $y^* = \frac{y}{\delta_c}$, where δ_t and δ_c represent characteristic thicknesses for the 'temperature boundary layer' and the 'concentration boundary layer' respectively, that is, regions where conductive and diffusive effects are important. Thus, we find

$$\frac{\delta_t}{L} = \frac{1}{\sqrt{ReP}}, \quad \frac{\delta_c}{L} = \frac{1}{\sqrt{ReS}}, \quad \text{and} \quad \frac{\delta_c}{\delta_t} \sim \sqrt{L} \tag{8.23}$$

The values of P and S (and L) are thus characteristic of the thickness of the various boundary layers: velocity boundary layer (or dynamic boundary layer), temperature boundary layer (or thermal boundary layer), and concentration boundary layer. These values also give the order of magnitude of their relative importance. When $P = S = 1 (= L)$, all boundary layers have a comparable thickness. Furthermore, the concept of the boundary layer becomes meaningless when $Re \to 0$ (rarefied gases, for example; Appendix 6.6).

8.5.2 General equations: two-dimensional flows

The concept of boundary layer enables us to simplify the Navier–Stokes equations. These simplifications are made below for steady plane and axisymmetric flows, which represent characteristic examples.

Plane flow

Starting again from the Navier–Stokes equations, we assume that the body along which the boundary layer is developing does not present any discontinuity and that its curvature is relatively weak, i.e. the curvature radius K^{-1} is large compared with the characteristic boundary layer thickness δ. In an intrinsic coordinate system x, y (Fig. 25), the differential elements $dx' = (1 + Ky)\, dx$ and dy' are then approximately equal to the Cartesian orthogonal elements dx and dy, respectively.

Applying the previous assumptions related to the boundary layer concept $\left(\frac{\delta}{L} = \varepsilon \ll 1 \text{ and } Re \sim \frac{1}{\varepsilon^2}\right)$, at each point of the boundary layer we have $x \gg y$, $u \gg v$, $\frac{\partial}{\partial y} \gg \frac{\partial}{\partial x}$, which means that the transverse dimensions and velocities are small compared with the longitudinal dimensions and velocities respectively, and that, as discussed previously, the transverse gradients are much more important than the longitudinal gradients.

The Navier–Stokes equations are written in a dimensionless form as in Chapter 7 by taking the Eulerian quantities defined at $y = 0$ as characteristic quantities assumed known (subscript e). Neglecting the terms of order ε and higher-order terms, we find simplified expressions for the components of the momentum equation along x and y axes, that is:

$$\rho u \frac{\partial u}{\partial x} + \rho v \frac{\partial u}{\partial y} = -\frac{\partial p}{\partial x} + \frac{\partial}{\partial y}\left(\mu \frac{\partial u}{\partial y}\right) \tag{8.24}$$

$$\frac{\partial p}{\partial y} = 0 \rightarrow p = p_e(x) \text{ (see Eqn. (7.52))} \tag{8.25}$$

Therefore, the pressure is constant through the boundary layer and has the same value as in the outer (Eulerian) flow, thus depending only on x.

As we also have $\frac{1}{ReP} \sim \frac{1}{ReS} \sim \varepsilon^2$ and $E \sim 1$, we obtain for the energy and species conservation equations the following expressions, respectively:

$$\rho u \frac{\partial h}{\partial x} + \rho v \frac{\partial h}{\partial y} = \frac{dp_e}{dx} + \mu \left(\frac{\partial u}{\partial y}\right)^2 + \frac{\partial}{\partial y}\left(\lambda \frac{\partial T}{\partial y}\right) + \frac{\partial}{\partial y}\left(\rho D \sum_p h_p \frac{\partial c_p}{\partial y}\right) \tag{8.26}$$

$$\rho u \frac{\partial c_p}{\partial x} + \rho v \frac{\partial c_p}{\partial y} = \frac{\partial}{\partial y}\left(\rho D \frac{\partial c_p}{\partial y}\right) + \dot{w}_p \tag{8.27}$$

The continuity equation remains unchanged and may be written in the following form:

$$\frac{\partial \rho u}{\partial x} + \frac{\partial \rho v}{\partial y} = 0$$

Axisymmetric case: general equations

For an axisymmetric flow, we can use the spatial coordinates r (transverse) and z (longitudinal). Thus, if u and v are the components of the velocity V on z and r, respectively, the continuity equation becomes

$$\frac{\partial \rho u}{\partial z} + \frac{\partial \rho v r}{\partial r} = 0 \qquad (8.28)$$

When the curvature of the wall is small, we may assume $\frac{\partial}{\partial r} \sim \frac{\partial}{\partial y}$, $\frac{\partial}{\partial z} \sim \frac{\partial}{\partial x}$, and $r = r_0$, where r_0 represents the radius of the body cross section.

A general form of the continuity equation, valid in plane and axisymmetric cases, is then

$$\frac{\partial \rho u r^j}{\partial x} + \frac{\partial \rho v r^j}{\partial y} = 0 \qquad (8.29)$$

with $j = 0$ (plane case) and $j = 1$ (axisymmetric case).

The other equations remain unchanged (Eqns (8.24)–(8.27)).

Global and interfacial quantities

When the quantities inside the boundary layer are determined, we can calculate the interfacial quantities, characteristic of the exchanges with the background (C_f, Nu, or St). We can also calculate the global quantities characteristic of the 'losses' due to the presence of a boundary layer. They are generally expressed as characteristic thicknesses; thus, the assumed Eulerian flow would undergo a mass loss rate calculated by writing that the mass flow rate through the boundary layer is equal to the mass flow rate of an Eulerian flow (with $u_e = $ const.) along a fictitious wall located at a distance δ^* from the real wall, so that

$$\int_0^\delta \rho u \, dy = \rho_e u_e (\delta - \delta^*) \qquad (8.30)$$

then

$$\delta^* = \int_0^\delta \left(1 - \frac{\rho u}{\rho_e u_e}\right) dy = \int_0^\infty \left(1 - \frac{\rho u}{\rho_e u_e}\right) dy \qquad (8.31)$$

Thus, δ^*, called displacement thickness, represents a mathematically and physically well-defined quantity contrary to the characteristic thickness δ. It is of course smaller than δ, but it has the same order of magnitude if, as is generally the case, δ is located at an ordinate y such that $u/u_e = 0.99$.

In the same way, other thicknesses may be defined, such as the momentum rate loss θ, equal to

$$\theta = \int_0^\infty \frac{\rho u}{\rho_e u_e}\left(1 - \frac{u}{u_e}\right) dy \tag{8.32}$$

These thicknesses are different, but they have the same order of magnitude ($P \sim S \sim 1$), so that in general we use only one of them, the simplest one, δ^*, corresponding to a virtual modification of the body geometry for the outer flow.

Vibrational non-equilibrium case

Here, if the momentum and continuity equations are unchanged (Eqns (8.24), (8.25), and (8.29)), the energy and relaxation equations for a pure gas may be written respectively as

$$\rho \frac{dh}{dt} = \frac{dp_e}{dx} + \frac{\partial}{\partial y}\left(\frac{\mu}{P}\frac{\partial h}{\partial y}\right) + \frac{\partial}{\partial y}\left[\frac{\mu}{P}(L-1)\frac{\partial e_V}{\partial y}\right] + \mu\left(\frac{\partial u}{\partial y}\right)^2$$

$$\rho \frac{de_V}{dt} = \rho \dot{e}_V + \frac{\partial}{\partial y}\left(\frac{L}{P}\mu\frac{\partial e_V}{\partial y}\right)$$

where P and L are frozen Prandtl and Lewis numbers, respectively.

Appendix 8.1 Method of characteristics

One-dimensional or quasi-one-dimensional unsteady flows, along with steady two-dimensional flows, may be computed by the method of characteristics, which is directly derived from the principles presented in Section 8.3.3.

As a simple example, we consider the case of the isentropic one-dimensional unsteady flow of an ideal gas presented above. In the x, t plane, we start from an initial data line, for example at $t = t_0$, where all flow quantities are known, and we write the equations $\frac{dx}{dt} = u \pm a$, $\frac{dx}{dt} = u$, $\frac{d_+ P}{dt} = 0$, $\frac{d_- Q}{dt} = 0$ in a finite difference form in order to compute the flow at time $t_0 + \Delta t$.

Thus, in Fig. 26, the P wave starting from point (1) intersects the line $t_0 + \Delta t$ at point (3) where a trajectory ($\frac{dx}{dt} = u$) and a Q wave may be defined, coming from points (4) and (2) respectively. After successive interpolations and iterations, the coordinates and state of point (3) are determined. It is of course necessary to ensure that the CFL (Courant–Friedrich–Levy) criterion[86] is satisfied and therefore $\Delta t \le \Delta x/a$.

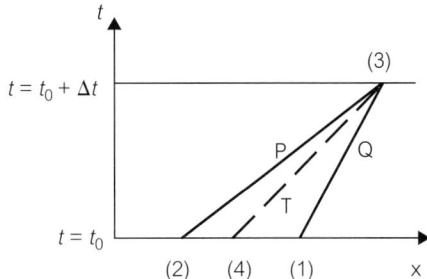

Figure 26. Characteristic mesh (current point).

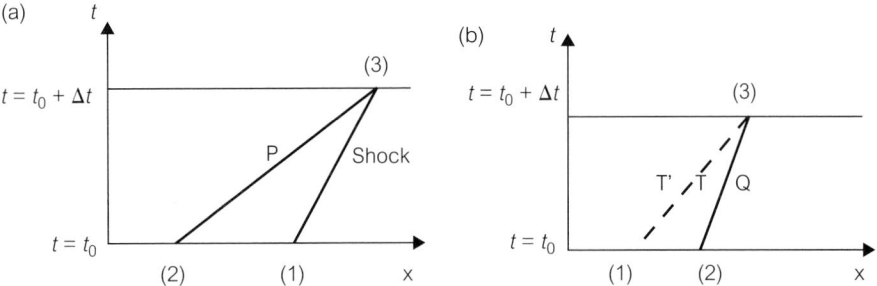

Figure 27. (a) Characteristic mesh (shock wave); (b) Characteristic mesh (contact surface).

This computation applies to current points of the field. If this field is limited by a moving shock wave or a contact surface, as is common, the type of mesh is different close to these boundaries.

Thus, for a moving shock wave, we have meshes represented in Fig. 27(a), where the Rankine–Hugoniot relations are used at point (3) on the shock trajectory. In the same way, for point (3) along a contact surface, we use the fact that the pressure and velocity are the same on each side of the contact surface (trajectories T and T′), and we also use the properties of a Q wave (Fig. 27(b)).

This method has been generalized to non-isentropic flows and non-equilibrium flows; it can also take into account various boundary conditions (geometry, physical conditions, and so on).

Appendix 8.2 Fundamentals of supersonic nozzles

A gas contained in a reservoir at a pressure p_0 is expanded in a convergent–divergent nozzle up to a tank where the pressure is equal to p_S ($p_S < p_0$) (Fig. 28).

In the convergent part of the nozzle, the flow is subsonic, and the pressure decreases up to a value p_C at the throat. If the exit pressure p_S is sufficiently

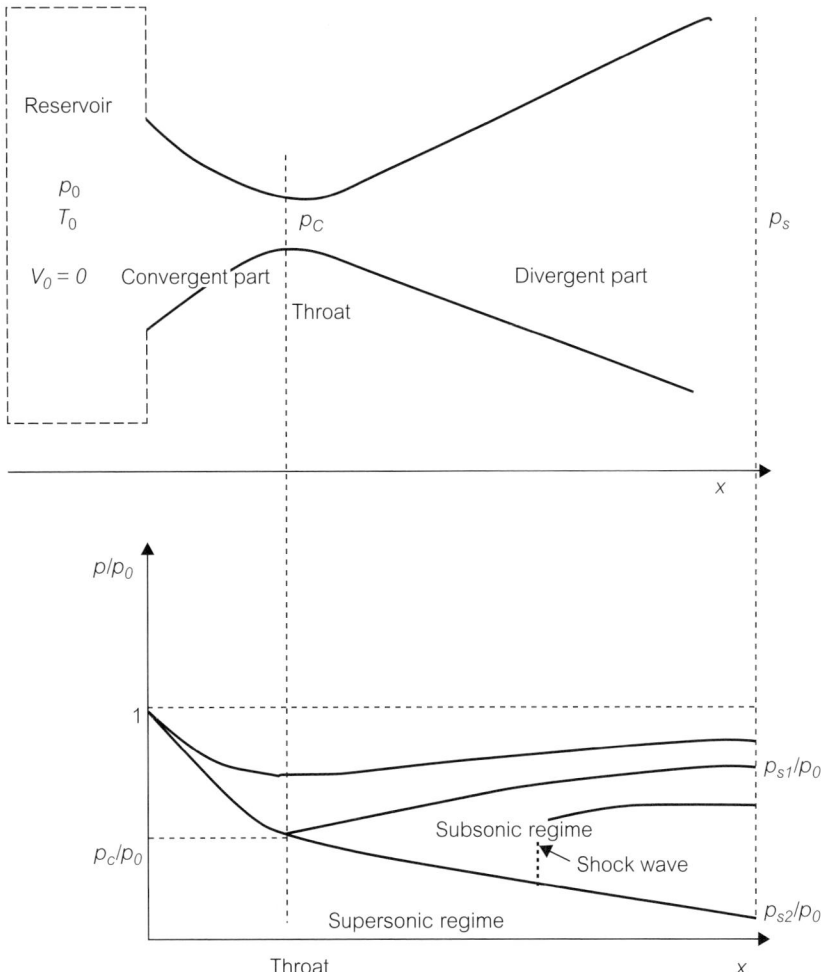

Figure 28. Scheme and running of a supersonic nozzle.

low, ($p_S < p_{S1}$) we obtain sonic conditions at the throat, i.e. $M_C = 1$ and $p_C/p_0 = (2/\gamma - 1)^{\gamma/\gamma-1}$. Then, the mass flow rate is maximum, the regime is supersonic in the divergent part of the nozzle, and the relative exit pressure p_{S2}/p_0 depends only on the cross section ratio S/S_C.

If $p_{S2} < p_S < p_{S1}$, the flow, first supersonic in the divergent part, then becomes subsonic through a stationary shock wave.

We can easily verify that a maximum velocity is attained when the total stagnation enthalpy (available energy) is transformed into kinetic energy. It should also be noted that these results remain qualitative, because of dissipative and real gas effects (Chapter 10).

Appendix 8.3 Shock waves: configuration and kinematics

Only a few general results are presented here. They are derived from the solution of the equations given above; difficulties which might be encountered generally arise from the particularities of each specific problem.[87–88]

Stationary shock waves

Flows around obstacles

The shock waves appearing in front of obstacles placed in supersonic flows may take different shapes and structures, depending on the type of obstacle and flow conditions.

If the flow of an ideal gas undergoes only a deviation of a constant angle α (wedge or cone, for example), the corresponding shock wave forms at a constant angle β to the free stream direction (Fig. 29).

The shock wave is said to be 'oblique', and the parameters of flow region 2 may be easily deduced from the free stream parameters (region 1) by applying mass and momentum conservation equations to the part of the flow normal to the shock wave. The tangential component of the velocity remains constant across the shock wave, as does the stagnation enthalpy. Thus, we have

$$\rho_1 u_1 \sin \beta = \rho_2 u_2 \sin (\beta - \alpha)$$
$$u_1 \cos \beta = u_2 \cos (\beta - \alpha)$$
$$p_1 + \rho_1 u_1^2 \sin^2 \beta = p_2 + \rho_2 u_2^2 \sin^2 (\beta - \alpha)$$
$$h_1 + \frac{u_1^2}{2} = h_2 + \frac{u_2^2}{2} \tag{8.33}$$

After solving this system, we see that, given a free stream Mach number, there exists a maximum angle α beyond which there is no solution, that is, the shock

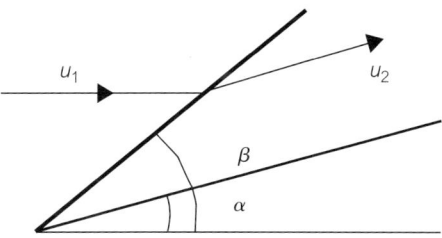

Figure 29. Attached shock wave.

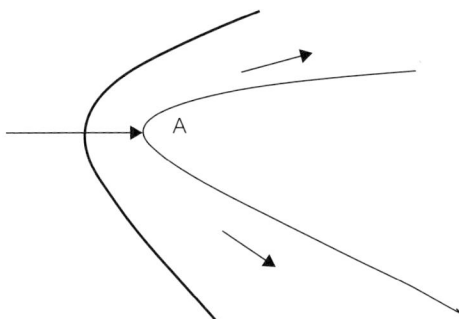

Figure 30. Detached shock wave.

Figure 31. (a) Regular reflection; (b) Mach reflection.

wave is detached from the model. Similarly, given an angle α, the shock wave is detached beyond a maximum value of the free stream Mach number.

In the case of a 'blunted' model, the shock wave is always detached (Fig. 30), and in general, the flow must be computed numerically. The flow separates into two parts on each side of stagnation point A and is subsonic in the neighbourhood of this point. A few examples of more complex geometries are presented in Chapters 9 and 10.

Shock wave reflection

A stationary incident shock wave I is reflected by a wall and becomes a shock wave R. The flow, after having crossed this wave system, is parallel to the wall, as can be seen in Fig. 31(a). The application of the relations of Eqn. (8.33) is sufficient to know the parameters of the emerging flow (regular reflection). When angle α is larger than a certain value for a given free stream, the problem has no solution of this type. However, a configuration similar to Fig. 31(b) represents a possible solution confirmed by experiment. We thus have a 'Mach reflection' that includes a quasi-normal wave N, a triple point, a reflected wave R, and an interface (or contact surface SC) separating the gas that has crossed the wave N from the gas that has crossed the waves I and R. This contact surface represents a discontinuity for temperature and velocity (and entropy) but not for pressure; therefore, in reality, it is a zone of important vorticity.

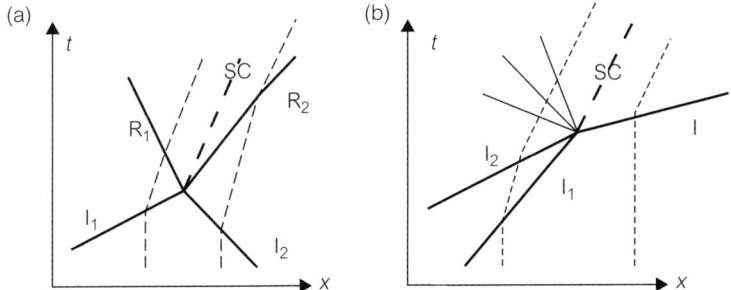

Figure 32. (a) Shock wave collision; _ _ _: Contact surface; (b) Shock wave merging. ---: Fluid trajectories.

Unsteady shock waves

One-dimensional case: interactions

The processes are generally visualized in an x, t diagram. A few examples are presented below.

Shock wave collisions

The problem is how to determine the location and intensity of the two reflected shock waves R_1 and R_2 (and downstream conditions) in relation to the incident shock waves I_1 and I_2. The entropy value is different for the fluid particles that have respectively crossed the shock waves I_1, R_1 or I_2, R_2; therefore, a contact surface develops from the intersection point of the incident shock waves (Fig. 32(a)).

This contact surface represents a discontinuity for temperature and density but not for pressure and velocity.

Merging of two shocks

After merging, the shock waves I_1 and I_2 constitute a unique shock wave I more intense than I_1 and I_2. The interaction process is completed by a simple wave system centred at the intersection point, and by a contact surface (Fig. 32(b)).

Simple wave catching a shock wave

The rarefaction waves R weaken the shock wave I, and the flow is no longer isentropic (Fig. 33(a)). Of course, compression waves would reinforce it (see example in Chapter 11).

Shock wave–interface interaction

A shock wave I_1 catches a contact surface SC_1; there is a transmitted shock I_2. The coherence of downstream conditions is operated by a contact surface SC_2 and

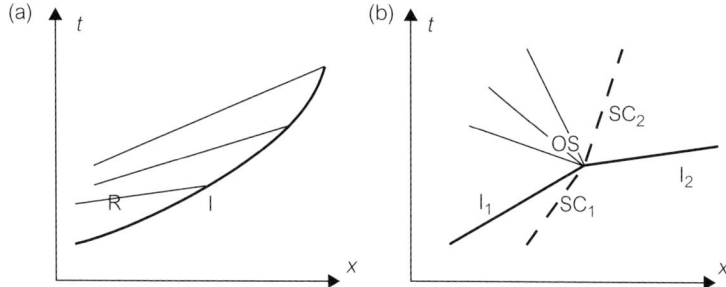

Figure 33. (a) Simple wave catching a shock wave; (b) Shock–interface interaction.

either by a simple wave OS (Fig. 33(b)) or by a shock wave (see example in Chapter 11).

Two-dimensional case

Many interaction problems (shock–shock, shock–model, shock–boundary layer, and so on) give rise to specific, and often complex, wave structures; examples are given in Chapter 11.

Appendix 8.4 Generalities on the boundary layer

The boundary layer considered as a perturbation

The equations of the boundary layer are second-order equations with respect to y and therefore require two boundary conditions along a normal to the wall, while they are first-order equations with respect to x and require only one longitudinal boundary condition. This last condition is generally given at $x = 0$, the starting point of the boundary layer (stagnation point, for example).

Along y, we generally have known boundary conditions (or assumed known) on the body ($y = 0$). For example, the velocity is zero on a solid wall, the temperature or the heat flux may be imposed, or the species concentrations may be known from the catalytic nature of the wall. The second condition is given by the 'outer flow' considered as a non-dissipative flow and also assumed known. From a mathematical point of view, this flow provides the boundary conditions for $y \to \infty$. However, this Eulerian flow varies along y, and the 'correct' value for the various quantities must be chosen. The solution is given by the concept of boundary layer itself, which, considered as a narrow region, may also be considered as a perturbation of the outer flow and thus may be determined by a 'perturbation method'.[89,90] Therefore, we proceed as follows:

1. An Euler flow is calculated around the model (or along the wall) with corresponding boundary conditions. Thus, we have no condition for velocity (except geometrical) or for temperature and concentrations at the wall.

2. In the natural coordinate system chosen for the boundary layer, from the preceding calculation we deduce the values of the flow parameters at $y = 0$, which depend only on x, i.e. $u_e(x)$, $h_e(x)$, $c_{pe}(x)$. These values are then considered to be boundary conditions $(y \to \infty)$ for the quantities inside the boundary layer, at least for those requiring two conditions along y.

3. The boundary layer flow is then computed generally with a method presented below. The 'losses' (mass, momentum, energy, and so on) due to the boundary layer are also calculated, and the Euler flow is recalculated by taking into account these losses (or corresponding thicknesses). Several iterations may be necessary, depending on the specific flows under consideration.

4. Pressure and transverse velocity have a particular behaviour, since they appear as first-order derivatives in the boundary-layer equations. Thus, for the transverse component of the velocity, there is no matching with the outer flow, but its value is always low. As for the pressure, it is known from the Eulerian flow, since

$$\frac{\partial p}{\partial x} = \frac{\partial p_e}{\partial x} = -\rho_e u_e \frac{du_e}{dx} \tag{8.34}$$

Therefore, the wall pressure measurements do not depend on the presence of the boundary layer and are representative of the outer pressure field.

More detailed methods providing a better matching with the outer flow are also available.[91] Curvature effects may also be taken into account.

Method of solution

Despite the fact that today, with modern computers, the complete Navier–Stokes equations may generally be solved, economic considerations or physical interest often induce engineers or researchers to solve the boundary-layer equations coupled with the Eulerian outer system.

These equations are of parabolic type and are strongly coupled. They also present mathematical and physical singular points (for example, stagnation points). Thus, in the past, coordinate transformations have been proposed and developed. They may suppress the singular points and simplify the integration field.[70,90] All these transformations may be reduced to a unique one, that is the LLD (Levy–Lees–Dorodnitsyn) transformation presented below. The main points of this transformation are the following:

244 CHAPTER 8 ELEMENTS OF GAS DYNAMICS

- The starting point of the boundary layer (singular point) at $x = 0$ is transformed into a starting line.

- The parabolic domain is transformed into a quasi-rectangular domain where the thickness of the boundary layer is quasi-constant.

- The transformation is independent of the dimensional character of the boundary layer (plane or axisymmetric).

- Obvious simplifications appear in the equations for the treatment of particular cases (for example, self-similar solutions).

New coordinate system: momentum equation

Now, we must find the transformation $(x, y) \rightarrow (\xi, \eta)$ with the above characteristics.

At $x = 0$, we must have a differential equation depending on only one variable, for example η (starting line), so that at this point we must also have $\xi = 0$, which leads to the following change:

$$\xi = \xi(x) \quad \text{and} \quad \eta = \eta(x, y) \tag{8.35}$$

Thus, ξ becomes a longitudinal variable. Moreover, in order to have a quasi-rectangular integration domain, η must be a quasi-self-similar variable ($\sim y/\delta$).

For any flow parameter F, we may write from Eqn. (8.35):

$$\frac{\partial F}{\partial x} = \frac{\partial F}{\partial \eta}\frac{\partial \eta}{\partial x} + \frac{\partial F}{\partial \xi}\frac{d\xi}{dx} \quad \text{and} \quad \frac{\partial F}{\partial y} = \frac{\partial F}{\partial \eta}\frac{\partial \eta}{\partial y} \tag{8.36}$$

It is also convenient to define a stream function ψ, so that the continuity equation is satisfied, that is:

$$\rho u r^j = \frac{\partial \psi}{\partial y} \quad \text{and} \quad \rho v r^j = -\frac{\partial \psi}{\partial x} \tag{8.37}$$

Setting

$$\frac{u}{u_e} = \frac{\partial \psi}{\partial \eta} F \tag{8.38}$$

we have

$$\frac{\partial \eta}{\partial y} = \rho u_e r^j F \tag{8.39}$$

where $F = F(x, y)$ is an unknown function. We therefore have

$$\eta = \rho_e u_e r^j \int_0^y \frac{\rho}{\rho_e} F\, dy \tag{8.40}$$

Now, in order to obtain explicit relations between (x, y) and (ξ, η), we need two equations which, in principle, may be defined arbitrarily. Equation (8.40) provides the motivation to define one of these. If η has to be a kind of self-similar variable, the function F must depend only on x (or ξ). Then, we easily verify that, in the incompressible case, we have $\eta/y = G(x)$; in the compressible case, we obtain a similar relation $\eta/Y = H(x)$ after having applied the transformation $Y = \int_0^y \frac{\rho}{\rho_e} dy$ (von Mises). Therefore, we have

$$\eta = \rho_e u_e r^j F(\xi) \int_0^y \frac{\rho}{\rho_e} dy \tag{8.41}$$

where the terms $\rho_e u_e r^j F(\xi)$ depend only on x (or ξ). Now, for convenience we set

$$\frac{u}{u_e} = \frac{\partial f}{\partial \eta} = f'(\xi, \eta) \tag{8.42}$$

where, conventionally, the differentiation is with respect to η.

Equation (8.38) then becomes

$$\psi = \frac{f(\xi, \eta)}{F(\xi)} \tag{8.43}$$

Now, in the momentum equation (8.24), we replace the variables u and v by the variables f and F with the help of Eqns (8.37), (8.41), and (8.43), and we set $C = C(\xi, \eta) = \rho\mu/\rho_e\mu_e$.

Then the momentum equation may be written in the following form:

$$\frac{1}{u_e}\frac{du_e}{d\xi}\left(\frac{\rho_e}{\rho} - f'^2\right) - ff'' \frac{1}{F}\frac{dF}{d\xi} + \rho_e\mu_e u_e r^{2j}\left(\frac{d\xi}{dx}\right)^{-1} F^2 (Cf'')'$$
$$= f'\frac{\partial f'}{\partial \xi} - f''\frac{\partial f}{\partial \xi} \tag{8.44}$$

Finally, we must choose the explicit dependence of F on ξ. Equation (8.44) hints at defining F such that $\frac{1}{F^3}\frac{dF}{d\xi} = K = \text{const}$.

If, arbitrarily, we choose $K = -1$, a possible form for F is $F = \frac{1}{\sqrt{2\xi}}$.

Finally, another arbitrary condition must be found in order to completely define the transformation: taking into account Eqn. (8.42), we can set $\rho_e\mu_e u_e r^{2j}\left(\frac{d\xi}{dx}\right)^{-1} = K'$. Moreover, choosing $K' = 1$, we at last obtain the following relations for ξ and η:

$$\xi = \int_0^x \rho_e \mu_e u_e r^{2j} dx$$

$$\eta = \frac{\rho_e u_e r^j}{\left(2 \int_0^x \rho_e \mu_e u_e r^{2j} dx\right)^{1/2}} \int_0^y \frac{\rho}{\rho_e} dy \qquad (8.45)$$

These equations define the LLD transformation.

The transformed momentum equation is then written as follows:

$$(Cf'')' + ff'' + \frac{2\xi}{u_e} \frac{du_e}{d\xi} \left(\frac{\rho_e}{\rho} - f'^2\right) = 2\xi \left(f' \frac{\partial f'}{\partial \xi} - f'' \frac{\partial f}{\partial \xi}\right) \qquad (8.46)$$

This third-order equation enables us to determine f. Equation (8.46) is, however, coupled with the energy equation (with ρ and μ). Furthermore, three boundary conditions along η are necessary. Without mass transfer, we generally have

$$f = f' = 0 \text{ at } \eta = 0 \qquad (8.47)$$

$$f' \to 1 \text{ when } \eta \to \infty \qquad (8.48)$$

We may also verify that, at $\xi = 0$, we must solve an ordinary differential equation (variable η).

Transformation of the other equations

Instead of using the energy equation in the form of Eqn. (8.26), we generally use Eqn. (7.14), which, in the boundary layer, may be written

$$\rho u \frac{\partial h_0}{\partial x} + \rho v \frac{\partial h_0}{\partial y} = \frac{\partial}{\partial y}\left(\frac{\mu}{P} \frac{\partial h_0}{\partial y}\right) + \frac{\partial}{\partial y}\left[\mu\left(1 - \frac{1}{P}\right) \frac{\partial}{\partial y}\left(\frac{u^2}{2}\right)\right]$$

$$+ \frac{\partial}{\partial y}\left[\frac{\mu}{S}\left(1 - \frac{1}{L}\right) \sum_P h_p \frac{\partial c_p}{\partial y}\right] \qquad (8.49)$$

It may be noted that in this equation only one transport coefficient, μ, explicitly appears, and this coefficient is not very sensitive to non-equilibrium (Chapter 3). The other transport terms appear in their usual dimensionless form also weakly sensitive to temperature and non-equilibrium (see examples in Fig. 21).

Now, the application of the LLD transformation (Eqn. (8.45)) to Eqn. (8.49) leads to the following:

$$\left(\frac{C}{P}g'\right)' + fg' + E_{0e}\left[\left(1 - \frac{1}{P}\right)Cf'f''\right]' + \left[\frac{C}{S}\left(\frac{1}{L} - 1\right)\sum_p \frac{h_{pe}c_{pe}}{h_{0e}}z'_p\right]'$$
$$= 2\xi\left(f'\frac{\partial g}{\partial \xi} - g'\frac{\partial f}{\partial \xi}\right) \tag{8.50}$$

where

$$g = g(\xi, \eta) = \frac{h_0}{h_{0e}}, \quad z_p = z_p(\xi, \eta) = \frac{c_p}{c_{pe}}, \quad \text{and}$$

$$E_{0e} = \frac{u_e^2}{h_{0e}} \text{ (Eckert number)}$$

with g and $z_p \to 1$ when $\eta \to \infty$, and with wall conditions ($\eta = 0$) depending on the specific problem (see, for example, Chapter 10 and Appendix 8.4).

The species conservation equation (8.27) is transformed into the following:

$$\left(\frac{C}{S}z'_p\right)' + fz'_p - \frac{2\xi}{c_{pe}}\left(f'z_p\frac{dc_{pe}}{d\xi} - \frac{\dot{w}_p}{\rho\rho_e\mu_e u_e^2 r^{2j}}\right) = 2\xi\left(f'\frac{\partial z_p}{\partial \xi} - z'_p\frac{\partial f}{\partial \xi}\right) \tag{8.51}$$

Of course, the vibrational relaxation equations may also be transformed if necessary (see, for example, Chapter 10).

These strongly coupled equations are generally solved numerically, and the inverse transformation gives the flow parameters in the physical plane (x, y) as well as the global quantities defined above. Simple cases, more or less classical, are presented in Appendix 8.5.

Appendix 8.5 Simple boundary layers: typical cases

Self-similar solutions

As used to be common, we may try to find solutions depending only on the variable η, that is, self-similar solutions valid in the whole boundary layer. These solutions, however, exist only for precise cases in which the terms depending on ξ are negligible.[70]

Here we examine the case of pure gas flows without chemical reaction or vibrational non-equilibrium. The case of reactive boundary layers is presented

CHAPTER 8 ELEMENTS OF GAS DYNAMICS

in Chapter 10. Thus, if there are self-similar solutions for f and g, we must solve the following two equations:

$$\left(Cf''\right)' + ff'' + \beta\left(\frac{T}{T_e} - f'^2\right) = 0$$

$$\left(\frac{C}{P}g'\right)' + fg' = E_{0e}\left[\left(\frac{1}{P} - 1\right)Cf'f''\right]' \quad (8.52)$$

with

$$\beta = \beta(\xi) = \frac{2\xi}{u_e}\frac{du_e}{d\xi} \quad (8.53)$$

When the functions f and g are found, we may compute the dimensionless interfacial quantities C_f and Nu from the following relations:

$$C_f = \frac{1}{\frac{1}{2}\rho_e u_e^2}\left(\mu\frac{\partial u}{\partial \eta}\right)_w\left(\frac{\partial \eta}{\partial y}\right)_w = \frac{2r^j\mu_e}{\sqrt{2\xi}}C_w f_w''$$

$$Nu = \frac{x}{T_{0e} - T_w}\left(\lambda\frac{\partial T}{\partial \eta}\right)_w\left(\frac{\partial \eta}{\partial y}\right)_w = \frac{r^j u_e x}{\sqrt{2\xi}}\rho_w\frac{g_w'}{1 - g_w} \quad (8.54)$$

We can generally assume that the outer flow is isoenergetic ($h_{0e} = $ const.; Chapter 7), but C, β, P, u_e, and T_e may depend on ξ. Taking into account the previous results, we assume that P is constant. It remains to analyse the influence of the external conditions resulting from the configuration of the flows based on these last two assumptions.

Flow without pressure gradient

Two-dimensional case: equations and results

This is the classical flow along a flat plate or along a wedge in the supersonic regime. Thus, we have

$$\beta = 0, \quad u_e = \text{const.}, \text{ and } T_e = \text{const.}$$

The momentum equation (8.46) becomes

$$\left(Cf''\right)' + ff'' = 0 \quad (8.55)$$

It is possible to modify the variable η (Eqn. (8.45)) so as to include C. Thus, we have

$$\eta = \frac{\rho_e u_e r^j}{\sqrt{2C\xi}}\int_0^y \frac{\rho}{\rho_e}dy \quad (8.56)$$

Then, Eqn. (8.55) reduces to

$$f''' + ff'' = 0 \quad \text{(Blasius equation)} \tag{8.57}$$

This equation has often been solved with the boundary conditions of Eqns (8.47) and (8.48). Thus, we obtain for f_w'' a value equal to 0.47. The skin-friction coefficient is therefore equal to

$$C_f = 0.66 \, (Re_x)^{-1/2} \tag{8.58}$$

A posteriori, we see that the variable η has indeed an order of magnitude of y/δ.

In this case, the energy equation (8.50) may easily be solved for a constant wall temperature T_w and with the assumption of the ideal gas model ($C^p = \text{const.}$). Then, we have

$$g'' + Pfg' + E_{0e}\,(P-1)\,(f'f'')' = 0 \tag{8.59}$$

First assuming $P = 1$, we see the proportionality between the stagnation temperature distribution (or the static temperature for a low Mach number) and the velocity distribution. This constitutes the 'Reynolds analogy'; this is of course not surprising, since the corresponding boundary layers have similar thicknesses. Another consequence is found for the Nusselt number, i.e. $Nu_x = \frac{C_f Re_x}{2}$. We obtain the same result when $E_{0e} \sim 0$ (low velocity or stagnation point).

In the more realistic case where the Prandtl number is constant but different from 1, the integration of the energy equation leads to a quasi-analytic expression for the temperature distribution as a function of velocity distribution. For the wall heat flux, we obtain

$$q_w = -\frac{\lambda_w \rho_w u_e}{\sqrt{2C_w \xi}} \frac{f_w''^P}{I(P)} (T_w - T_{wr}) \tag{8.60}$$

where $T_{wr} = T_e \left(1 + \frac{u_e^2}{C^p T_e} PJ(P)\right)$ is the 'recovery temperature' (i.e. wall temperature for a zero heat flux). Approximate expressions are often used, such as $\frac{f_w''^P}{I(P)} \simeq f_w'' P^{1/3}$ and $r = 2PJ(P) \simeq P^{1/2}$ (recovery factor). Finally, we find

$$Nu_x = 0.33 P^{1/3} Re_x^{1/2} \tag{8.61}$$

For $P = 1$, we have $T_{wr} = T_{0e}$, which means that the conduction effects compensate the dissipation effects.

Physical interpretations

When there is no wall heat flux, the boundary layer is an isolated system since the heat flux is also zero at the external boundary. At every boundary-layer cross

CHAPTER 8 ELEMENTS OF GAS DYNAMICS

section, there is a total energy rate equal to $C^p T_{0e}$. For $P > 1$, we have $T_{wr} > T_{0e}$, and the energy dissipated by viscosity in the external part of the boundary layer between δ_d and δ_t (where $T = T_e$) is recovered by conduction close to the wall. The opposite is the case for $P < 1$ (the present case of gases).

When the wall temperature is constant, the system is no longer isolated, and depending on the relative values of T_w, T_e, and T_{0e}, the temperature distribution may be non-monotonic. This is the case with high-speed flow arising from the expansion of a compressed hot gas ($T_{0e} \gg T_e \sim T_w$). The gas in the boundary layer tends to recover the stagnation temperature, and this tendency is increased by conduction, since $\delta_t > \delta_d$. However, close to the wall, the influence of the wall heat flux and viscous dissipation tend to decrease the temperature. Therefore, there is a maximum for the temperature, which sometimes may be of importance (Chapter 10).

Axisymmetric case: cone without incidence

In the supersonic regime, if θ is the semiangle of the cone, we have

$$\xi = \rho_e \mu_e u_e \frac{x^3}{3} \sin^2 \theta$$

with $r = x \sin \theta$, and where u_e, p_e, T_e are constant. We also have

$$\eta = \left(\frac{3\rho_e \mu_e}{2 C u_e x}\right)^{1/2} \int_0^y \frac{\rho}{\rho_e} dy \qquad (8.62)$$

We may then obtain the following values for the skin friction and the heat flux:

$$\tau_w = \sqrt{3} \tau_{wpp}, \quad \text{and} \quad q_w = \sqrt{3} q_{wpp} \qquad (8.63)$$

where τ_{wpp} and q_{wpp} represent the corresponding values of the flat plate with the same outer conditions.

Flows with pressure gradient

Stagnation point

For this, we have $x = \xi = 0$, $u_e = 0$, $\frac{du_e}{dx} = \text{const.} = K$, $\beta = \frac{1}{j+1}$, and $r \sim x$

Therefore, self-similar solutions may be found for the transformed equations, that is:

$$f''' + ff'' = \frac{1}{j+1}\left(f'^2 - \frac{\rho_e}{\rho}\right)$$

$$g'' + Pfg' = 0 \qquad (8.64)$$

APPENDIX 8.5 SIMPLE BOUNDARY LAYERS: TYPICAL CASES

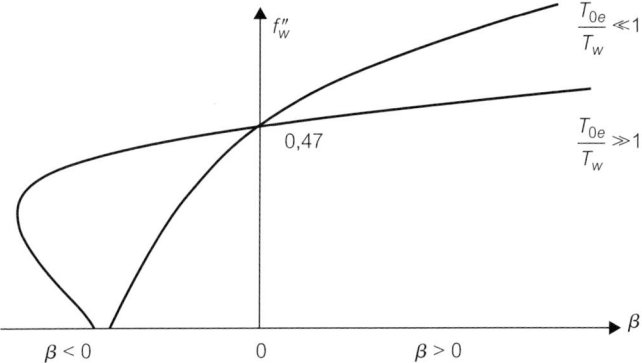

Figure 34. Variation of f_w'' with β.

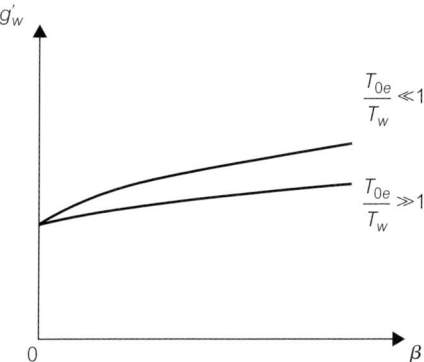

Figure 35. Variation of g_w' with β.

General case

The numerical solutions obtained[92] for $\beta \neq 0$ lead to values for f_w'' (proportional to the skin friction) and for g_w' (proportional to the wall heat flux) shown in Figs 34 and 35.

When the pressure gradient is 'adverse' (compression $\beta < 0$), the boundary layer becomes unstable and separates when $f_w'' = 0$ (Fig. 34); the separation is favoured by a 'hot' wall.

For a given value of β, f_w'' and g_w' are strongly influenced by the value of T_{0e}/T_w. For a 'cold' wall, however, they do not depend significantly on the pressure gradient (particularly g_w' and therefore the heat flux). In many cases, this may constitute the basis of simplifications.

We may also note that, for low-speed flows, we find the Falkner–Skan equation for the momentum conservation equation, i.e.

$$f''' + ff'' + \beta\left(1 - f'^2\right) = 0 \qquad (8.65)$$

Appendix 8.6 The turbulent boundary layer

Important problems related to turbulence are beyond the scope of this book. The phenomenon of turbulence is characterized by the emergence of instability in flows when the deterministic process of the streamlines of 'laminar' flows seems to be destroyed. Vortex structures arise and diffuse in the flow, so that the various quantities seem to show a random variation in time and space and appear to reproduce the molecular agitation at the macroscopic level. It also seems that statistically distinct average values could be discerned for these flow quantities. However, the structure of turbulent flow depends on the 'history' of the flow and generally is not a 'local' phenomenon,[93] so that the turbulence may be considered as a non-equilibrium phenomenon.

However, in a boundary layer where the transverse gradients are important, a relative homogenization occurs, and the boundary layer may be considered in near equilibrium and dominated by production–dissipation processes. Furthermore, in high-speed gas dynamics, the turbulence is generally not fully developed. Here, therefore, we restrict ourselves to a brief analysis of this type of turbulence.

Among the instability factors, the values of the pressure gradient and the stagnation temperature (β and T_{0e}/T_w) can play an important role in the laminar–turbulent transition. However, the Reynolds number remains the main factor, so that when its value becomes sufficiently high, the transition may occur in various forms; one example of which is given in Chapter 11.

Statistical analysis of homogeneous turbulence

Therefore, assuming random fluctuations, we may consider turbulent flow as the superposition of a mean flow and an unsteady fluctuating flow. Then, each scalar flow quantity F (density, temperature, and so on) may be written as

$$F = \overline{F} + F'' \qquad (8.66)$$

with

$$\overline{F''} = 0$$

The mean value $\overline{F}$ is defined in a timescale long enough to be significant but short enough to exclude the eventual unsteadiness of the mean flow. Moreover, if we take into account the compressibility of the flow, the mean velocity is a mass-weighted-averaged quantity (such as for a mixture in kinetic theory).[94,95]

Thus, for a component u_i, we have

$$\tilde{u}_i = \frac{\overline{\rho u_i}}{\overline{\rho}} \qquad (8.67)$$

with

$$u_i = \tilde{u}_i + u'_i \text{ and } \overline{\rho u'_i} = 0$$

Similarly, we can define the other quantities F such that

$$F = \tilde{F} + F'$$

Of course, the engineer is essentially interested in knowing the mean flow (deterministic description), which, however, depends on correlations between fluctuating quantities.

In order to describe this mean flow, we can use the Navier–Stokes equations that remain locally valid, and replacing each quantity by a sum given either by Eqn. (8.66) or Eqn. (8.67), we take the time average of the terms appearing in the resulting equations, which in fact represent moment equations of the Navier–Stokes equations.[95] Thus, we obtain mean continuity, mean momentum, mean energy, and mean species conservation equations in the following forms for a two-dimensional steady boundary layer:

$$\frac{\partial}{\partial x}\left(\overline{\rho}\,\overline{u}r^j\right) + \frac{\partial}{\partial y}\left(\overline{\rho v}r^j\right) = 0$$

$$\overline{\rho}\,\overline{u}\frac{\partial \overline{u}}{\partial x} + \overline{\rho v}\frac{\partial \overline{u}}{\partial y} = -\frac{d\overline{p}}{dx} + \frac{\partial}{\partial y}\left(\overline{\mu}\frac{\partial \overline{u}}{\partial y} - \overline{\rho u'v'}\right)$$

$$\overline{\rho}\,\overline{u}\frac{\partial \overline{h_0}}{\partial x} + \overline{\rho v}\frac{\partial \overline{h_0}}{\partial y}$$

$$= \frac{\partial}{\partial y}\left[\overline{\mu}\left(1 - \frac{1}{P}\right)\overline{u}\frac{\partial \overline{u}}{\partial y} + \frac{\overline{\mu}}{P}\frac{\partial \overline{h_0}}{\partial y} - \overline{\rho h'_0 v'} + \frac{\overline{\mu}}{S}\left(1 - \frac{1}{L}\right)\sum_P \overline{h_p}\frac{\partial \overline{c_p}}{\partial y}\right]$$

$$\overline{\rho}\,\overline{u}\frac{\partial \overline{c_p}}{\partial x} + \overline{\rho v}\frac{\partial \overline{c_p}}{\partial y} = \frac{\partial}{\partial y}\left(\frac{\overline{\mu}}{S}\frac{\partial \overline{c_p}}{\partial y} - \overline{\rho c'_p v'}\right) \qquad (8.68)$$

These equations take into account the usual hypotheses of the boundary layer. They present a formal similarity with those of the laminar boundary layer (Eqns (8.24)–(8.29)). However, 'turbulent transport' terms appear: They are added to the molecular transport terms and generally have much higher values. They must be determined or modelled in order to close the equation system (Eqn. (8.68)).

Examples of boundary-layer modelling

Closure of the system of Eqn. (8.68) may be operated by assuming a functional dependence between the turbulent transport terms $\overline{u'v'}$, $\overline{h'_0 v'}$ and $\overline{c'_p v'}$ and the parameters of the mean flow. Thus, by analogy with kinetic theory, we can express these terms in the form of products of mean quantity gradients and 'turbulent transport coefficients' (Boussinesq), i.e.

$$\overline{\rho u'v'} = -\mu_t \frac{\partial \overline{u}}{\partial y}$$

$$\overline{\rho h'_0 v'} = -\frac{\mu_t}{P_t} \frac{\partial \overline{h_0}}{\partial y}$$

$$\overline{\rho c'_p v'} = -\frac{\mu_t}{S_t} \frac{\partial \overline{c_p}}{\partial y} \quad (8.69)$$

Here, $\mu_t, \lambda_t = \frac{P_t}{C^P \mu_t}$ and D_t represent viscosity, conductivity, and diffusion turbulent coefficients respectively, and P_t and S_t are the corresponding Prandtl and Schmidt numbers. However, the above represents only a first step in the modelling, since these coefficients are unknown.

Algebraic models

If we continue the analogy with kinetic theory, for which the viscosity coefficient μ may be expressed as a function of the mean free path λ and the mean velocity U (Chapters 1 and 3), i.e. $\mu = \rho \lambda U$, we may assume

$$\mu_t = \overline{\rho} l_m V_t \quad (8.70)$$

where l_m is a characteristic 'mixing length' (Prandtl) and V_t a characteristic fluctuating velocity, both depending on the flow structure. In the framework of the same analogy, we assume

$$V_t = l_m \frac{\partial \overline{u}}{\partial y}$$

Therefore

$$\mu_t \simeq \overline{\rho} l_m^2 \frac{\partial \overline{u}}{\partial y} \quad (8.71)$$

The dependence of l_m on y remains to be defined, but a linear dependence is often sufficient.

Generally, the other coefficients are modelled in their form of turbulent Prandtl and Schmidt numbers, which strongly depend on the flow configurations.

Models with one or several moment equations

In the boundary layer, experiments have shown that the mixing-length hypothesis is valid and could explain the boundary-layer structure. The more sophisticated models are marginally better but more complex to implement. However, the non-local nature of the turbulence may be taken into account with evolution equations of characteristic quantities (moment equations).

Thus, if V_t^2 represents an average value of the turbulent kinetic energy, that is $V_t^2 \sim \overline{k} = \overline{(\frac{1}{2}u'^2 + \frac{1}{2}v'^2 + \frac{1}{2}w'^2)}$, we can obtain a transfer equation for $\overline{k}$ from the moment equations for u'^2, v'^2, w'^2. Thus, with the usual assumptions, we have

$$\overline{\rho}\,\overline{u}\frac{\partial \overline{k}}{\partial x} + \overline{\rho}\,\overline{v}\frac{\partial \overline{k}}{\partial y} = -\overline{\rho u'v'}\frac{\partial \overline{u}}{\partial y} - \frac{\partial}{\partial y}\left(\overline{\rho k v'} + \overline{p'v'}\right) - \overline{\mu}\overline{\left(\frac{\partial u'}{\partial y}\right)^2} \qquad (8.72)$$

$$\uparrow \qquad \qquad \uparrow \qquad \qquad \uparrow \qquad \qquad \uparrow$$

$$\text{Convection} \qquad \text{Production} \qquad \text{Diffusion} \qquad \text{Dissipation}$$

The terms of this equation require development; for example, they may be modelled as gradient terms or analysed by dimensional considerations.

We can also use other moment equations as an evolution equation of the dissipation term $\overline{\varepsilon} = \overline{\mu}\overline{\left(\frac{\partial u'}{\partial y}\right)^2}$. The system of Eqn. (8.68) and the evolution equations for $\overline{k}$ and $\overline{\varepsilon}$ constitute the $k - \varepsilon$ model largely used in engineering calculations.

Appendix 8.7 Flow separation and drag in MHD

As discussed in Appendix 7.6, ionized gas flows can be modified by electromagnetic forces proceeding from external magnetic and electric fields.[96,97] In the following, we analyse a very simple MHD flow around an obstacle with the aim of 'improving' the flow, that is, delaying (or suppressing) the boundary-layer separation and, as far as possible, reducing the drag of the obstacle.

Thus, we consider the steady incompressible plane flow of an ionized gas around an insulating cylinder (radius R) with an applied constant magnetic field $\boldsymbol{B}_A$ normal to the flow[98] (Fig. 36).

Assuming that the induced magnetic field may be neglected ($R_\sigma \ll 1$) and that we can divide the flow into an inviscid part and a boundary layer, the MHD equations are strongly simplified and, in the inviscid flow, may be written in the

256 CHAPTER 8 ELEMENTS OF GAS DYNAMICS

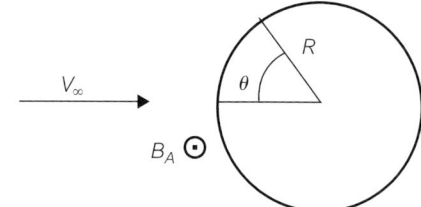

Figure 36. MHD flow around a circular cylinder.

following dimensionless form:

$$\frac{\partial \cdot V^*}{\partial r^*} = 0 \tag{8.73}$$

$$V^* \cdot \frac{\partial V^*}{\partial r^*} = -\frac{\partial p^*}{\partial r^*} - IV^* \tag{8.74}$$

where

$$r^* = \frac{r}{R}, \quad V^* = \frac{V}{V_\infty}, \quad p^* = \frac{p}{\rho V_\infty^2}, \quad B^* = \frac{B}{B_A}$$

and where I is an 'interaction parameter', the ratio of the Laplace force to the inertial force, that is:

$$I = \frac{\sigma B_A^2 R}{\rho V_\infty}$$

Thus, fluid particles are slowed down by the electromagnetic force, and assuming as usual that the inviscid flow is irrotational, the Laplace force field derives from a potential ϕ, i.e.

$$\frac{\partial \phi^*}{\partial r^*} = IV^*$$

The velocity field is not modified, but the momentum equation (8.74) becomes

$$p + \frac{1}{2}\rho V^2 + \phi = K$$

where K is a constant. This is the generalized Bernoulli equation (see Eqn. (7.13)).

If we compute the pressure coefficient on the cylinder, that is, $C_p = \frac{p - p_\infty}{1/2 \rho V_\infty^2}$, we find

$$C_p = C_{p0} - 4I \cos \theta$$

where $C_{p0} = 1 - 4 \sin^2 \theta$ is the pressure coefficient without a magnetic field, so that $C_p > C_{p0}$ on the windward side, and $C_p < C_{p0}$ on the leeward side. The pressure decreases from the stagnation point up to a point corresponding to $\theta = \text{Arc} \cos (I/2)$, which is located on the leeward side, instead of $\theta = \frac{\pi}{2}$ without a magnetic field. If $I = 2$, the pressure gradient remains negative up to the downstream stagnation point. As a consequence, we expect the boundary layer to remain attached all along the cylinder (Appendix 8.4).

This conclusion must be confirmed by a computation of the boundary layer. Thus, the corresponding dimensionless equations may be written:

$$\frac{\partial u^*}{\partial x^*} + \frac{\partial v^*}{\partial y^*} = 0$$

$$u^* \frac{\partial u^*}{\partial x^*} + v^* \frac{\partial u^*}{\partial y^*} = -\frac{dp^*}{dx^*} + \frac{1}{Re} \frac{\partial^2 u^*}{\partial x^2} - I u^*$$

with

$$-\frac{dp^*}{dx^*} = u_e^* \frac{du_e^*}{dx^*} + I u_e^*$$

and where $x^* = x/R$, $y^* = y/R$, $u^* = u/V_\infty$, $v^* = v/V_\infty$.

After having applied the usual LLD transformation, the above boundary equations are solved, and the skin-friction coefficient C_f (Eqn. (7.32)) is computed, with the following result:

$$C_f = \frac{1}{\sqrt{Re}} \frac{4 \sin^2 \theta}{\sqrt{1 - \cos \theta}} f_w''$$

The quantity $C_f \sqrt{Re} = F(\theta)$ is plotted in Fig. 37 for three values of the interaction parameter I; thus, skin friction cancels out at the downstream stagnation point for values of I larger than 2.5, for which no separation therefore occurs. Without a magnetic field, separation takes place for $\theta \simeq 110°$.

Of course, these results are approximate because of the interaction between the inviscid flow and the boundary layer, which becomes relatively thick on

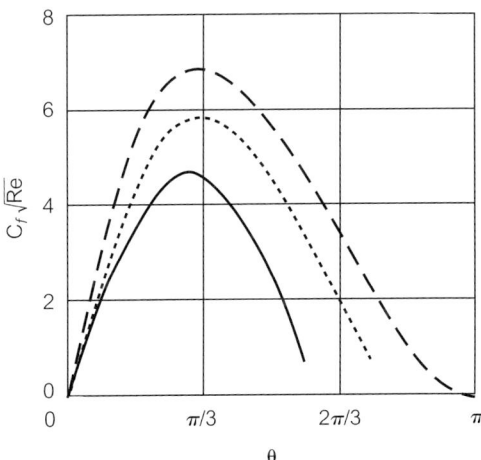

Figure 37. Skin-friction coefficient along a cylinder. ———: $I = 0$, $\cdots$: $I = 1$, $---$: $I = 2.5$.

the leeward side. Thus, when taking this interaction into account, or making a complete Navier–Stokes calculation, we find the separation point at $\theta \simeq 80°$ without a magnetic field (value confirmed by experiment), so that values for I larger than 2.5 are necessary to avoid any separation.

As for the drag, the viscous part is indeed very small, but contrary to the case without a magnetic field, there is a non-null pressure drag equal to $2\pi \mu V_\infty R_h^2$, where $R_h = \sqrt{IRe}$ is the Hartmann number. This drag may take important values, so that the application of an 'appropriate' electric field should be necessary for reducing (or suppressing) this drag.

NINE
Reactive Flows

9.1 Introduction

In this chapter, we examine the interaction of the chemical processes taking place in gaseous media with the flow parameters. We first consider separately the equilibrium regimes ($Da \to \infty$) and the non-equilibrium regimes ($Da \sim 1$). Frozen cases, in which the influence of chemical reactions is negligible ($Da \sim 0$), correspond to flows analysed with methods for traditional gas dynamics, a few examples of which were presented in Chapter 8.

The influence of vibrational kinetics on flow is also discussed when it represents the only process at a moderate temperature or when chemical reactions occur at the same time (Chapter 5).

The developments presented in this chapter are not entirely new,[83,99] and they include a summary of the main elements necessary for understanding reactive gas flows. Examples of simple flow modelling are also presented. These are essentially Eulerian flows behind relatively intense shock waves or expanding Eulerian flows in supersonic nozzles, exemplified here to point out the influence of 'real gas effects' on corresponding flows. Hypersonic flows around models are analysed in Chapter 10 in the framework of dissipative regimes.

9.2 Generalities on chemical reactions

The problem of reactive collisions was discussed in Chapter 2 in connection with the Boltzmann equation, and definitions of binary reaction-rate constants and equilibrium constants were also given. In this chapter, these notions are generalized to any type of reaction but are applied only to the macroscopic level.

In fact, we must derive the source term $\dot{w}_p$ included in the species conservation equations (7.60). We therefore consider the reactions of the following type taking place in a homogeneous gaseous phase:

$$\sum_p v_p A_p \rightleftarrows 0 \qquad (9.1)$$

where ν_p represents the stoichiometric coefficient of component p in the reaction, and A_p is the symbolic notation for species p. Conventionally, $\nu_p < 0$ and $\nu_p > 0$ correspond to the 'reactants' and 'products' of the reaction, respectively.

Consequently, there exists a similarity relation between the values of the molar concentrations $X_p = n_p/N$ of the components, that is:

$$\frac{dX_1}{\nu_1} = \frac{dX_2}{\nu_2} = \cdots = \frac{dX_p}{\nu_p} = \cdots = dZ \qquad (9.2)$$

The entropy variation dS_c due to the reactions (in TRV equilibrium) may be deduced from Eqn. (7.18), that is:

$$dS_c = -\frac{1}{T} \sum_p \mu_p dc_{pc} \qquad (9.3)$$

where dc_{pc} represents the concentration change of species p due to the reactions.

9.3 Equilibrium flows

9.3.1 Law of mass action: chemical equilibrium constant

For $Da \to \infty$, we have $\dot{w}_p = 0$ and $dS_{pc} = 0$.

The species conservation equation is (formally) uncoupled from the other conservation equations, so that the species concentrations must be independently found. These concentrations, however, depend on local conditions governed by the other equations.

As we have $c_p = X_p M_p/\rho$, we can deduce from Eqns (9.2) and (9.3) the relation $\sum_p \nu_p \mu_p M_p dZ = 0$. With the definition for the chemical potential per mole of the species p, $\hat{\mu}_p = \mu_p M_p$, we have

$$\sum_p \nu_p \hat{\mu}_p = 0 \qquad (9.4)$$

This relation (Eqn. (9.4)) constitutes the 'law of mass action', from which we can obtain another relation involving the concentrations. Thus, taking into account the expressions (Eqns (7.16) and (7.10)) for S_p and h_p respectively, we have

$$\mu_p = \mu_p^0 + R_p T \operatorname{Log} p_p \qquad (9.5)$$

where μ_p^0 is a function of T only. Now, applying Eqn. (9.4), we find

$$\prod_p p_p^{\nu_p} = K_p \qquad (9.6)$$

where K_p depends only on T. From Eqn. (9.6), we can deduce the partial pressures of the components. Then, since $p_p = X_p \mathcal{R} T$, we can obtain a relation between the molar concentrations, that is:

$$\prod_p X_p^{\nu_p} = K_c \tag{9.7}$$

where $K_c = K_p/(\mathcal{R}T)^{\sum_p \nu_p}$ also depends only on temperature. Here, K_c is the equilibrium constant of the reaction. We can then calculate the enthalpies h_p and h. However, the relations giving c_p and N_p (molar fraction equal to X_p/ρ) also depend on the pressure.

The reactions are endothermic or exothermic according to whether they absorb or give energy (heat of reaction) to the surroundings. This is equivalent to an enthalpy variation per mole equal to $\Delta \hat{H} = \sum_p \nu_p h_p$: this represents the enthalpy balance of the reaction, which may be easily related to K_p by the following relation (Van't Hoff):

$$\frac{d}{dT}(\text{Log } K_p) = \frac{\Delta \hat{H}}{\mathcal{R}T^2} \tag{9.8}$$

As $\Delta \hat{H}$ varies little with T, we can obtain an approximate expression for K_p:

$$K_p \simeq A \exp\left(-\Delta \hat{H}/\mathcal{R}T\right) \tag{9.9}$$

where A is a constant.

Therefore, the value of the equilibrium constant can be deduced from a measurement of ΔH.

9.3.2 Examples of reactions

Example 1: Single reaction

We consider the following bimolecular reaction:

$$A + B \rightleftarrows C + D \ (\text{or: } C + D - A - B \rightleftarrows 0) \tag{9.10}$$

We can represent this reaction in an energetic diagram (Fig. 38 for the exothermic case).

An important application concerns the dissociation of diatomic molecules: for high temperatures, molecular collisions may give rise to the following dissociation reaction:

$$A_2 + A_2 \rightleftarrows 2A + A_2 \tag{9.11}$$

The dissociation is an endothermic process (and the recombination is exothermic). We have $\Delta \hat{H} = \hat{E}_{DA_2}$, where $\hat{E}_{DA_2}$ is the dissociation energy per mole. We can therefore calculate the

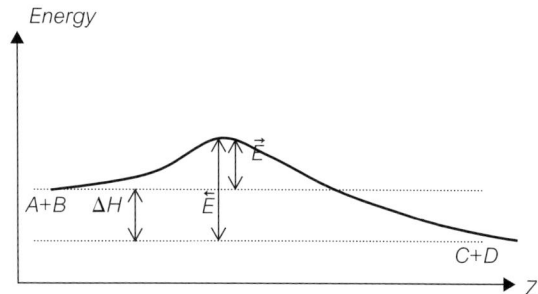

Figure 38. Energetic diagram of the reaction $A + B \leftrightarrows C + D\vec{E}$: Direct activation energy; $\vec{E}$: Inverse activation energy ($\Delta H = \vec{E} - \vec{E}$)

equilibrium concentrations of the components A_2 and A, the enthalpy of the mixture, and its equation of state.

As usual, we set $c_A = \alpha$ (dissociation rate) and $c_{A_2} = 1 - \alpha$.

With the law of mass action, we have

$$\frac{p_A^2}{p_{A_2}} = K_p \text{ and consequently } \frac{4\alpha^2}{1-\alpha} = \frac{K_p}{p} = \frac{K_c \mathcal{R} T}{p} \tag{9.12}$$

that is

$$\alpha = \alpha(p, T)$$

With the approximate relation (Eqn. (9.9)), we have $K_p \simeq A \exp(\theta_{DA2}/T)$, where $\theta_{DA_2} = \hat{E}_{DA2}/R_{A2}$ is the characteristic temperature of the dissociation of molecule A_2. Thus, for oxygen we have $\theta_{DO_2} \simeq 59\,500$ K, and for nitrogen $\theta_{DN_2} \simeq 113\,000$ K. It is clear that α begins to be important at much lower temperatures. In fact, the dissociation of oxygen is significant from 2000 K.

From Eqns (9.9) and (9.12), we may find a universal (but approximate) relation for diatomic molecules at a given pressure,[83] that is:

$$\alpha = \alpha(\theta_{DA2}/T)$$

We also obtain the expression of the enthalpy of the mixture from its definition, that is:

$$h = \alpha h_A + (1 - \alpha) h_{A_2} \tag{9.13}$$

A reference enthalpy must be chosen; here, as usual, we choose the molecular state at $T = 0$, so that, in the definition of enthalpy, we must take into account the formation enthalpy of atoms $h_A^0 = \hat{E}_{DA_2}/M_{A_2} = R_{A_2}\theta_{DA_2}$. Therefore, for h, we have

$$h = \alpha \left(h_A^0 + \int_0^T C_A^p dT \right) + (1 - \alpha) \int_0^T C_{A_2}^p dT \tag{9.14}$$

9.3 EQUILIBRIUM FLOWS

The continuation of the calculation depends on the chosen physical model and on the temperature range. Thus, for example, when $T \gg \theta_{VA_2}$, we also obtain a universal relation for the reduced enthalpy, that is:

$$\frac{h}{R_{A_2}\theta_{DA_2}} = F\left(\frac{\theta_{DA_2}}{T}\right)$$

The equation of state for the mixture may be written in the following form:

$$p = \rho RT = (1+\alpha)\rho R_{A_2} T \quad (9.15)$$

At higher temperature, the atoms and molecules may be ionized by the following process:

$$A + A \rightleftarrows A^+ + A + e \quad (9.16)$$

This type of reaction may also be considered a particular case of the general reaction (Eqn. (9.1)) with an equilibrium constant, a characteristic ionization temperature, and so on. The energies involved are generally expressed in electronvolts (eV), where 1 eV corresponds to about 11600 K.

Example 2: Mixture of reactive gases

A typical example is that of air at high temperature. For temperature in the range 7–8000 K, we may assume that there are three dissociation–recombination reactions and one ionization reaction,[7] that is:

$$O_2 + M \rightleftarrows O + O + M \left(\Delta \hat{H} = 5.1 \text{ eV}\right)$$

$$N_2 + M \rightleftarrows N + N + M \text{ (9.8 eV)}$$

$$NO + M \rightleftarrows N + O + M \text{ (6.5 eV)}$$

$$NO + M \rightleftarrows NO^+ + e + M \text{ (9.25 eV)} \quad (9.17)$$

Here, M represents any species of the mixture, that is, O_2, N_2, O, N, NO, NO^+, or e.

Thus, we have a mixture with seven species. The concentrations of these species may be deduced from the law of mass action (Eqn. (9.7)), that is:

$$\frac{X_O^2}{X_{O_2}} = K_{c1} \quad \text{and} \quad \frac{X_N^2}{X_{N_2}} = K_{c2}$$

$$\frac{X_N X_O}{X_{NO}} = K_{c3} \quad \text{and} \quad \frac{X_e X_{NO^+}}{X_{NO}} = K_{c4} \quad (9.18)$$

To these four relations, we can add two conservation equations for the N and O atoms (molar concentrations per unit mass of the mixture), that is:

$$2N_{O_2} + N_O + N_{NO} + N_{NO^+} = 2(N_{O_2})_0$$
$$2N_{N_2} + N_N + N_{NO} + N_{NO^+} = 2(N_{N_2})_0 \quad (9.19)$$

where $N_p = X_p/\rho$, and where $(N_p)_0$ is the initial mole number (assumed known) of the species p per unit mass of the mixture.

A seventh relation is deduced from the electric neutrality of the mixture, that is:

$$N_{NO^+} = N_e \tag{9.20}$$

From these seven equations, we may obtain the mixture composition as a function of temperature and pressure. At higher temperature, more reactions must be taken into account (Appendix 9.2).

The enthalpy and the other thermodynamic functions of the mixture may then be computed, as well as the equation of state.

9.3.3 Examples of equilibrium flows

Flow behind a shock wave

The high temperatures attained behind shock waves give rise to processes of excitation of internal modes, dissociation, and chemical reactions. Fluid particles, after crossing a shock wave, tend to an equilibrium situation which may be determined a priori from the upstream conditions but also from the shape and the intensity of the shock wave. For a straight shock wave, the Rankine–Hugoniot conditions (Eqn. (8.12)) give the downstream equilibrium conditions if we use adequate expressions for the enthalpy h.

Thus, for a pure diatomic gas, with vibrational excitation and dissociation, we use Eqn. (9.14). Figures 39 and 40 show the variation of the density and temperature ratios ρ_2/ρ_1 and T_2/T_1 with the shock wave Mach number M_s and with upstream conditions p_1, T_1, for oxygen and nitrogen; they are compared to the corresponding values obtained for an ideal gas[100] (Eqns (8.13) and (8.14)).

Flow in supersonic and hypersonic nozzles

When the reservoir conditions include high values for pressure and temperature, the rapidly expanding flow in a nozzle cannot generally preserve an equilibrium state; the corresponding equilibrium solution therefore represents only a theoretical reference. As in the case for a shock wave, a realistic expression for h must be used in the conservation equations.

For the quasi-one-dimensional case, a rough calculation may be carried out, as with the ideal gas model, i.e. with Eqns (7.47), (7.48), and (7.40), that is:

$$\rho u S = \text{const.}, \quad h_0 = \text{const.}, \quad \text{and} \quad \rho u du + dp = 0$$

If p, ρ, and T decrease along the nozzle because of the transformation of the various types of energy into kinetic energy, M increases, and u tends to a limiting

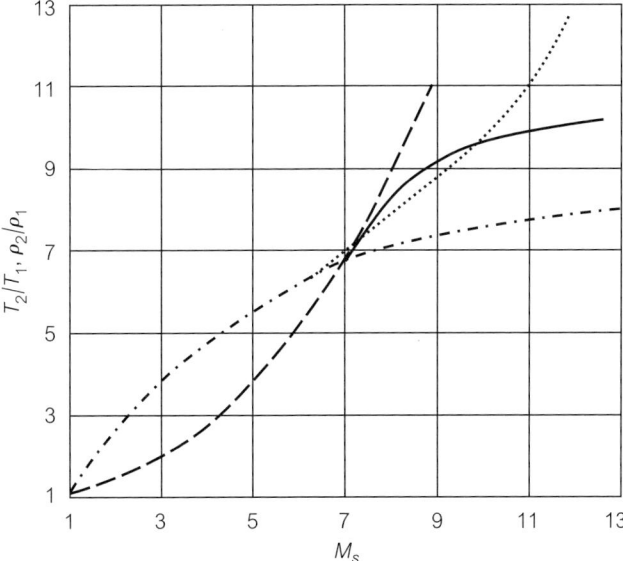

Figure 39. Temperature and density ratios (straight shock wave) for oxygen. Temperature: ———: equilibrium, — — —: ideal; Density: ····: equilibrium, — · — · —: ideal; $(T_1 = 300\text{ K}, p_1 = 10^3\text{ Pa})$.

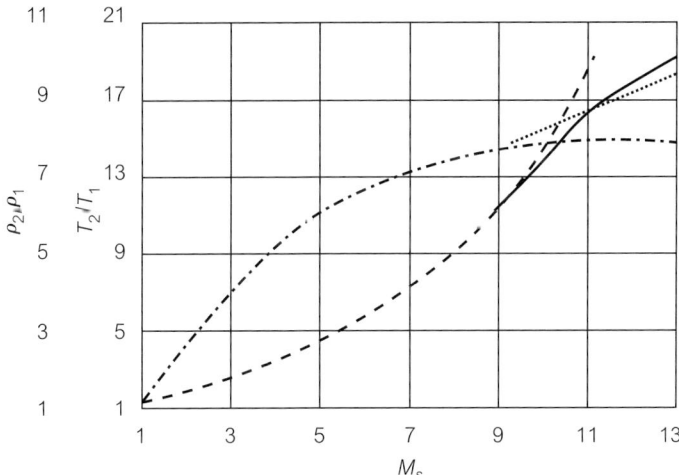

Figure 40. Temperature and density ratios (straight shock wave) for nitrogen (Notation as Fig. 39).

value. The temperature, however, remains higher than in the ideal case because the chemical system tends to adapt itself to the local conditions and because the recombination favoured by the expansion is an exothermic process. We also note that the flow at the throat is sonic as for the ideal gas model.

9.4 Non-equilibrium flows

Flows in chemical and/or vibrational non-equilibrium were analysed from a fundamental statistical point of view in the first part of this book. Here, we generalize or apply the results obtained in the first part to concrete and complex situations requiring a macroscopic approach. As in the previous sections, we first examine chemical non-equilibrium situations and vibrationally relaxing flows separately and then the situations where both processes occur simultaneously.

Applications to characteristic Euler flows are also presented in order to emphasize the importance of real gas effects and the influence of the physical models.

9.4.1 Chemical kinetics

Regarding equilibrium flows, we must develop the source term $\dot{w}_p$, but from the complete species conservation equation (7.60). With $Da \sim 1$, we have either a high value for ReS, in which case we use the Euler system, or a low value, for which we use the Navier–Stokes system. In either case, all equations must be solved simultaneously.

Assuming TRV equilibrium, we consider the following general chemical reaction, involving only non-ionized species:

$$\sum_p v'_p A_p \rightarrow \sum_p v''_p A_p \qquad (9.21)$$

where v'_p and v''_p are positive coefficients in contrast to the form of Eqn. (9.1).

The molar rate of disappearance of the species p per unit volume is related to the reaction rate dZ by Eqn. (9.2), that is:

$$\dot{X}_p^d = v'_p \frac{dZ}{dt} \qquad (9.22)$$

From Section 9.3.1 (law of mass action), it is clear that we have

$$\frac{dZ}{dt} = k_f \prod_p X_p^{v'_p} \qquad (9.23)$$

where the proportionality factor k_f is called the rate constant of the reaction (Eqn. (9.21)), similar to the rate constant defined in Chapter 2. We then have

$$\dot{w}_p = k_f v'_p M_p \prod_p X_p^{v'_p} \qquad (9.24)$$

The rate constant $k_f = k_f(T)$ is proportional to the total collision rate, which, at moderate temperature, can be considered the elastic collision rate

9.4 NON-EQUILIBRIUM FLOWS

(Appendix 2.2). Among these collisions, only those that have a rate constant proportional to $\exp(-\hat{E}/\mathcal{R}T)$ are energetic enough to start a reaction (Appendix 9.3), so that we may generally write (Arrhenius):

$$k_f = C_f T^n \exp\left(\frac{-\hat{E}}{\mathcal{R}T}\right) \qquad (9.25)$$

Similarly, a rate of appearance of the species p may be defined, i.e.

$$\dot{X}_p^a = k_f v_p'' M_p \prod_p X_p^{v_p''} \qquad (9.26)$$

For a reversible reaction such as $\sum_p v_p' A_p \underset{k_f}{\overset{k_r}{\rightleftharpoons}} \sum_p v_p'' A_p$, the balance $\dot{w}_p$ is equal to

$$\dot{w}_p = M_p \left(v_p'' - v_p'\right) \left(k_f \prod_p X_p^{v_p'} - k_r \prod_p X_p^{v_p''}\right) \qquad (9.27)$$

In that case, for equilibrium conditions ($\dot{w}_p = 0$), we again find the relation of Eqn. (9.7) as long as the TRV equilibrium is maintained:

$$\prod_p X_p^{v_p'' - v_p'} = \frac{k_f}{k_r} = K_c(T) \qquad (9.28)$$

From Eqns (9.25) and (9.28), we deduce that the rate constant of the reverse reaction of Eqn. (9.21) has the following form:

$$k_r = C_r T^m \qquad (9.29)$$

Example 1: Dissociation of a diatomic gas

This example is typical (Chapter 5), and we have

$$A_2 + M \underset{k_r}{\overset{k_f}{\rightleftharpoons}} 2A + M \qquad (9.30)$$

Thus we find

$$\dot{X}_A = 2(k_f X_{A_2} X_M - k_r X_A^2 X_M) \qquad (9.31)$$

with

$$k_f = C_f T^n \exp\left(\frac{\theta_D}{T}\right) \qquad (9.32)$$

and finally

$$\dot{w}_A = 2 M_A k_f X_M \left(X_{A_2} - \frac{1}{K_c} X_A^2\right) \qquad (9.33)$$

Example 2: Reacting gas mixture

One example of this is air at high temperature.

Considering a mixture of n species p involving r reactions, we have for the mth reaction:

$$\sum_{p=1}^{n} v'_{pm} A_p \underset{k_{bm}}{\overset{k_{fm}}{\rightleftharpoons}} \sum_{p=1}^{n} v''_{pm} A_p \qquad (9.34)$$

The total balance of the species p due to all r reactions is

$$\dot{X}_p = \sum_{m=1}^{r} (\dot{X}_p)_m = \sum_{m=1}^{r} \left(v''_{pm} - v'_{pm} \right) \left(k_{fm} \prod_{p=1}^{n} X_p^{v'_{pm}} - k_{rm} \prod_{p=1}^{n} X_p^{v''_{pm}} \right) \qquad (9.35)$$

and in TRV equilibrium, we have

$$\dot{X}_p = \sum_{m=1}^{r} \left(v''_{pm} - v'_{pm} \right) k_{fm} \left(\prod_{p=1}^{n} X_p^{v'_{pm}} - \frac{1}{K_{cm}} \prod_{p=1}^{n} X_p^{v''_{pm}} \right) \qquad (9.36)$$

so that

$$\dot{w}_p = M_p \dot{X}_p$$

9.4.2 Vibrational kinetics

In principle, the fundamental problem is how to determine vibrational populations in various situations involving vibrational relaxation and their influence on flow quantities. To do this, it is necessary first of all to know or to determine the individual TV and VV transition probabilities. Fortunately, it is not always absolutely necessary to make such calculations, and 'global' methods based on realistic models are often used (Chapter 2). These may give sufficiently accurate results, particularly when only macroscopic flow parameters are required. Sometimes, it is also necessary to make further assumptions, particularly in complex situations where several reactive components are involved. Examples of such calculations in specific cases (shock waves, expansion flows) are given below.

Global models

Many models derived with global methods have been presented and used for relaxation times τ_{TV} (pure gases and mixtures) and τ_{VV} (mixtures) in the general framework of the harmonic oscillator model.[11] Thus in Fig. 41, particular results for TV relaxation times (N_2/N_2, O_2/O_2, O/O), deduced from SSH (Schwarz–Slavsky–Herzfeld) modelling,[15] are shown. As indicated in Chapter 2, they show

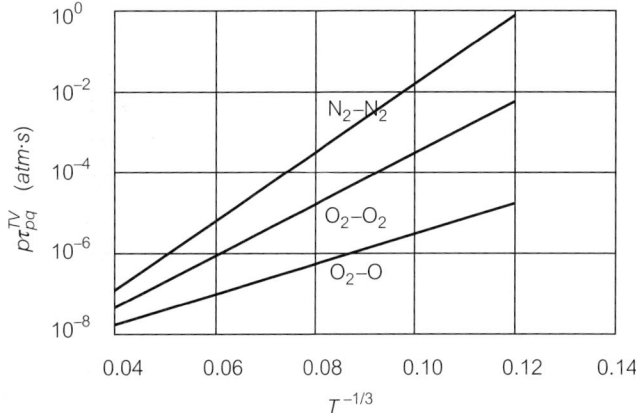

Figure 41. Examples of TV relaxation times.

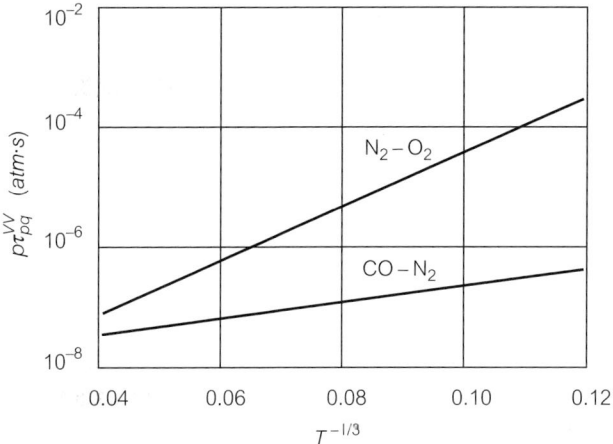

Figure 42. Examples of VV relaxation times.

a linear dependence of the product $p\tau_{TV}$ with $T^{1/3}$. In particular, we point out the efficiency of the O_2/O collisions. However, in view of the experimental results, the model may be questioned both at high temperature, because in these conditions polyquantum transitions may occur, and at low temperature for which the long-range effects of the interaction potential may be important.

In Fig. 42, values for τ_{VV} in mixtures N_2/O_2 and CO/N_2, deduced from the same model, are also shown.

In addition, this model is not completely correct for VV transitions (Chapter 12). However, qualitatively, it shows the different characteristics of these two mixtures, essentially due to the quasi-resonant CO/N_2 collisions: the spacing between the vibrational levels is nearly the same for these two molecules, in contrast to the N_2/O_2 case. Thus, the VV relaxation time τ_{VV} is relatively small for the CO/N_2 case, and this mixture relaxes essentially as a pure gas.

For mixtures composed of more than two components, the harmonic oscillator model enables us to simplify the general relaxation equation (2.39). In particular, we may take into account all TV collisions by using the following barycentric formula:

$$\frac{1}{\tau_p^{TV}} = \sum_q \frac{\xi_q}{\tau_{pq}^{TV}} \qquad (9.37)$$

where τ_p^{TV} represents an 'average' relaxation time for the species p.

STS Models

In principle, the determination of the individual collision rates $a_{i,j}^{k,l}$, called the State to State (STS) approach, gives a detailed description of vibrational nonequilibrium flows and does not depend on approximate global models. The STS method must be used when particular exchanges are preferential, as is the case for gas-dynamic lasers (CO_2/N_2, $\cdots$). However, when we need to know only the macroscopic flow parameters, the complexity of the STS method may not justify its use.

STS models are available for TV and VV exchanges.[17,101–103] Thus, Figs 43 and 44 show the variation of a few collision rates with the vibrational quantum number for N_2 and O_2, respectively. As pointed out in Chapter 2, we observe that, for moderate temperatures, the VV transitions are dominant for the lowest levels and the TV transitions for the highest levels. Moreover, as the resonant transitions represent the most probable transitions, we therefore validate the definition of a

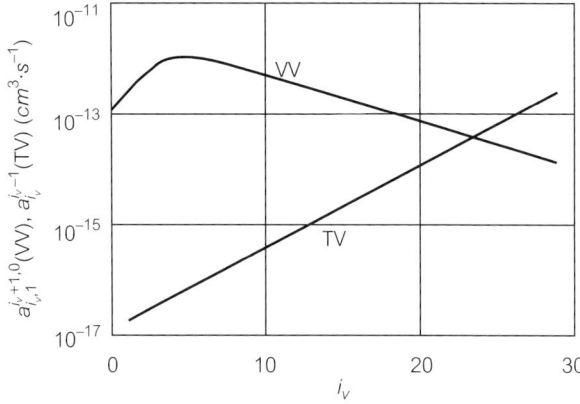

Figure 43. TV and VV collision rates ($N_2 - N_2$, T = 1000 K).

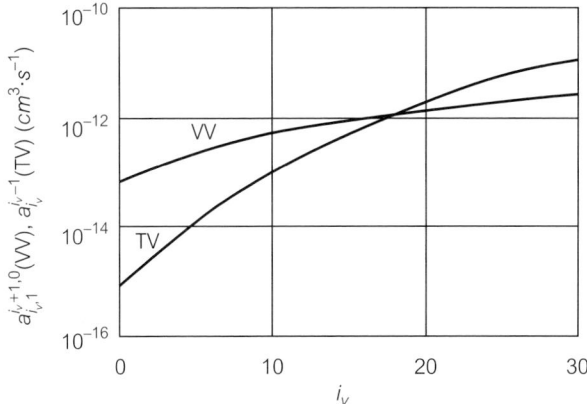

Figure 44. TV and VV collision rates (O_2–O_2, T = 1000 K).

'vibrational temperature', at least for not too high temperatures and for situations where the vibrational population is essentially concentrated in low levels.

9.4.3 General kinetics

As already discussed, vibrational kinetics is generally faster than chemical kinetics (Figs 9–10). Vibrational and chemical non-equilibriums may, however, coexist, for example in situations where non-equilibrium processes start simultaneously, such as behind strong shock waves. A statistical analysis of the resulting interaction between both processes was presented in Chapter 5. Thus, the chemical rate constants depend on the vibrational populations, and therefore, the relaxation equations are modified by the chemical reactions. The application of the corresponding models analysed in Chapter 5 is carried out below within the framework of specific examples of non-equilibrium flows.

9.5 Typical cases of Eulerian non-equilibrium flows

9.5.1 Flow behind a straight shock wave

Vibrational kinetics: pure gases

For relatively weak shock waves, there are generally no chemical reactions, only vibrational excitation, and the corresponding equilibrium is attained after a number of collisions, depending on density and temperature. Therefore, behind the shock wave there exists a vibrational non-equilibrium region generally governed by the Euler equations, closed by the (level by level) balance equations

(Eqn. (2.15)). For one-dimensional flow behind a straight shock wave moving at constant speed, we can use the Rankine–Hugoniot relations (Eqn. (8.12)) and the Landau–Teller equation written in a shock wave fixed coordinate system, that is:

$$u\frac{de_V}{dx} = \frac{\bar{e}_V - e_V}{\tau_V} \tag{9.38}$$

The relaxation time τ_V represents the TV exchanges (examples of which are shown in Fig. 41). At $x = 0$ (just behind the shock wave), the flow quantities are vibrationally frozen (Rankine–Hugoniot relations for an ideal gas), and the vibrational energy is the same as that upstream from the shock wave. Then, after solving the Rankine–Hugoniot system and Eqn. (9.38), we obtain the flow parameters $T(x)$, $e_V(x)$, and so on. For increasing values of x, the equilibrium state is approached. These equilibrium conditions of course correspond to the intensity of the shock wave and upstream conditions, so that they can be a priori determined independently of the relaxation (Section 9.3.3). Moreover, if we assume that the resonant collisions are dominant, we can define a vibrational temperature $T_V(x)$, that evolves from its frozen value to the equilibrium value when $T_V = T$. An example[104] of the evolution of temperatures T and T_V is presented in Fig. 45, where we observe that they vary significantly, mainly of course T_V. As the pressure varies little in the relaxation zone (which is also the case for the velocity), density ρ increases significantly, so that many experimental determinations of τ_V are based on the measurement of this parameter as well as on e_V measurements (Chapter 11).

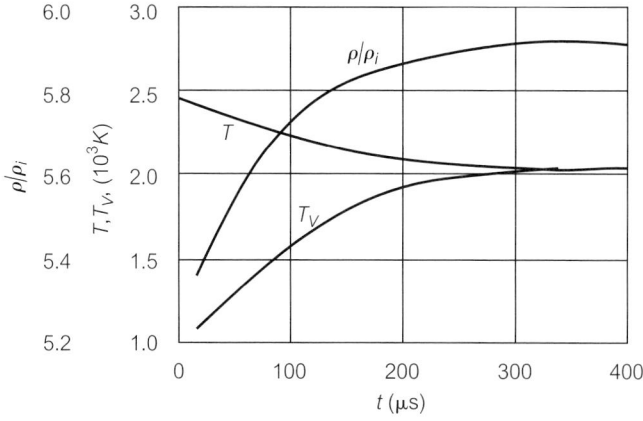

Figure 45. Evolution of temperatures and density ratio behind a shock wave (nitrogen, $M_s = 6.12$, $p_i = 3947$ Pa, $T_i = 295$ K).

Vibrational kinetics: gas mixtures

In the case of a binary mixture, when the gap between the vibrational levels of the components is rather different (N_2/O_2 mixture, for example), the VV exchanges are difficult, and both components tend to evolve independently to equilibrium with two distinct vibrational temperatures T_{VN_2} and T_{VO_2}. In contrast, as discussed in Section 9.4.2, if the VV exchanges are easy (N_2/CO case for which $\theta_{VN_2} \sim \theta_{VCO}$, for example), the coupling is important, and both gases evolve to equilibrium quasi-simultaneously $(T_{VN_2} \sim T_{VCO})$: the TV relaxation times are close, and $\tau_{CO-N_2}^{VV}$ is small (Chapter 12).

Thus, an example of the variation of temperatures for a N_2/O_2 mixture (air) behind a straight shock wave is shown in Fig. 46(a) with and without VV

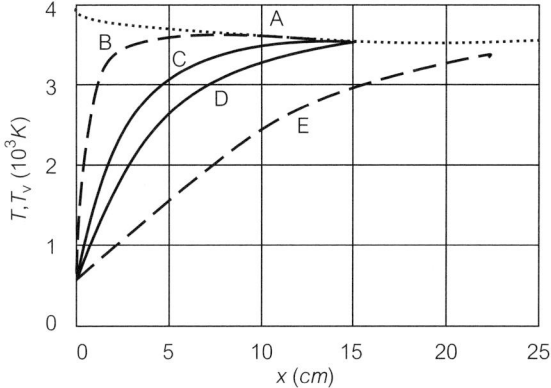

Figure 46. (a) Spatial variation of temperatures behind a shock wave in air ($M_S = 8$, $p_1 = 102$ Pa, $T_1 = 271$ K). A: Temperature T, B: T_{VO2} (without VV), C: T_{VO2} (with VV), D: T_{VN2} (with VV), E: T_{VN2} (without VV).

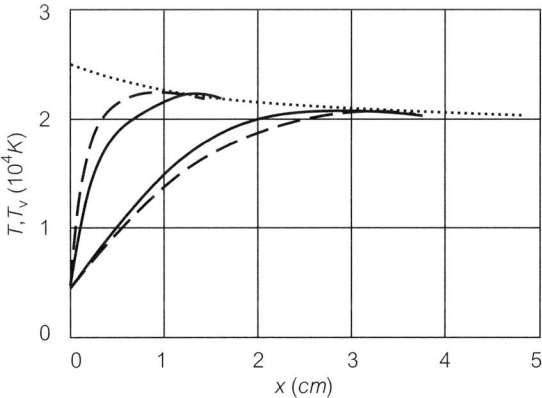

Figure 46. (b) Spatial variation of temperatures behind a shock wave in air ($M_S = 25$, $p_1 = 102$ Pa, $T_1 = 205$ K). (Notation of Fig. 46(a)).

collisions. We observe that, for relatively moderate temperatures (shock wave Mach number equal to 8), the evolution of the vibrational temperatures for O_2 and N_2 is different ($\tau^{TV}_{N_2} \gg \tau^{TV}_{O_2}$). However, at higher temperatures (Fig. 46(b)), the dominant influence of the TV collisions, and therefore their efficiency, tends to accelerate the process of relaxation, and the evolution of the vibrational temperatures with and without VV collisions are comparable. We also observe that, because of the VV collisions, the relaxation of N_2 is accelerated while the relaxation of O_2 is slowed down.

We may also note that the use of STS methods does not bring important differences in the evolution of the macroscopic flow quantities, even if the vibrational population does not have a Boltzmann distribution. An example of the evolution of vibrational populations behind a shock wave, with and without dissociation, is presented in Appendix 9.1.

Chemical kinetics

If we consider that, at high pressure and temperature, vibrational equilibrium is attained just behind the shock wave (at $x = 0$), chemical reactions take place in TRV equilibrium conditions, and therefore, the chemical rate constants depend only on the temperature common to the three modes (Arrhenius) from the chemically frozen case.

As an example, we consider the case of a shock wave in an oxygen–nitrogen mixture (21–79%). In a Mach number range 10–25 (characteristic of the re-entry of a space shuttle into Earth's atmosphere), the following chemical reactions must be taken into account:[105–106]

$$N_2 + M \leftrightarrows N + M$$
$$O_2 + M \leftrightarrows 2O + M$$
$$NO + M \leftrightarrows N + O + M$$
$$N_2 + O \leftrightarrows NO + N$$
$$O_2 + N \leftrightarrows NO + O$$
$$O_2 + N_2 \leftrightarrows 2NO \quad (9.39)$$

with $M = N_2, O_2, NO, N, O$. Thus, there are 15 dissociation–recombination reactions and 3 exchange reactions that include the rate constants k_f and k_r of the same type as those of Eqns (9.25) and (9.29), which, in principle, are known.[105]

By solving the Rankine–Hugoniot system and the species conservation equations $u \frac{dc_p}{dx} = \dot{w}_p$, we can determine the variation of the concentrations as

9.5 TYPICAL CASES OF EULERIAN NON-EQUILIBRIUM FLOWS

well as that of the flow quantities from the frozen state ($x = 0$) to the equilibrium state.

In fact, for re-entry conditions in the high-altitude atmosphere (high temperature, relatively low pressure), the vibrational relaxation of N_2, O_2, and NO and the vibration–chemistry interaction must be taken into account. Furthermore, for the highest Mach numbers, the molecular and atomic components may be ionized (Appendix 9.2).

With regard to the vibration–chemistry interaction, the dissociation-rate constants are modified according to Eqn. (5.39), and the recombination and exchange-rate constants by similar formulae. However, for recombination reactions involving only atoms as reactants, it is difficult to know the vibrational level of recombined molecules, so that we consider that the corresponding reaction-rate constants k_r depend only on the translation temperature of the atoms, that is $\bar{k}_r(T)$ (Section 9.5.2): this corresponds to generally used assumptions.[34,37] We may also note that this hypothesis involves only 4 reactions among the 36 reactions of the system of Eqn. (9.39) and that, moreover, the recombination reactions become really important only close to the equilibrium state.

An example of the modification of the dissociation-rate constants due to vibrational relaxation is shown in Fig. 47 (a) and (b) for nitrogen and oxygen respectively behind a shock wave in air.[107,108] These values are compared to their corresponding values without interaction[106] (Arrhenius) and to those deduced

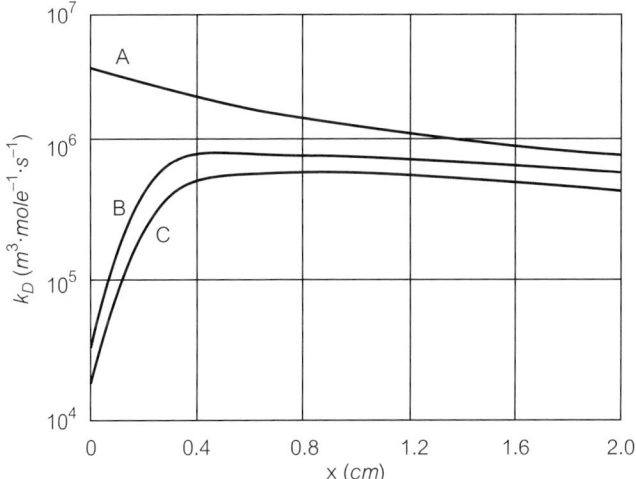

Figure 47. (a) Dissociation-rate constants of nitrogen behind a shock wave in air ($M_S = 25$, $p_1 = 8.5$ Pa, $T_1 = 205$ K); A: k_D Arrhenius, B: k_D with interaction, C: k_D empirical.[35]

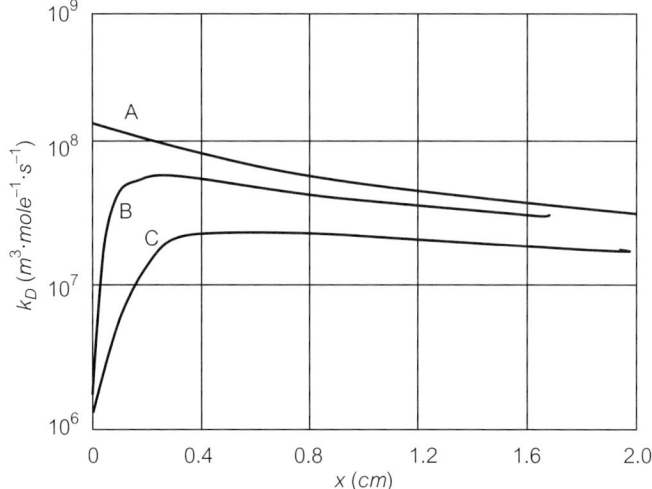

Figure 47. (b) Dissociation-rate constants for oxygen behind a shock wave in air (Conditions and notation of Fig. 47(a)).

from an empirical model widely used.[36] We easily understand that the equilibrium Arrhenius values are deeply modified when we observe that the vibrational temperatures vary from about 300 K to 6500 K in the chosen example.

The variation of the temperatures and species concentrations corresponding to this example is represented in Fig. 48 (a) and (b), respectively. These results are deduced from the solutions of the Rankine–Hugoniot system, the species conservation equations, and the relaxation equations in which the vibration–chemistry interaction, as described above, has been taken into account.

This interaction contributes to slowing down the chemical reactions and to a decrease of the temperature. We may also note the existence of a minimum for T_{VO_2}, which is due to an important dissociation rate of O_2 from the highest vibrational levels and then compensated by exchanges with the other components of the mixture. This minimum does not appear when the interaction is neglected or in the case of pure gases.

The temperature T_{VN_2} exhibits a maximum because of the relatively slow relaxation of N_2, whereas the concentration of N atoms increases monotonously. The concentration of O atoms also presents a maximum owing to the rapid relaxation and dissociation of O_2, then decreases following the decrease in T. The maximum observed in the evolution of the NO concentration is due to the successive phases of formation and dissociation.

As for possible ionization, in the Mach number range considered above (10–25), the ionization rate is about 1%, so that this process can be considered as

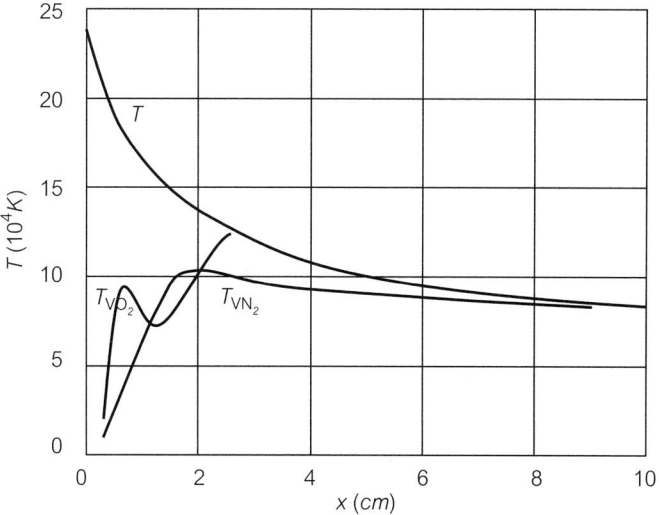

Figure 48. (a) Spatial variation of temperatures behind a shock wave in air (Conditions of Fig. 47(a)).

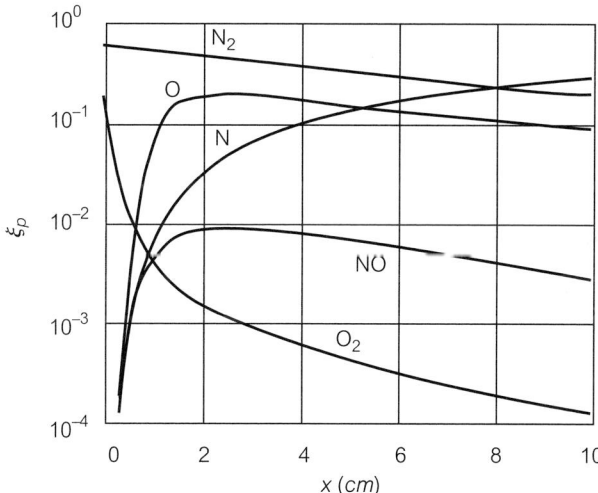

Figure 48. (b) Spatial variation of concentrations behind a shock wave in air (Conditions of Fig. 47(a)).

a perturbation. Thus, concentrations of the ionized species may be determined from the conditions given by the preceding calculation. However, for practical applications related to communications with re-entry vehicles, it is important to determine these concentrations (Appendix 9.2).

9.5.2 Flow in a supersonic nozzle

This type of flow also represents a relatively simple example of non-equilibrium flow. The main aerodynamic features of this flow were presented in Chapter 8.

If we assume equilibrium reservoir conditions (gas at rest with high pressure and temperature), the temperature and density of the gas expanding into the nozzle quickly decrease, the initially dissociated gases tend to recombine, and the vibrational energy tends to decrease. However, if the characteristic recombination times and/or the vibrational relaxation times have the same order of magnitude as the characteristic flow time in the nozzle, the flow is out of equilibrium. Thus, there is a 'freezing' of the concentrations as well as of the vibrational energy, corresponding to values higher than those for local equilibrium conditions.

An estimation of the freezing processes may be obtained from the quasi-one-dimensional equations (7.47), (7.48), and (7.40), closed by the species conservation equations and by the vibrational relaxation equations.

Example 1: Vibrational kinetics

A simple example is shown in Fig. 49 (a) and (b), concerning a non-dissociated nitrogen flow in a nozzle of small dimensions. The macroscopic flow quantities (temperatures, Mach number) along the nozzle are determined by the classical Landau–Teller equation (Fig. 49(a)): we observe that the vibrational energy freezes at a high value.

In Fig. 49(b), this frozen value (curve A) is compared to the corresponding value deduced from a computation of the probabilities of vibrational transitions, taking into account the

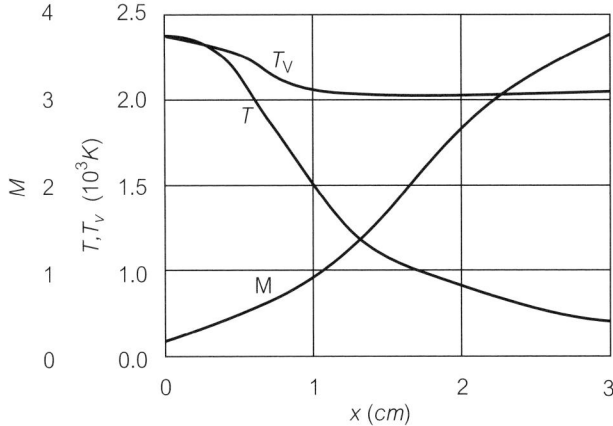

Figure 49. (a) Spatial variation of temperatures along a supersonic nozzle $[S/S_C = 1 + tg^2\alpha \, (x - x_C/R_C)^2]$. (Nitrogen, $T_0 = 2400$ K, $p_0 = 10^7$ Pa, $x_C = 0.5$ cm, $R_C = 0.1$ cm, $\alpha = 8°$).

9.5 TYPICAL CASES OF EULERIAN NON-EQUILIBRIUM FLOWS

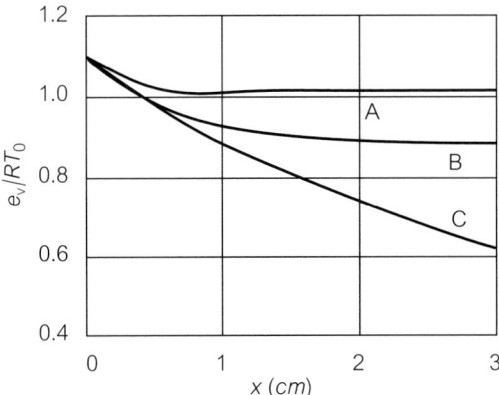

Figure 49. (b) Spatial variation of vibrational energy along a supersonic nozzle. A: Landau–Teller, B: Non-spherical interaction, C: Equilibrium (Conditions of Fig. 49(a)).

anharmonicity of the molecules and the non-sphericity of the intermolecular potential (curve B; Appendix 1.3).[109] In this last case, the freezing level is clearly lower (about 10%) than in the Landau–Teller case, because the vibrational exchanges, especially the VV exchanges, tend to re-equilibrate the system. Curve C of Fig. 49(b) would correspond to the local equilibrium.

Example 2: Vibrational and chemical kinetics

As another example,[101] we consider a partially dissociated nitrogen flow, with reservoir conditions $T_0 = 10\,000$ K, $p_0 = 28 \times 10^6$ Pa, $\alpha = 0.4$. Vibrational energy and N-atom concentration quickly freeze in the divergent part of the nozzle. When the vibration–recombination interaction is not taken into account, curve A (Fig. 50(a)) represents the evolution of the atom concentration along the nozzle (with Arrhenius constants and Landau–Teller model). The MS model (Section 6.6) derived from the $(SNE)_C+(WNE)_V$ model gives, for the atom concentration, a freezing level that is slightly lower (curve B, Fig. 50(a)), because the recombination-rate constant of reaction 1 (Eqn. (5.44)) is more important (high level of vibrational energy), but also because no coupling has been assumed for reaction 2 (Eqn. (5.52)). The $(SNE)_C+(SNE)_V$ model, defined also in Chapter 5, cannot be applied without further hypothesis, such as by assuming that the relation $K_c = k_R/k_D$ remains valid.

Similarly, the STS approach does not greatly modify the freezing level of the atom concentration (curve C, Fig. 50(a)) or the flow quantities. However, the spatial variation of the vibrational populations may be determined in this way (Fig. 50(b)). Thus, we observe a Boltzmann distribution close to the throat because, in the convergent part of the nozzle, the flow is practically in equilibrium. Downstream, the lowest levels preserve a quasi-Boltzmann distribution (in fact, close to a Treanor distribution) and freeze at a high temperature level. In

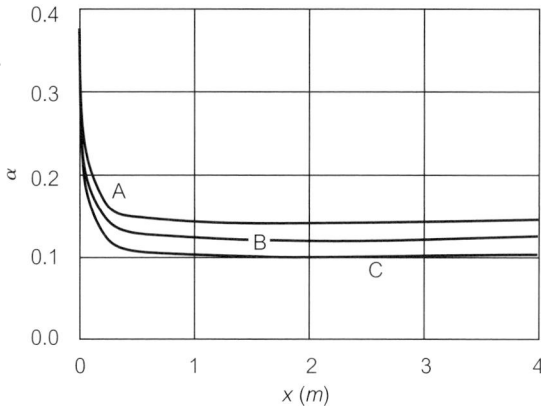

Figure 50. (a) Dissociation rate along a supersonic nozzle (nitrogen, $R_C = 0.5$ cm, $S_S/S_C = 4489$, $L = 3.42$ m, $T_0 = 10\,000$ K, $p_0 = 28 \times 10^6$ Pa). (A: Landau–Teller–Arrhenius, B: MS model, C: STS model).

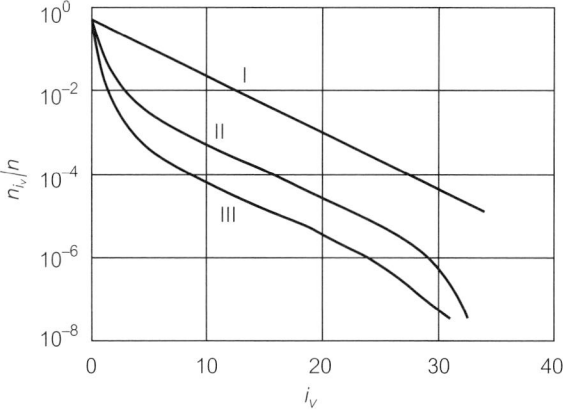

Figure 50. (b) Vibrational population distribution in a supersonic nozzle I: $x = 0.0015$ m, II: $x = 0.50$ m, III: $x = 3.42$ m (Conditions of Fig. 50(a)).

contrast, if the population of the high levels also preserves a quasi-Boltzmann distribution, this corresponds to a much lower temperature level—in fact, close to the local equilibrium.

In summary, therefore, the assumption $k_R = k_R(T)$ seems credible and is validated by experimental results (Chapter 12).

We may also think that the energy arising from the recombination reactions cannot be found in rotational or vibrational form in the recombined molecules without these molecules undergoing a new dissociation.[99] We cannot find this

9.5 TYPICAL CASES OF EULERIAN NON-EQUILIBRIUM FLOWS

energy in a radiative form, because the spontaneous transition characteristic time is much longer than the vibration period. Furthermore, as ternary collisions are rare, we may assume a two-step scheme of following type:

$$A + A \rightleftarrows A_2^{**} \quad \text{then} \quad A_2^{**} + M \rightleftarrows A_2^* + M$$

Therefore, the M particle possesses an important kinetic energy.

Most experimental nozzles are either conical or 'contoured' (that is, a priori computed with aerodynamic and physical models). Therefore, a two-dimensional calculation of the flow is necessary. This does not question the preceding results, which remain qualitatively valid in the central part of the nozzle. A two-dimensional Euler computation, however, gives a more detailed description[110] of the flow but requires boundary-layer corrections (or a complete Navier–Stokes computation must be made, as reported in Chapter 10). The Euler computations are nonetheless useful, and they are relatively fast and inexpensive, particularly for complex mixtures. A typical example is presented below.

Example 3: Airflow in an axisymmetric nozzle

An Euler computation of the airflow expanding into a nozzle that has a conical divergent part is thus carried out, starting from equilibrium reservoir conditions for which the dissociation rate of oxygen is important. An example of isomass flux lines[111] close to the throat (where the boundary layer is very thin) is shown in Fig. 51, pointing out the 'source effect' of the throat.

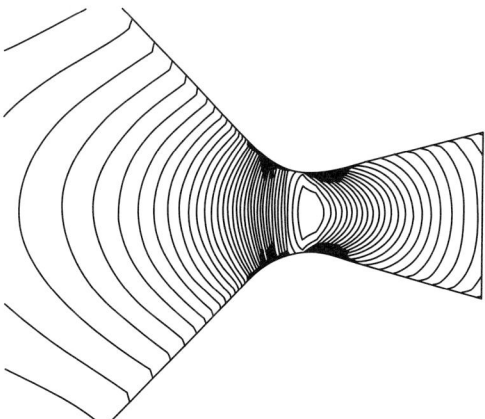

Figure 51. Airflow close to the throat of a nozzle (isomass flux lines). (Flow from left to right) ($R_C = 0.175$ cm, $\alpha = 15°$, $T_0 = 6000$ K, $p_0 = 10^7$ Pa, $\xi_O = 0.16$, $\xi_{NO} = 0.09$, $\xi_N = 0.01$, $\xi_{O_2} = 0.02$).

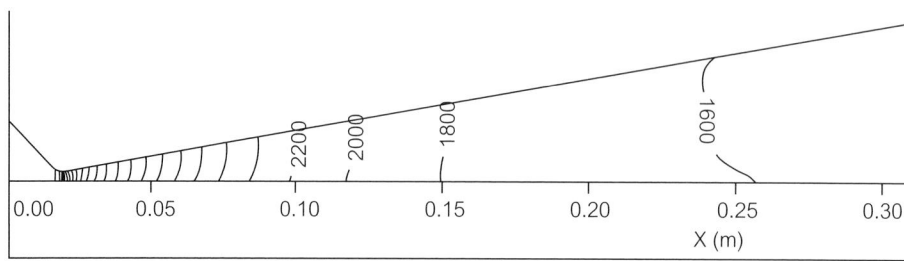

Figure 52. Oxygen vibrational temperature in a nozzle ($R_C = 0.3$ cm, $\alpha = 10°$, air, $T_0 = 6500$ K, $p_0 = 1.53 \times 10^6$ Pa).

From a similar computation, Fig. 52 shows isothermal lines for the vibrational temperature of oxygen in the nozzle.[110] However, the freezing level of 1500 K given by this Euler computation is not realistic, because of the presence of a non-negligible boundary layer in the divergent part. Thus, in Fig. 73 of Chapter 10, for the same nozzle, we compare the frozen values for T_{VO_2} and T_{VN_2} obtained in the Euler and Navier–Stokes regimes.[112]

9.5.3 Flow around a body

The Euler computations of high-speed flows around bodies cannot give a quantitative description of all parameters, nor can they in particular solve the important problem of exchanges between the flow and the body (Chapter 10). However, as for the internal flows, they do give correct general features of the real gas effects.

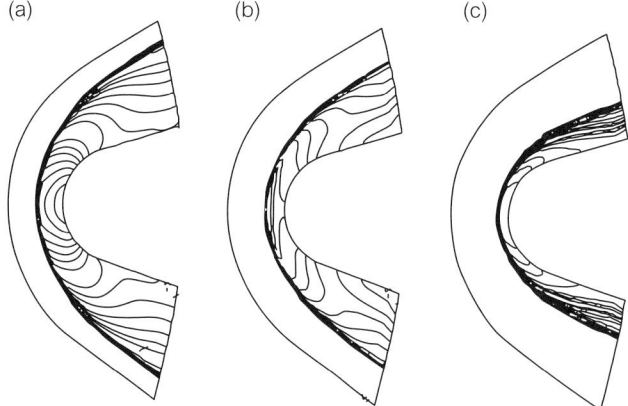

Figure 53. Isothermal lines of hypersonic flow around a hemisphere-cone body ($M_\infty = 17.9$, $T_\infty = 231$ K, $p_\infty = 10$ Pa).

Thus, shown in Fig. 53 are the isothermal lines of a hypersonic airflow around a hemisphere-cone body, deduced from an Euler calculation. In particular, these lines show the location of the detached shock wave in front of the body when assuming frozen flow (a), non-equilibrium flow (b), and equilibrium flow (c), for the same upstream free flow.[113] Thus, important differences may be observed.

Appendix 9.1 Evolution of vibrational populations behind a shock wave

An example of the evolution of the vibrational populations of a diatomic gas (oxygen) behind a straight shock wave is analysed. Two situations are considered: the first concerns a temperature range 600–2000 K for which dissociation may be neglected, and the second has a temperature range 2000–5000 K in which vibrational excitation and dissociation occur simultaneously.[114]

An STS method is used with the values of individual transition probabilities for the 'bound–bound' transitions[115] and dissociative transition 'bound–free' probabilities using a preferential model discussed in Chapter 5.

In the first case, the relaxation equations are given by Eqn. (2.17) or, taking the relative populations n_{i_v}/n as variables, by the following equations:

$$\frac{d}{dt}\left(\frac{n_{i_v}}{n}\right) = Z(O_2 - O_2) \sum_{j_v,k_v,l_v} \left(\frac{n_{k_v} n_{l_v}}{n^2} P^{i_v,j_v}_{k_v,l_v} - \frac{n_{i_v} n_{j_v}}{n^2} P^{k_v,l_v}_{i_v,j_v}\right) \quad (9.40)$$

In the second case, taking into account the presence of O atoms, we write the population evolution equations as follows:

$$\frac{dc_{i_v}}{dt} = \frac{Z(O_2 - O_2)}{c_{O_2}} \left[\sum_{j_v,k_v,l_v} \left(c_{k_v} c_{l_v} P^{i_v,j_v}_{k_v,l_v} - c_{i_v} c_{j_v} P^{k_v,l_v}_{i_v,j_v}\right)\right.$$

$$+ \sum_{j_v,l_v} \left(2c_O^2 c_{l_v} P^{i_v,j_v}_{D,l_v} - c_{i_v} c_{j_v} P^{D,l_v}_{i_v,k_v}\right)\right]$$

$$+ Z(O_2 - O) \left[\sum_{k_v} \left(c_{k_v} P^{i_v}_{k_v} - c_{i_v} P^{k_v}_{i_v}\right) + 2c_{O_2} P^{i_v}_D - c_{i_v} P^D_{i_v}\right] \quad (9.41)$$

where the mass concentrations are the unknowns, and where the index O_2 has been omitted for the concentrations of molecular-level populations.

These equations are coupled with the Rankine–Hugoniot (RH) relations and with a species conservation equation, for example O species, that is:

$$\frac{dc_O}{dt} = \frac{Z(O_2 - O_2)}{c_{O_2}} \left[\sum_{i_v, j_v, k_v, l_v} (c_{i_v} c_{j_v} P_{i_v,k_v}^{D,l_v} - 2c_O^2 c_{l_v} P_{D,l_v}^{i_v,k_v}) \right]$$
$$+ Z(O_2 - O) \left[\sum_{i_v} \left(c_{i_v} P_{i_v}^D - 2c_O^2 P_D^{i_v} \right) \right] \quad (9.42)$$

The evolution of vibrational populations deduced from these equations is compared to that obtained from global models, such as the Landau–Teller model (model A) or the Treanor model (model B). Moreover, in order to appreciate the relative influence of the TV and VV transitions, a particular computation is carried out with Eqns (9.41) and (9.42), but excluding the VV transitions (model C).

9.1.1 Evolution without dissociation

For conditions without dissociation, Fig. 54 (a) and (b) show the evolution of the populations of the fourth and sixteenth vibrational levels respectively behind a shock wave in oxygen ($M_s = 6$, $T_f \simeq 2200$ K, and $T_e \simeq 1900$ K).

Behind the shock wave, at $x = 0$, 99% of the total population is in level 0, but in equilibrium conditions, the population of level 0 represents only 68%, whereas the population of the first level n_1 increases from 0.32% to 21%. We also observe that the populations of the lowest levels (such as level 4) computed with various models are nearly the same, except with the Landau–Teller model. This confirms that we have a non-Boltzmann distribution in the non-equilibrium

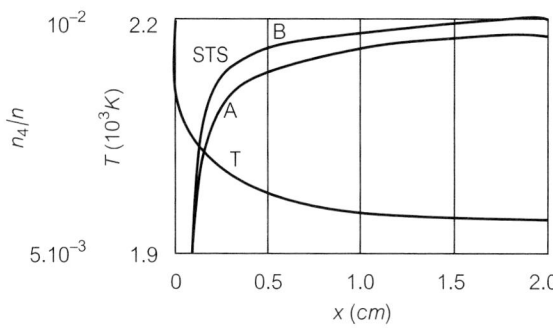

Figure 54. (a) Relative vibrational population of the fourth level behind a shock wave (oxygen, $M_s = 6$, $T_1 = 278$ K, $p_1 = 140$ Pa; A: Landau–Teller, B: Treanor, STS: state to state, T: temperature).

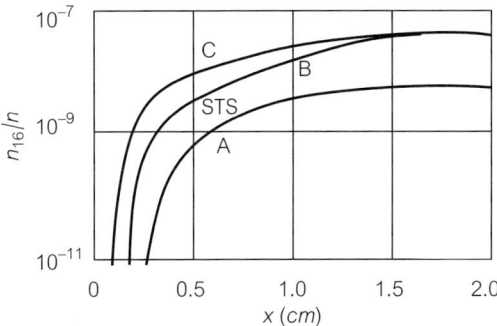

Figure 54. (b) Relative vibrational population of the sixteenth level behind a shock wave (Conditions of Fig. 54(a); A: Landau–Teller, B: Treanor, C: without VV collisions, STS: state to state, T: temperature).

region, which is particularly true for the high levels. However, the differences between the models increase with increasing level.

These results also show that the Treanor distribution is very close to the STS model, while model C (no VV collisions) overestimates the populations, thus pointing out the 'regulating' role of the VV collisions.

9.1.2 Evolution with dissociation

For conditions with dissociation, Fig. 55 (a) and (b) represent examples of the evolution of populations of the fourth and seventeenth levels respectively behind a shock wave in oxygen ($M_s = 10$, $T_e \simeq 3000$ K).

In these conditions, vibrational relaxation is very rapid compared to chemical evolution ($\tau_V \ll \tau_D$), so that both processes are clearly separated: there is an initial phase of a rapid increase in the vibrational populations of those levels that are equal to or higher than 1, practically identical for all levels and corresponding to a chemical 'incubation phase' (vibrational relaxation), followed by a slow decrease in populations up to the equilibrium values (chemical phase). In contrast, as expected, the population of level 0 decreases during the first phase and then slowly increases up to its equilibrium value.

For the low levels (the fourth is represented in Fig. 55(a)), all models are equivalent, whereas for the high levels (the seventeenth is represented in Fig. 55(b)), models A, B, and C overestimate the populations in comparison with the STS model, which once more points out the importance of the VV collisions reducing the effect of the TV collisions.

As already discussed, if the populations (especially those of the high levels) are sensitive to the models, the macroscopic quantities depend very little on them.

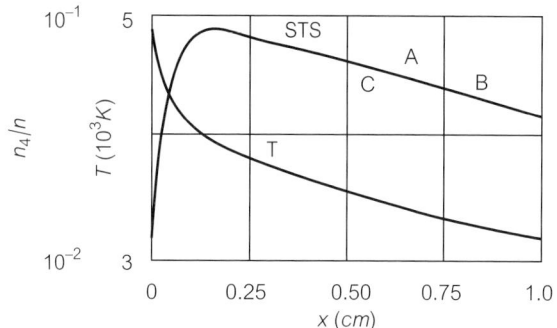

Figure 55. (a) Relative vibrational population of the fourth level behind a shock wave (oxygen, $M_s = 10$, $T_1 = 271$ K, $p_1 = 390$ Pa; notation of Fig. 54(b)).

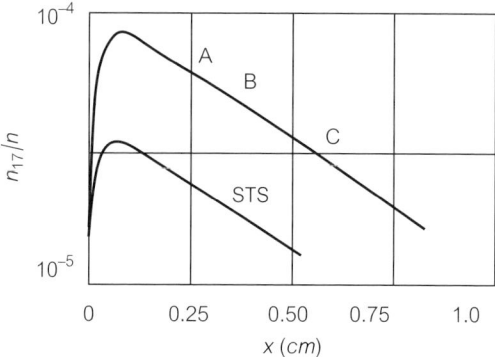

Figure 55. (b) Relative vibrational population of the seventeenth level behind a shock wave (Conditions of Fig. 55(a); notation of Fig. 54(b)).

Appendix 9.2 Air chemistry at high temperature

9.2.1 Air chemistry in equilibrium conditions

As indicated in Section 9.5.1, in the Mach number range 10–25, we must take into account 18 reactions (9.39), with equilibrium constants having the following form:

$$K_c = CT^c \exp(-\theta_c/T) \tag{9.43}$$

where $\theta_c = \theta_D$ for the dissociation reactions.

This important number of reactions does not lead to a number of equations higher than unknown quantities. For example, the equilibrium constant K_{c18} is

APPENDIX 9.2 AIR CHEMISTRY AT HIGH TEMPERATURE

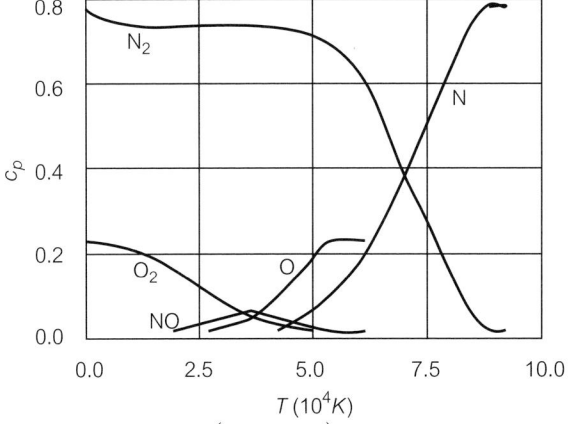

Figure 56. (a) Equilibrium air composition $\left(p = 10^5 \text{ Pa}\right)$.

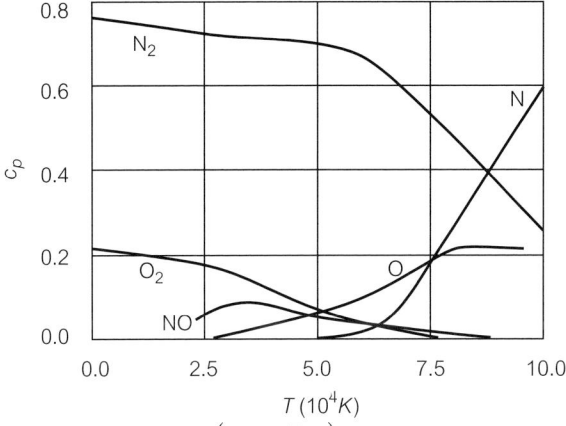

Figure 56. (b) Equilibrium air composition $\left(p = 10^7 \text{ Pa}\right)$.

a function of preceding reaction constants, that is:

$$K_{c18} = \frac{N_{NO}^2}{N_{O_2} N_{N_2}} = \frac{K_{c1} K_{c2}}{K_{c3}^2} \tag{9.44}$$

The air composition for equilibrium conditions is represented in Fig. 56(a) and (b) for two pressures. Of course, for better accuracy, other species (Ar, CO_2, and so on) should be taken into account.

9.2.2 Ionization phenomena

Behind a straight shock wave, in the same Mach number range as before (10–25), the only source of electrons is from the reaction of Eqn. (9.17) and reactions of associative type,[105] the more efficient of which are the following:

$$O + N \rightleftarrows NO^+ + e^-$$
$$O + O \rightleftarrows O_2^+ + e^-$$
$$N + N \rightleftarrows N_2^+ + e^- \qquad (9.45)$$

If we consider the different peculiar velocities of heavy species and electrons, the rate constants of the forward reactions depend on T, and those of backward reactions on T_e.

The cross sections of these reactions are small, so that the electronic density n_e increases slowly behind the shock wave. However, as soon as the number of electrons becomes 'significant', new reactions start (reactions by impact of electrons), the most important of which are

$$O + e^- \rightleftarrows O^+ + 2e^-$$
$$N + e^- \rightleftarrows N^+ + 2e^- \qquad (9.46)$$

Then, there is a steep increase in the electron density, known as the 'avalanche process'.

As the density of ionized species simultaneously increases, charge exchange reactions may take place. The most important of these are

$$O + O_2^+ \rightleftarrows O_2 + O^+$$
$$N_2 + N^+ \rightleftarrows N_2^+ + N$$
$$O + NO^+ \rightleftarrows O^+ + NO$$
$$N_2 + O^+ \rightleftarrows N_2^+ + O$$
$$N + NO^+ \rightleftarrows N^+ + NO$$
$$O_2 + NO^+ \rightleftarrows O_2^+ + NO$$
$$NO^+ + N \rightleftarrows N_2^+ + O \qquad (9.47)$$

These reactions may be sensitive to vibrational non-equilibrium as long as the corresponding relaxation is not over.

It is of course necessary to know the evolution of the electronic temperature T_e, which can be deduced from the corresponding energy balance equation. If we neglect the dissipative processes, charge separation effects, and radiative

APPENDIX 9.2 AIR CHEMISTRY AT HIGH TEMPERATURE

phenomena,[116–118] we can write this equation as follows:

$$\frac{d}{dx}\left(\frac{3}{2} n_e T_e u\right) + n_e T_e \frac{du}{dx} = 3 n_e \left(\sum_s \frac{m_e}{m_s} Z_{es} + \sum_c \frac{m_e}{m_c} Z_{ec}\right)(T - T_e)$$

$$+ \frac{3}{2} \sum_p n_e \frac{T - T_e}{\tau_{erp}}$$

$$+ \sum_p \frac{\theta_{Vp}^2}{T_{Vp} T_e} n_p \sum_{i_v} a_0^{i_v} i_v^2 (T_{Vp} - T_e) - \sum_c I_c \frac{\dot{n}_{ec}}{k}$$

(9.48)

The first three terms on the right-hand side of Eqn. (9.48) represent the balance of elastic collisions with neutral species s and with charged particles c, the balance of inelastic collisions with molecules p, involving rotational then vibrational energy; the meaning of various quantities involved is given in the nomenclature at the start of this part of the book.

An example[116] of the spatial variation of electronic temperature T_e, electronic density n_e, and concentrations of the main ionized species behind a shock wave is given in Fig. 57 (a) and (b) and Fig. 58 respectively: the maximum observed in the evolution of n_e may be easily explained by successive phases of the creation and disappearance of electrons.

Sometimes it may be necessary to take into account the electronic excitation of neutral particles. Then, we generally suppose[117] that there is equilibrium between the distribution of the excited electronic states and that of free electrons, so that we assume that there 'exists' only one temperature T_e; a common electronic energy, e_e, of free and bound electrons is therefore defined such that

$$\rho e_e = \sum_q \rho_e e_{eq}$$

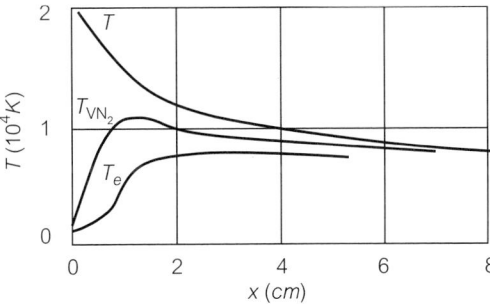

Figure 57. (a) Spatial variation of temperatures behind a shock wave in air ($M_S = 25$, $T_1 = 250$ K, $p_1 = 8.5$ Pa).

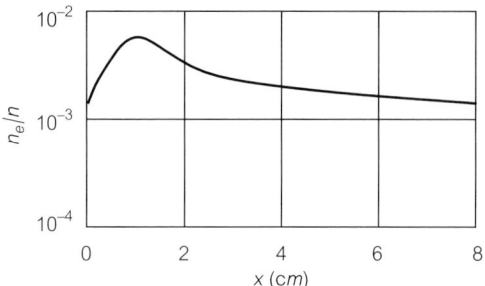

Figure 57. (b) Spatial variation of electronic density behind a stock wave in air (Conditions of Fig. 57(a)).

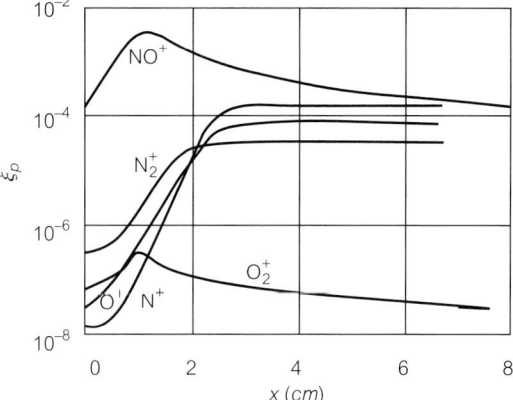

Figure 58. Spatial variations of ionized species concentrations behind a shock wave in air (Conditions of Fig. 57(a)).

where q includes all species and electrons.

Equation (9.48) is then slightly modified.

Another consequence of the presence of electrons is the possibility of energy exchange between the electrons and the vibrational mode of the molecules, ionized or not.[102] Then for example, in the Landau–Teller equation (2.19), a new term must be added on the right-hand side, which can be written in the following simplified form: $\dfrac{E_{vp}(T_e) - E_{vp}}{\tau_{evp}}$

Appendix 9.3 Reaction-rate constants

If we again consider the example of Chapter 2 (Section 2.8.1), we have the following bimolecular reaction:

$$P + Q \rightarrow P' + Q'$$

APPENDIX 9.3 REACTION-RATE CONSTANTS

and

$$\frac{\dot{w}_p}{m_p} = -k_f n_p n_q$$

where, neglecting the internal modes, we have

$$k_f = \int_{\Omega, \boldsymbol{v}_p, \boldsymbol{v}_q} \frac{f_p f_q}{n_p n_q} I_{p,q}^{p',q'} g_{pq} d\Omega \, d\boldsymbol{v}_p d\boldsymbol{v}_q = \int_{\boldsymbol{v}_p, \boldsymbol{v}_q} \frac{f_p f_q}{n_p n_q} C_{p,q}^{p',q'} g_{pq} d\boldsymbol{v}_p d\boldsymbol{v}_q$$

Independently of the determination of the cross sections, which constitutes an important difficulty, we can obtain a general expression for k_f which may be used for the correlation or representation of the experimental results.[99]

The reactive collision rate k_f represents one part of the total collision rate, assimilated to the elastic collision rate k_{el}, that is:

$$k_{el} = \int_{\Omega, \boldsymbol{v}_p, \boldsymbol{v}_q} \frac{f_p f_q}{n_p n_q} g_{pq} d\Omega \, d\boldsymbol{v}_p d\boldsymbol{v}_q$$

Among these collisions, we consider to be reactive collisions those that correspond to values for relative velocity that are higher than a value denoted g_*.

With a Maxwellian distribution and a rigid sphere model (Appendix 2.2), we obtain for k_{el}:

$$k_{el} = \left(\frac{d_p + d_q}{2}\right)^2 \left(\frac{8\pi kT}{m_r}\right)^{1/2}$$

The rate of collisions for which $g_{pq} \geq g_*$, is equal to

$$k_* = k_{el}\left(1 + \frac{m_r g_*^2}{2kT}\right) \exp\left(-\frac{m_r g_*^2}{2kT}\right)$$

It is more significant to consider that the collision is reactive when the component of the relative velocity parallel to r is higher than a value g_{**}; then we simply have

$$k_{**} = k_{el} \exp\left(-\frac{m_r g_{**}^2}{2kT}\right)$$

We can relate the quantity $\frac{m_r g_{**}^2}{2}$ to an activation energy per molecule ε_a; therefore

$$k_f \sim k_{**} \sim k_{el} S \exp\left(-\frac{\varepsilon_a}{kT}\right)$$

The exponential term may be considered the probability of a reactive collision. Moreover, the term S may take into account the relative orientation of the colliding molecules (steric factor). However, in consideration of the restrictive hypotheses of the present analysis, a more general formula of the type of Eqn. (9.25) is often used.

Appendix 9.4 Nozzle flows

In Chapters 3 and 4, linearized expressions (WNE) for vibrational non-equilibrium were derived (Eqn. (3.66) for pure gases, and Eqns (4.51) and (4.52) for binary mixtures). If we apply these results to a particular nozzle flow, we obtain an evolution of the vibrational energy along the nozzle,[119] shown in Fig. 59, where it is compared to the evolution given by a global model (SNE, Landau–Teller).

It is not surprising to observe that the WNE freezing level is lower than the SNE freezing level; moreover, this level is not really a plateau but rather an inflexion point, so that, further along the nozzle, the vibrational energy tends to its local equilibrium value. There is indeed some competition in the expression of E_V between the increasing term $\frac{\tau_V}{C_{TRV}^2}$ and the decreasing term $C_V T \frac{du}{dx}$, which becomes dominant ($\frac{du}{dx} \to 0$).

For smaller angles of the conical divergent part of the nozzle, the difference between the freezing levels is less important, but for angle values higher than 4–5°, the WNE freezing values oscillate about the corresponding SNE value. Then, we must be aware of the conditions for applying the WNE method, that is $\frac{\tau_V}{\theta} \ll 1$. In a nozzle, we have $\theta \sim \left(\frac{du}{dx}\right)^{-1}$, so that close to the throat we have $\frac{\tau_V}{\theta} \sim 10^{-2}$, but we rapidly attain the limit of the validity of the WNE method in

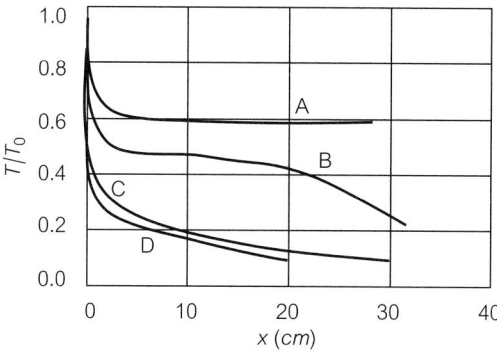

Figure 59. Spatial variation of temperatures along a nozzle (nitrogen, $T_0 = 2500$ K, $p_0 = 12 \times 10^6$ Pa, $\alpha = 4°$). A: T_V/T_0(SNE), B: T_V/T_0(WNE), C: T/T_0(WNE), D: T_{TR}/T_0(SNE).

the divergent part. From a practical point of view, in usual supersonic nozzles, the divergence angles are 8–10° or higher.

We may also note that the choice of an oscillator model is not very important, and the freezing level remains practically the same for a harmonic or an anharmonic model (for the latter, it is 1–2% higher). This is not true for the molecular interaction model, which influences this level significantly (Fig. 49(b)). Similarly, the choice of a model for the relaxation time itself may also be important for the freezing level, and its value may vary significantly (Chapter 12).

TEN
Reactive Flows in the Dissipative Regime

10.1 Introduction

In the preceding chapter, the basics of reactive flows were presented and applied to simple situations to highlight the main features of the interaction between physical–chemical phenomena and aerodynamic processes. The determination of more complex flow fields constitutes the subject of the present chapter, with a view to examining realistic and practical situations. Even if the geometrical configurations of these flows are simple, dissipative effects are taken into account, as are the correlative exchanges with the background: thus, the Navier–Stokes or boundary-layer equations constitute the essential means to obtaining an accurate description of these flows.

We successively examine dissipative flows in chemical equilibrium (flat plate, stagnation point), and then non-equilibrium flows for which the boundary conditions, particularly chemical conditions, play an important role: this is the case for catalytic phenomena which govern wall heat flux processes.

Vibrational non-equilibrium boundary layers, which present specific features and influence chemical kinetics, are studied in the second part of the chapter. In the final part, typical examples of dissipative supersonic and hypersonic flows in nozzles, around various bodies, or in mixing zones are presented.

Some specific problems of importance, such as accommodation and exchange phenomena in Knudsen layers, generalized Rankine–Hugoniot relations behind intense shock waves, and quantitative data on particular transport coefficients, are discussed in the Appendices. The details of the kinetics of vibrational exchanges in gas-dynamic lasers are also presented here.

Finally, this chapter would be incomplete without giving the principles and general features of numerical methods often used to solve the Navier–Stokes equations.

10.2 Boundary layers in chemical equilibrium

We first consider the simple case of flows in complete equilibrium (TRV and chemical equilibrium) and particularly the flow of a dissociating pure diatomic gas A_2 (Chapter 9). The most important problem is the determination of the wall heat flux and its comparison with the corresponding frozen heat flux (without dissociation; Chapter 8).

To this purpose, we consider the boundary-layer equations ((8.24)–(8.26) and (8.29)), and the species conservation equation $\dot{w}_p = 0$, which, in this case, becomes $\alpha p = F(T)$ (Eqn. (9.12)), where α is the dissociation rate. Enthalpy h is given by Eqn. (9.14), and the component of the heat flux along an axis normal to the wall, q_y, which includes conduction and diffusion fluxes, may be written as follows:

$$-q_y = \lambda \frac{\partial T}{\partial y} + \rho D \left(h_A - h_{A_2}\right) \frac{\partial \alpha}{\partial y} \tag{10.1}$$

since

$$\frac{\partial p}{\partial y} = 0$$

As we have

$$h_A^0 \gg \int_0^T \left(C_A^p - C_{A_2}^p\right) dT$$

we can write

$$-q_y \simeq (\lambda + \lambda_R) \frac{\partial T}{\partial y} = \lambda_e \frac{\partial T}{\partial y} \tag{10.2}$$

where $\lambda_R = \rho D h_A^0 \frac{d\alpha}{dT}$ is called the 'reaction conductivity', and $\lambda_e = \lambda + \lambda_R$ is the apparent conductivity.

Now, we consider the typical cases of the flat plate and the stagnation point.[70]

10.2.1 The flat plate

Here, we have $u_e = A$, $p = p_e = B$, and $\alpha = \alpha(T)$, where A and B are constants. As μ, D, and λ_e depend only on T, we have

$$dh = C^p dT \tag{10.3}$$

with

$$C^p = C^p(T) \simeq C_f^p(T) + h_A^0 \frac{d\alpha}{dT} \tag{10.4}$$

We can also define an 'apparent' Prandtl number P_e such that

$$P_e = \frac{C^p \mu}{\lambda + \lambda_R} \simeq P \left[\frac{C_f^p}{C^p}(1-L) + L \right]^{-1} \tag{10.5}$$

where L is the frozen Lewis number equal to $\frac{\rho D C_f^p}{\lambda}$, and P is the frozen Prandtl number equal to $\frac{C_f^p \mu}{\lambda}$.

We then have a complete formal analogy between the equations of the dissociated boundary layer (in equilibrium) and the equations of the frozen boundary layer (Eqns (8.55) and (8.59)).

Therefore, assuming analogous hypotheses (P_e is constant), we find a skin-friction coefficient that is practically equal to Eqn. (8.58) and therefore not very sensitive to the dissociation, and a wall heat flux which, in view of Eqn. (8.60), may be written:

$$q_w = \left(\frac{\rho_w \mu_w u_e}{2x} \right)^{1/2} f_w'' P_e^{-2/3} (h_{wr} - h_w) \tag{10.6}$$

As P_e is not very different from P, the influence of the dissociation appears essentially in the term $h_{wr} - h_w \simeq h_e - h_w + r\frac{u_e^2}{2}$. Moreover, as the recovery factor r depends weakly on α, the dependence of the heat flux on the dissociation is mainly found in the term $h_e - h_w$, with

$$h_e - h_w = \int_{T_w}^{T_e} \left[\alpha C_A^p + (1-\alpha) C_{A_2}^p \right] dT + h_A^0 (\alpha_e - \alpha_w) \tag{10.7}$$

Thus, the last term of the right-hand side of Eqn. (10.7) represents the direct contribution of the dissociation to the wall heat flux. As this term is proportional to the difference of the dissociation rates in the outer flow and at the wall, it may be important for a significantly dissociated flow ($0 < \alpha_e < 1$) along a 'cold' wall where the atoms are recombining (catalytic wall, $\alpha_w \simeq 0$).

10.2.2 The stagnation point

$$x = \xi = 0, \quad u_e = 0, \quad \frac{du_e}{dx} = K \text{ (const.)}, \quad j = 1 \quad \text{(axisymmetric body)}$$

This case is an important example for the heat flux, which can take high values when the upstream flow is hypersonic (re-entry of space shuttles). Of course, we assume that $T_{0e} \gg T_w$ (cold wall).

10.2 BOUNDARY LAYERS IN CHEMICAL EQUILIBRIUM

It is clear (Chapter 8) that there are similarity solutions for the transformed boundary-layer equations (Eqn. (8.64)), if we include C in the transformation formula giving η (Eqn. (8.56)), and if we use Eqn. (9.12) giving α.

If we are essentially interested in the heat flux, we may neglect the pressure gradient term in Eqn. (8.64) in view of Fig. 35, so that, assuming $P_e = $ const., we have to solve the following system:[70]

$$f''' + ff'' = 0$$
$$g'' + P_e f g' = 0$$
$$\alpha = \alpha(T) \qquad (10.8)$$

This system is equivalent to the system used for the flat plate in Chapter 8 (with $E_{0e} = 0$). Therefore, we have the same expressions for f_w'' and g_w'. For the heat flux, taking into account the relation giving η, that is, $\left(\frac{\partial \eta}{\partial y}\right)_w = \left(\frac{2K\rho_w}{\mu_w}\right)^{1/2}$, we have

$$-q_w = (2K\rho_w \mu_w)^{1/2} f_w'' P_e^{-2/3} (h_{0e} - h_w) \qquad (10.9)$$

with

$$h_{0e} - h_w = \int_{T_w}^{T_{0e}} \left[\alpha C_A^p + (1-\alpha) C_{A_2}^p\right] dT + h_A^0 (\alpha_{0e} - \alpha_w) \qquad (10.10)$$

As $T_{0e} \gg T_w$, the heat flux may be very important if the stagnation enthalpy is high (and therefore the upstream velocity) and if the (cold) wall is catalytic.

This result suggests that we should consider the case of a frozen boundary layer for a dissociated gas at a stagnation point. In this case, the species conservation equation is reduced to a diffusion equation, that is:

$$z'' + Sfz' = 0 \qquad (10.11)$$

with $z = \frac{\alpha}{\alpha_e}$ (Chapter 8).

The corresponding momentum equation may be written as follows, neglecting the pressure gradient term as above:

$$f''' + ff'' = 0$$

And for the energy equation, we have

$$g'' + Pfg' = (1-L) z'' \qquad (10.12)$$

with the usual boundary conditions:

$$\eta \to 0, \quad f = f' = 0, \quad g = g_w, \quad z = z_w$$
$$\eta \to \infty, \quad f', g, z \to 1$$

These equations are easily solved, and assuming that for a cold wall we have $(h_A - h_{A_2})_w \simeq h_A^0$, the wall heat flux $-q_w = \left(\lambda \frac{\partial T}{\partial \eta} + \rho D \frac{\partial \alpha}{\partial \eta}\right)_w \left(\frac{\partial \eta}{\partial y}\right)_w$ may be written in the following approximate form:

$$-q_w = (2K\rho_w\mu_w)^{1/2} f_w'' P^{-2/3} \left[h_{0e} - h_w + h_A^0 \left(L^{2/3} - 1\right)(\alpha_e - \alpha_w)\right] \quad (10.13)$$

For a cold wall, it is clear that, quantitatively, this expression differs little from Eqn. (10.9), which corresponds to the equilibrium case; it is then apparent that in the non-equilibrium case the heat flux will have approximately the same value.[120,121]

Therefore, the heat flux is not very sensitive to the boundary layer 'chemical regime' but essentially to the temperature difference $T_{0e} - T_w$ and also to the dissociation rate difference $\alpha_{0e} - \alpha_w$. This last term may be important if, as discussed above, the wall is catalytic. With a view to eventually reducing this part of the heat flux, it is necessary to examine the catalytic properties of the wall for the recombination of the impinging atoms.

10.2.3 Reactive boundary layer and wall catalycity

When the boundary layer is frozen or in non-equilibrium, a layer of non-recombined atoms may be found close to the wall if the recombination reactions are inhibited by the non-catalytic properties of this wall, so that the heat flux may be reduced.

Assuming that the recombination process is fed by the diffusion of atoms towards the wall, we have[122]

$$\left(\rho D \frac{\partial \alpha}{\partial y}\right)_w = (k_R \rho \alpha)_w \quad (10.14)$$

We also assume that the wall is 'cold' enough to lay down a negligible atom concentration in the catalytic case.

Here, k_R is the recombination-rate constant, which is dependent on the gas–wall coupling ($k_R \to 0$: non-catalytic wall; $k_R \to \infty$: catalytic wall) and corresponds to the following reaction, taking place after the adsorption of atoms:

$$A + A_{-s} \to A_2 + s \quad \text{(Eley–Rideal process)}$$

or to the reaction

$$A_{-s} + A_{-s} \to A_2 + 2s \quad \text{(Langmuir–Hinshelwood process)}$$

where s is a site for an atom A_{-s} on the wall surface.

10.2 BOUNDARY LAYERS IN CHEMICAL EQUILIBRIUM

From Eqn. (10.14), we deduce the following expression:

$$z'_w = \left(\frac{\rho_w \mu_w}{2K}\right)^{1/2} \frac{k_{Rw}}{\rho_w D} z_w \qquad (10.15)$$

The solution of Eqn. (10.11) is still valid, so that

$$z'_w = f_w'' S^{1/3} (1 - z_w) \qquad (10.16)$$

We deduce from Eqns (10.15) and (10.16) an expression for z'_w, that is:

$$z'_w = f_w'' S^{1/3} \Phi \qquad (10.17)$$

with

$$\Phi = \left[1 + f_w'' \frac{(2K\mu_w)^{1/2}}{S^{2/3} \rho_w^{1/2} k_{Rw}}\right]^{-1} \qquad (10.18)$$

Here, $\Phi = 0$ and $\Phi = 1$ correspond to the non-catalytic and to the fully catalytic case, respectively.

Finally, for the wall heat flux, we find

$$-q_w = (2KC\rho_e\mu_e)^{1/2} f_w'' P^{-2/3} \left[h_{0e} - h_w + (L^{2/3}\Phi - 1) h_A^0 (\alpha_e - \alpha_w)\right] \qquad (10.19)$$

For $\Phi = 1$, we again find Eqn. (10.13).

For $\Phi = 0$, in Eqn. (10.19), we find the 'thermal contribution' only.

We may also define a recombination coefficient γ_A, representing the ratio of the net mass rate created at the wall to the incident mass flux, that is:[55]

$$\gamma_A = \frac{(k_R \rho_A)_w}{m_A N_A}$$

where N_A, incident atom flux, is equal to $n_{Aw}\sqrt{\frac{kT_w}{2\pi m_A}}$ (Appendix 6.5). Thus, we have

$$k_{Rw} = \gamma_A \sqrt{\frac{kT_w}{2\pi m_A}}$$

From measurements of γ_A carried out for various temperatures (Chapter 12), we can deduce the activation energy ε_A for the corresponding reaction, with

$$\gamma = \gamma_0 \exp\left(-\frac{\varepsilon_A}{kT_w}\right)$$

It is also possible that the energy arising from the recombination is not entirely transmitted to the wall in the form of a heat flux q_{cw}: a chemical accommodation coefficient β may then be defined as being equal to $q_{cw}/N_A E_{DA_2}$.

10.2.4 Boundary layer along a body

Numerical solutions are generally used in solving the complete Navier–Stokes equations between the detached shock wave and the body (see below).

However, approximate solutions may be found from the stagnation point along the body. Assuming, for example, a flat-plate similarity (quasi-independency of g'_w with the pressure gradient) and $L = 1$, we find an approximate expression for q_w along the body, that is:[123]

$$\frac{q_w}{q_{w0}} = \frac{1}{2} \frac{\rho_e \mu_e}{(\rho_e \mu_e)_0} u_e r^2 K^{-1/2} \left(\int_0^x \frac{\rho_e \mu_e}{(\rho_e \mu_e)_0} u_e r^2 dx \right)^{1/2} \quad (10.20)$$

where the subscript 0 is relative to the stagnation conditions.

Owing to the expansion process along a convex body, the heat flux decreases quickly along this type of body.

10.3 Boundary layers in vibrational non-equilibrium

Several examples of vibrationally relaxing boundary layers are presented below: they are representative of the interaction between non-equilibrium phenomena and dissipative processes. Moreover, in the third example, the consequences on the chemical rate constants are pointed out.

10.3.1 Example 1: boundary layer behind a moving shock wave

We consider the boundary layer developing along a flat plate behind a straight shock wave moving at constant speed[124] (Fig. 60). The one-dimensional inviscid flow, which was analysed in Chapter 9, is vibrationally relaxing. The boundary-layer equations are solved between this Eulerian flow and the wall, which is assumed to be at constant temperature and catalytic for the vibration. In the shock wave fixed coordinate system, the two-dimensional boundary layer is stationary and then described by Eqns (8.57) and (8.59) and a Landau–Teller relaxation equation, which, in the transformed coordinate system ξ, η, may be written for a non-dissociated pure gas in the following form:

$$\left(\frac{L_f}{P_f} C\varepsilon' \right)' + f\varepsilon' = 2\xi \left(\frac{\bar{\varepsilon} - \varepsilon}{\rho_e \mu_e u_e^2 \tau_V} + f' \frac{\partial \varepsilon}{\partial \xi} + \varepsilon' \frac{\partial f}{\partial \xi} \right) \quad (10.21)$$

with

$$\varepsilon = \frac{e_V}{\bar{e}_{Ve}}$$

10.3 BOUNDARY LAYERS IN VIBRATIONAL NON-EQUILIBRIUM

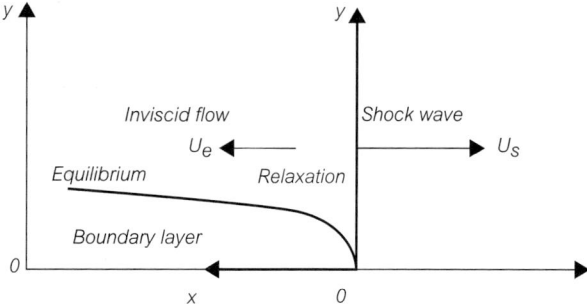

Figure 60. Boundary layer behind a moving shock wave ($x = U_s t - X$).

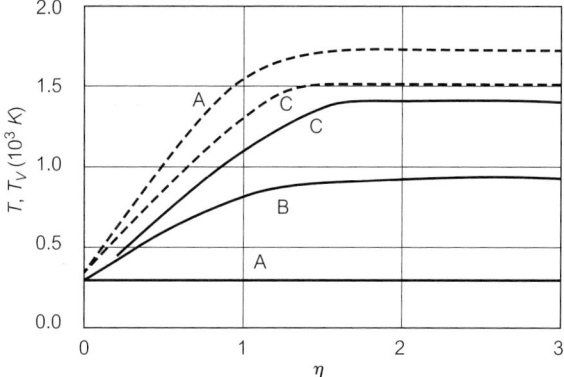

Figure 61. Temperature distribution across the boundary layer (nitrogen, $M_s = 5$, $T_i = 295$ K, $p_i = 10^4$ Pa). ----Translation–rotation temperature; ──Vibrational temperature. A: $x = 0$ cm, B: $x = 0.2$ cm, C: $x = 1$ cm.

The temperature distribution across the boundary layer may be deduced from these equations, and Fig. 61 shows distributions of $T = T(\eta)$ and $T_V = T_V(\eta)$ for increasing distances x from the shock wave. The evolution of the energetic vibrational non-equilibrium is also represented in Fig. 62. Thus, we observe that a maximum becomes visible while the inviscid flow is close to equilibrium: this is because of longer relaxation times in the boundary layer, where the temperature is lower than outside but the pressure is the same.

10.3.2 Example 2: boundary layer in a supersonic nozzle

Assuming that the boundary layer in the divergent part of a nozzle develops from the throat (Fig. 63), we can also calculate energetic vibrational non-equilibrium (for a non-dissociated pure gas) in this boundary layer:[125] as at the throat, the flow is generally in TRV equilibrium (Chapters 8 and 9), the flow entering the boundary layer has a high temperature. Assuming a cold catalytic wall, there

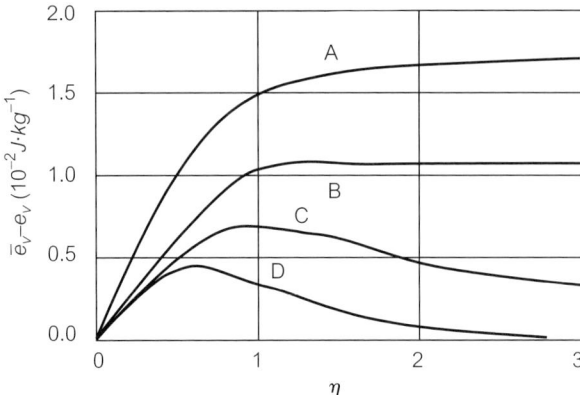

Figure 62. Energetic vibrational non-equilibrium across the boundary layer (Conditions and notation of Fig. 61).

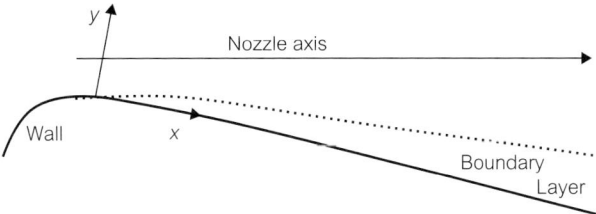

Figure 63. Scheme of boundary layer in a nozzle.

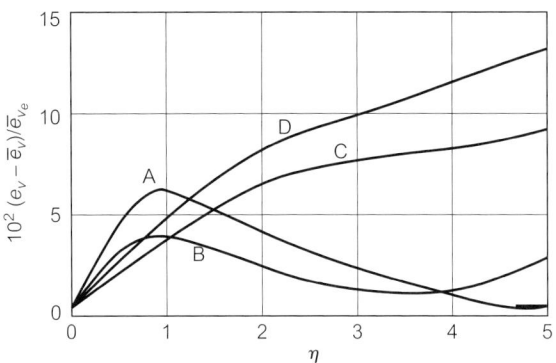

Figure 64. Energetic non-equilibrium distribution across the boundary layer of a supersonic nozzle (nitrogen, conditions of Fig. 49(a)). A: $x = 2$ cm, B: $x = 16$ cm, C: $x = 80$ cm, D: $x = 120$ cm.

is a rapid freezing of the vibrational energy (Fig. 64), while the inviscid flow is close to equilibrium. Thus, close to the throat, there is a maximum for the non-equilibrium across the boundary layer (curve A). Further in the nozzle, this maximum tends to decrease, while, as expected, the freezing becomes important outside; therefore, we have distributions along the nozzle represented by curves B, C, and D.

10.3.3 Example 3: boundary layer behind a reflected shock wave

Behind a reflected shock wave at the end wall of a tube (Fig. 65), the relaxation is more complex than it is behind an incident shock wave, since the reflected shock wave may propagate in a relaxing gas. If we do not take into account the presence of side-wall boundary layers, the inviscid flow behind the reflected shock wave may be computed by a method of characteristics: thus, an example of successive TR temperature distributions is shown in Fig. 66. These result from a competition between relaxation and attenuation effects (the shock slows down), and a minimum appears in the distributions. Close to the end wall, a boundary layer develops, essentially because of the important temperature difference between the gas and the wall, since the flow velocity is very small behind the reflected shock wave.

The computation of this boundary layer (Appendix 10.3), gives the time evolution of the temperature distributions T and T_V and of the energetic non-equilibrium[126] (Fig. 67). The unsteady inviscid conditions give a particular character to these distributions: thus, the non-equilibrium decreases first between the shock and the wall during the first instants following the shock reflection. Then, with increasing time, a minimum appears close to the wall, so that in the outer part of the boundary layer, we have $T_V < T$ as usual behind a shock, while in the inner part, we have a freezing zone where $T_V > T$ appears and develops until the inviscid flow is in equilibrium. This behaviour results from competition between a diffusion characteristic time of the excited

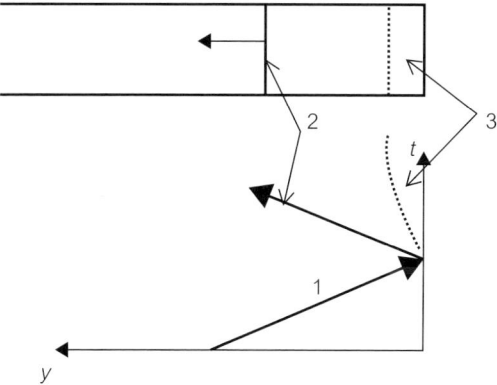

Figure 65. Scheme of boundary layer developing at the end wall of a tube (Physical plane and y, t diagram). 1: Incident shock, 2: Reflected shock, 3: Boundary layer.

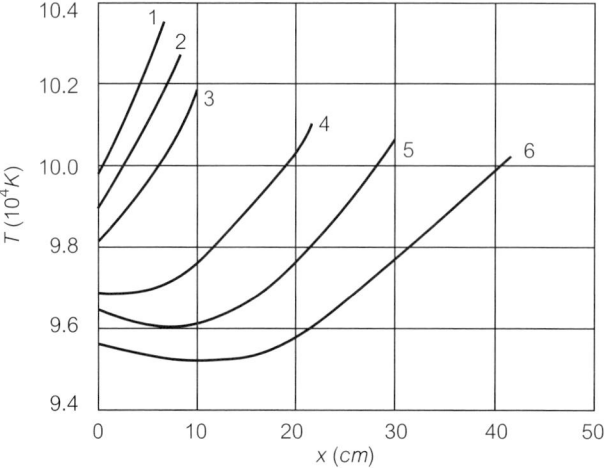

Figure 66. Temperature distributions behind a reflected shock wave (nitrogen, $M_s = 12, p_1 = 200$ Pa). Time after reflection: 1: 15 μs, 2: 25 μs, 3: 40 μs, 4: 400 μs, 5: 500 μs, 6: 600 μs.

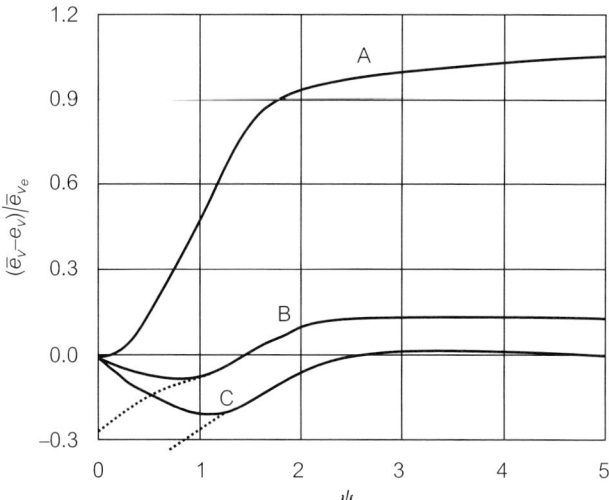

Figure 67. Vibrational non-equilibrium distributions across the boundary layer (Conditions of Fig. 66), $\psi = \left[2 \int_0^t \rho_e \mu_e dt \right]^{-1/2} \int_0^y \rho \, dy$. Time after reflection: A: 25 μs, B: 500 μs, C: 900 μs. Full line: Catalytic wall. Dotted line: Non-catalytic wall.

molecules going towards the end wall and a relaxation time that is longer in the colder zone. This occurs even when the wall is catalytic for the vibrational mode ($T_{Vw} = T_w$).

When the reflected shock wave is more intense, dissociation also takes place, but the vibrationally frozen zone is still present, so that, in the inner part of the boundary layer where $T_V > T$, the dissociation-rate constant k_D is higher

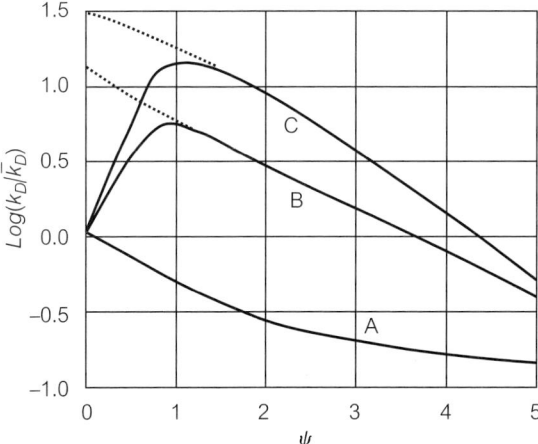

Figure 68. Dissociation-rate constants across the boundary layer (Conditions and notation of Figs 66 and 67).

than its local equilibrium value $\bar{k}_D$: thus, Fig. 68 represents an example of the variation of the ratio $k_D/\bar{k}_D$ across the boundary layer, and a maximum is clearly visible for a catalytic wall.[127]

If the wall is non-catalytic $\left(\left(\frac{\partial T_V}{\partial y}\right)_w = 0\right)$, the freezing zone is of course more important, the non-equilibrium does not present any minimum, and the vibrational non-equilibrium increases up to the wall. The same is of course true for the ratio $k_D/\bar{k}_D$ (Fig. 68). We may also verify that the rate constant itself, k_D, seems to behave in an anomalous way, since it increases up to the wall, while the temperature is decreasing.

10.4 Two-dimensional flows

The solutions of the Navier–Stokes equations, including chemical reactions and vibrational non-equilibrium, generally give realistic and complete descriptions of many flows, without the need for other assumptions than those related to physical and chemical processes (various models). These equations are therefore extensively used in more or less complex situations connected to practical problems. A few examples are presented below.

10.4.1 Hypersonic flow in a nozzle

Experimental simulations of hypersonic flows are generally carried out in nozzles requiring computation prior to any construction. In particular, the flow at the nozzle exit, where various bodies are placed, necessitates such computations.

The general characteristics of nozzle flows have already been described in the preceding chapters pointing out the vibrational and chemical non-equilibrium inherent to the hypersonic regime.

A typical example is presented below: a scheme of the nozzle is shown in Fig. 69, and the reservoir conditions are given in Table 3. The hypersonic flow at the exit of the nozzle is intended to reproduce conditions close to those encountered by a re-entry body in high atmosphere.[112]

A Navier–Stokes computation including the species conservation equations and vibrational relaxation equations enables us to determine the complete flow in the nozzle and especially at the exit. Various results are shown in Figs 70–73; thus, in Fig. 70, iso-Mach lines are represented with a clearly visible boundary layer.

In Fig. 71, the variation along the nozzle of transverse TR temperature distributions is shown: the first distribution, A, is close to the throat, the third, C, is at the exit, and an intermediate distribution is labelled B. In the same figure, the corresponding distributions of T_{VN_2} are also shown, in order to point out the evolution of the non-equilibrium. The temperatures T_{VO_2} are, of course, evolving more rapidly, but the trend is similar.

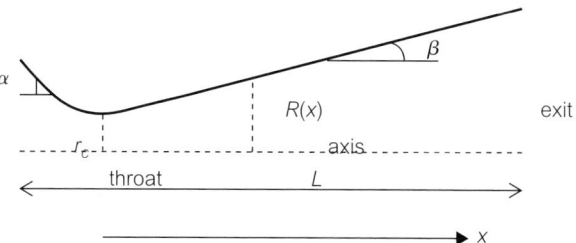

Figure 69. Scheme of the nozzle.

Table 3. General characteristics of the nozzle

		Geometrical characteristics			
α	β	r_c (cm)	L (m)	S_S/S_C	
45°	10°	0.3	1.13	4444	
		Reservoir conditions			
T_0 (K)	p_0 (Pa)	c_{N_2}	c_{O_2}	c_{NO}	c_N	c_O
6500	153×10^6	0.69	0.05	0.16	0.005	0.095
		Wall conditions			
	$T_W = 600$ K			Catalytic wall	

10.4 TWO-DIMENSIONAL FLOWS

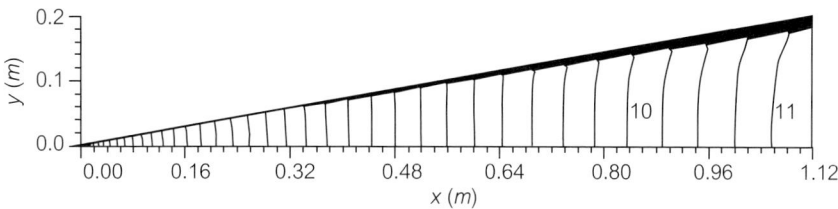

Figure 70. Iso-Mach lines in the nozzle ($\Delta M = 0.25$).

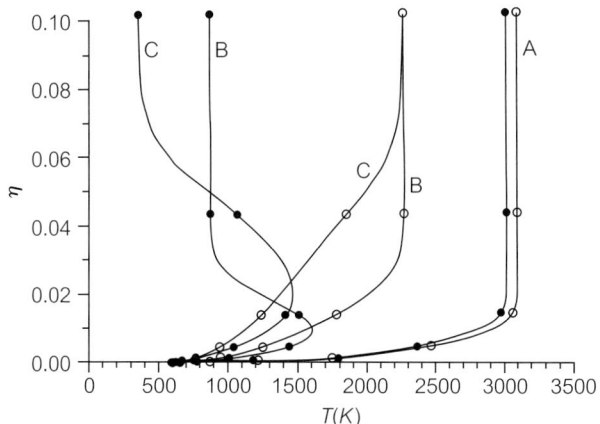

Figure 71. Spatial variation of temperatures along the nozzle (A: 0.04 m, B: 0.34 m, C: 1.12 m, •: T, o: T_{VN_2}).

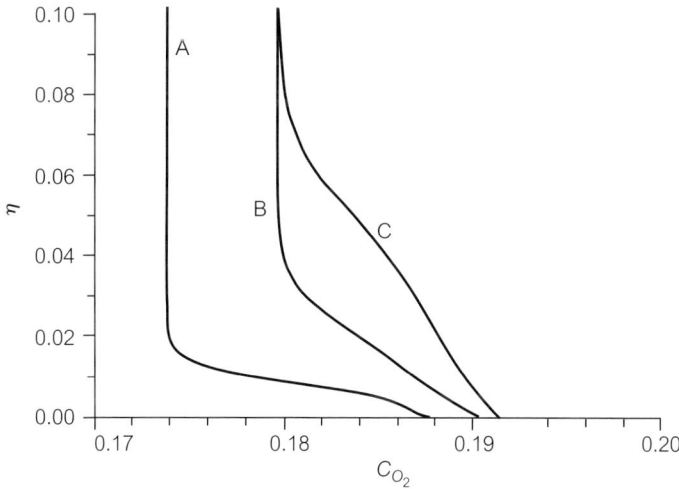

Figure 72. Spatial variation of O_2 concentration along the nozzle (Notation of Fig. 71).

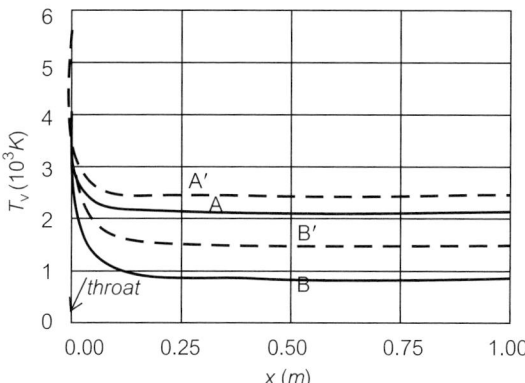

Figure 73. Spatial variations of vibrational temperatures along the nozzle axis (Full lines: Navier–Stokes, dotted lines: Euler). A, A': Nitrogen; B, B': Oxygen.

In Fig. 72, the evolution of the O_2 concentration is shown: this concentration continues to evolve in the boundary layer along the nozzle, while the frozen level is attained in the inviscid flow.

The evolution of the temperatures T_{VN_2}, T_{VO_2} along the axis of the nozzle is represented in Fig. 73 for comparison with the evolution of the same temperatures obtained with an Euler computation (Chapter 9): in that latter case, we observe that the freezing levels are higher, since the expansion is quicker.

10.4.2 Hypersonic flow around a body

Example 1

The proposed present example consists of a steady hypersonic airflow around a sphere-cone model[101] (nose radius R_C): the free stream is assumed to be in vibrational and chemical equilibrium, and the wall is non-catalytic (Table 4).

The distribution of mass concentrations along the stagnation line of the body is shown in Fig. 74. The results are consistent with what could be expected: we observe a decrease in concentrations of O_2 and N_2 from the shock towards the wall, then an increase for N_2 due to the dissociation of NO and to the non-catalycity of the wall; there is a maximum for the NO concentration (formation then dissociation).

From this computation, we can deduce the shock stand-off distance (distance between the shock and the body along the stagnation line). This quantity is sensitive to non-equilibrium (Chapter 9) and can be compared to experimental values (Chapter 12).

Example 2

Within the framework of another example (experimental plane X38) and for free stream conditions given in Table 5, the distribution of T and T_{VN_2} along the stagnation line is shown in Fig. 75, where a rapid relaxation may be observed.[101] The influence of the wall catalycity on the heat

Table 4. Example of hypersonic flow

R_C(cm)	T_∞(K)	p_∞(Pa)	V_∞(m/s)	M_∞	T_W(K)
3.5	811	192	5010	8.9	300

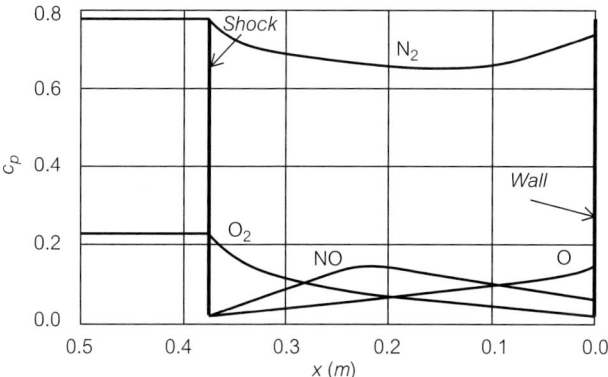

Figure 74. Distribution of mass concentrations along the stagnation line (Conditions of Table 3).

Table 5. Free stream conditions for X38

M_∞	V_∞(m/s)	T_∞(K)	p_∞(Pa)	Altitude (km)
20	6105	231	12	52
25	7310	205	4	73

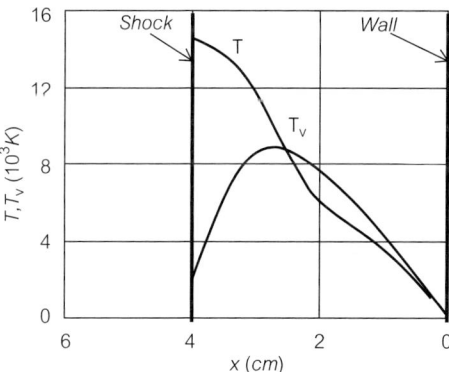

Figure 75. Distribution of temperatures T and T_{VN_2} along the stagnation line of X38 ($M_\infty = 20$).

flux may also be observed in Fig. 76, where the distribution of the N concentration along the stagnation line is shown. The use of a partially catalytic coating (curve NC) significantly reduces the heat flux in comparison with a fully catalytic wall (curve C).

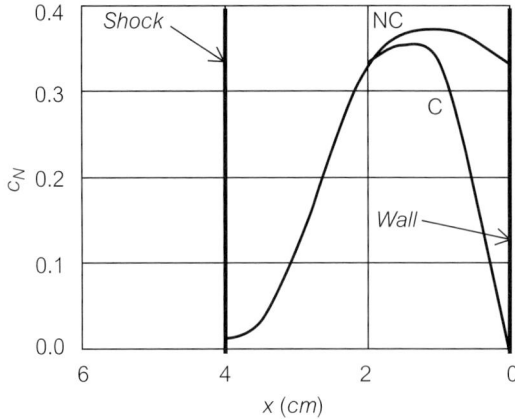

Figure 76. Distribution of N concentration along the stagnation line of X38 ($M_\infty = 25$, C: catalytic case, NC: non-catalytic case).

Table 6. Stagnation and free stream conditions for the shock tunnel[128]

T_0 (K)	P_0 (MPa)	M	T (K)	$T_V(N_2)$ (K)	$T_V(O_2)$ (K)
6000	18	7.25	420	2120	900

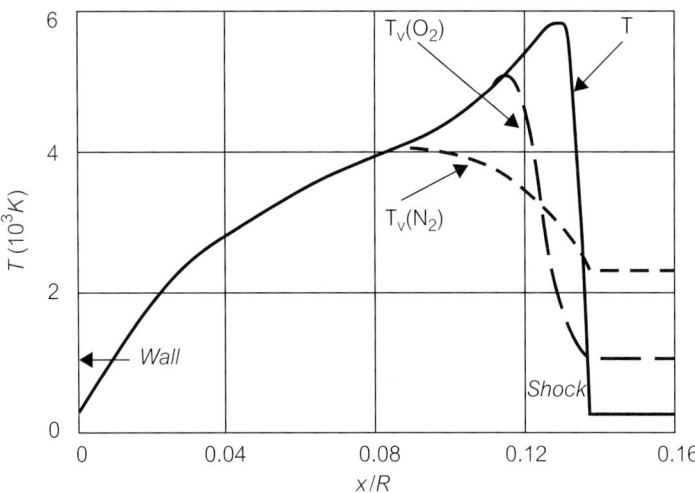

Figure 77. Temperature distributions along the stagnation line (Conditions of Table 6).

Example 3

A somewhat different example is given by the airflow around a hemisphere-cylinder (nose radius 5 cm) placed at the exit of a shock tunnel nozzle,[128] where, as we know, the free stream is in strong non-equilibrium (conditions in Table 6). The computation of the axisymmetric flow around the body is made by using the MS vibration–reaction model developed in Chapter 6.

The evolution of temperatures along the stagnation line (Fig. 77) shows that the non-equilibrium effects are strongly reduced because the vibrational temperature freezes at a high level in the free stream (Table 6).

10.4.3 Mixtures of supersonic reactive jets

Another type of reactive flow is represented by gas-dynamic lasers in which gaseous components are reacting and can create 'population inversions' in their internal energy modes (electronic, vibrational, and so on). As is well known, these particular non-equilibrium conditions are required to obtain a laser effect (Appendix 10.4). The experimental equipments used for this purpose are of various types because of the differing nature of the reacting media and that of the efficiency of the 'pumping' processes used to maintain the population inversion.[129]

The proposed present example is a gas-dynamic CO_2/N_2 laser using the population inversion of the vibrational modes of CO_2. This inversion is obtained from the interaction of a vibrationally excited nitrogen flow and a jet of carbon dioxide (mixed with other chemically neutral gases):[130] this interaction is characterized by a quasi-resonant VV transfer (Chapter 2) between the excited nitrogen molecules and the highest vibrational mode of the carbon dioxide, which therefore becomes 'overpopulated' (Appendix 10.4). The reactive mixture is then expanded (therefore frozen) in a laser cavity, where the laser effect is obtained from the de-excitation of the carbon dioxide.

It is therefore essential, in particular so as to increase the efficiency of the reacting mixture, to understand the flow field, for which a Navier–Stokes computation is necessary.

There are two steps in the process: first an excited nitrogen flow is expanded into a nozzle and mixed with a CO_2-expanded flow in a 'premixing nozzle' (Fig. 78), then this mixture is mixed again with another carbon dioxide flow in the laser cavity. Figure 78 shows a basic cell of the whole set-up, which includes a large number of similar cells. The geometry is quasi-three-dimensional because it also includes an expansion in the direction normal to the plane of the figure in order to preserve a partial freezing. In the cavity, the laser line corresponding to the vibrational transition $v_3 \rightarrow v_1$ and to the rotational transition corresponding to the P14 branch, i.e. 10.6 μm, is amplified.

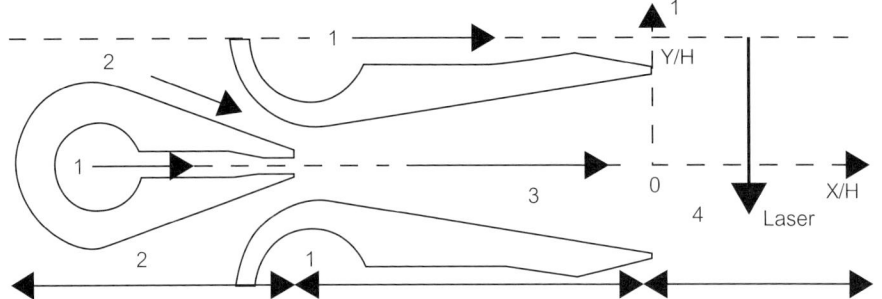

Figure 78. Scheme of gas-dynamic laser. 1: $CO_2/Ar/He/H_2$ nozzle ($T_0 = 500$ K, $p_0 = 10^5$ Pa), 2: N_2 nozzle ($T_0 = 500$ K, $T_{V0} = 3000$ K, $p_0 = 10^5$ Pa), 3: Premixing nozzle, 4: Laser cavity (H = 0.305 cm).

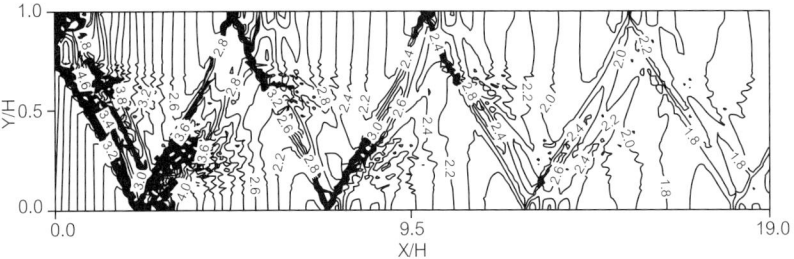

Figure 79. Isobars (10^{-2} Pa) in the laser cavity.

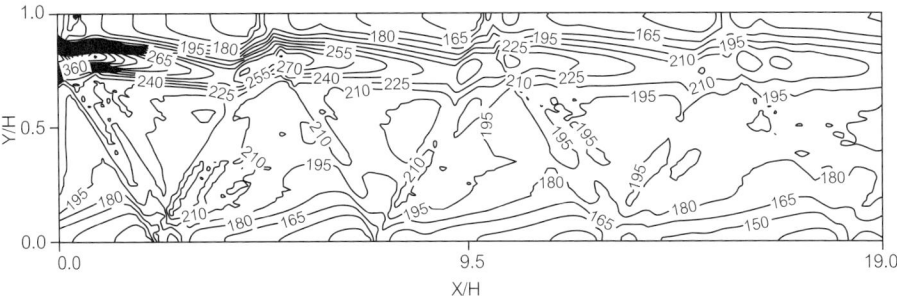

Figure 80. Isotherms (K) in the laser cavity.

A few results of the Navier–Stokes computation are presented in Figs 79 and 80. The flow-field isobars (Fig. 79) are characteristic of the mixtures of supersonic jets, including oblique shock waves at the exit of nozzles, reflecting on the symmetry axis and attenuating because of the divergence of the cavity. Similarly, the isotherms (Fig. 80) show the development of the boundary layers coming from the nozzles. The optical running of the cavity is presented in Appendix 10.4.

Appendix 10.1 Catalycity in the vibrational non-equilibrium regime

As in the case of dissociated flows, there is also a catalycity problem for the vibrational mode of the molecules impinging on a wall: thus, a vibrational catalycity may be defined with the extreme cases $T_{Vw} = T_w$ (catalytic case) and $\left(\frac{\partial T_V}{\partial y}\right)_w = 0$ (non-catalytic case). The problem, however, may be approached in a somewhat different way.

We consider the boundary layer of a pure gas in vibrational non-equilibrium and the corresponding Knudsen layer (Appendix 6.5). At the (macroscopic) level of the wall, vibrational exchanges may occur, either with the wall or with the TR mode of the gas molecules (considered as a whole). We can therefore define two accommodation coefficients specific to each mode (TR and V), that is, α_{TR} and α_V, and two exchange coefficients TR→V and V→TR, that is, γ_{TR} and γ_V. We also assume that the energy exchanged with the wall or between modes is proportional to the corresponding available energy. For example, the rate of TR→V energy exchange $Q_{TR \to V}$ is such that $Q_{TR \to V} = \gamma_{TR} (F_i - F_w)$ and so on. It is necessary of course to take into account the exchanged energy between the molecular modes when writing the balance of each type of energy of the gas with the wall.

Now if, as before (Appendix 6.5), we write the balance of molecule and energy fluxes through the Knudsen layer, we obtain the following expressions[54,131] for the TR and V 'temperature jumps' at the wall, ΔT and ΔT_V, that is:

$$\Delta T = T - T_w = \frac{1}{A} \left[F(\alpha_{TR}, \gamma_{TR}, \gamma_V) q_{TR} - G(\alpha_V, \gamma_{TR}, \gamma_V) q_V \right]_w \quad (10.22)$$

$$\Delta T_V = T_V - T_w = \frac{1}{B} \left[F(\alpha_V, \gamma_V, \gamma_{TR}) q_{TR} - G(\alpha_{TR}, \gamma_V, \gamma_{TR}) q_V \right]_w$$

$$(10.23)$$

with

$$A = n_w k \left(\frac{2kT_w}{\pi m}\right)^{1/2} \quad \text{and} \quad B = \frac{C_V}{2} n_w \left(\frac{2kT_w}{\pi m}\right)^{1/2}$$

$$F(a, b, c) = \frac{2(1-c) - a(1-b-c)}{2a(1-b-c)}$$

$$G(a, b, c) = \frac{b}{a(1-b-c)}$$

The coupling between ΔT and ΔT_V is pointed out in these expressions, which depend on both heat fluxes.

A simple application enables us to determine the influence of these coefficients on the wall heat flux. Thus, we consider the boundary layer in vibrational non-equilibrium developing at the end wall of a tube behind a reflected shock wave (Appendix 10.3). We apply to the corresponding equations the boundary conditions given by Eqns (10.22) and (10.23), and because of the vibrational freezing, we assume that $\gamma_V = 0$ (dominant $TR \to V$ exchanges).

Independently of the numerical solution of the equations, we can obtain a general idea of the relative importance of the three coefficients α_{TR}, α_V, and γ_{TR} by linearizing the equations and considering only the short times following the shock reflection: thus, we obtain an analytic expression for the total heat flux $q_w = (q_{TR} + q_V)_w$, that is:

$$\frac{q_w}{(q_w)_{\gamma_{TR}=0}} = \frac{1 + \gamma_{TR}\left(\frac{2\alpha_V}{2-\alpha_V} - 1\right)}{1 + \gamma_{TR}\frac{\alpha_{TR}}{2-\alpha_{TR}}} \quad (10.24)$$

where $(q_w)_{\gamma_{TR}=0}$ represents the heat flux without exchange.

Equation (10.24) is represented in Fig. 81.

Thus, it appears that the intermode coupling gives the (theoretical) possibility of obtaining a wall heat transfer more important than without exchange ($\gamma_{TR} = 0$) if α_V is greater than α_{TR}. The case of $\alpha_V \ll \alpha_{TR}$, however, seems more realistic and is experimentally confirmed (Chapter 12), so that it is similar to the case of a dissociated but frozen boundary layer on a non-catalytic wall. The importance of the heat flux then depends on the exchange coefficient γ_{TR} (Chapter 12).

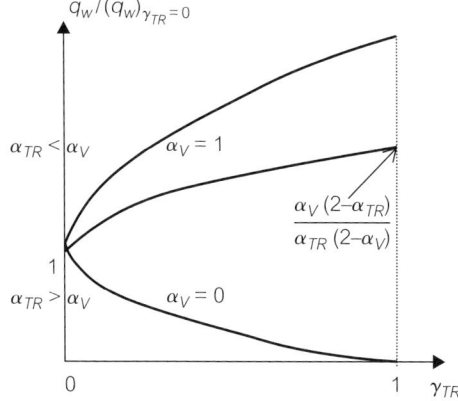

Figure 81. Influence of the exchange coefficient γ_{TR} on the wall heat flux behind a reflected shock wave.

Appendix 10.2 Generalized Rankine–Hugoniot relations

Behind intense shock waves, the gradients of various quantities are important enough to be taken into account at $x = 0$, and therefore, for the balance equations across the shock wave, it is necessary to use the Navier–Stokes equations (Chapman–Enskog distribution) rather than the Euler equations (Maxwellian distribution), which may then be used for $x > 0$. The corresponding relations or 'shock slip' conditions constitute the generalized Rankine–Hugoniot relations: in the most general case of vibrational and chemical non-equilibrium, they may be easily deduced from the general Navier–Stokes equations (Chapter 8) written in the following one-dimensional form, using only μ and η as dimensional transport variables:

$$\frac{d}{dx}(\rho u) = 0$$

$$\frac{d}{dx}(\rho u u) = -\frac{dp}{dx} + \frac{d}{dx}\left[(2\mu + \eta)\frac{du}{dx}\right]$$

$$\frac{d}{dx}(\rho u h_0) = \frac{d}{dx}\left\{\frac{2\mu + \eta}{P_f}\left[\frac{dh_0}{dx} + (P_f - 1)\frac{d}{dx}\left(\frac{u^2}{2}\right)\right]\right\}$$

$$+ \frac{d}{dx}\left[\sum_p (L_f - 1)\frac{2\mu + \eta}{P_f} h_p \frac{dc_p}{dx}\right]$$

$$+ \frac{d}{dx}\left[\sum_p (\Gamma_p - 1)\frac{2\mu + \eta}{P_f}\frac{de_{Vp}}{dx}\right]$$

$$\frac{d}{dx}(\rho u c_p) = \dot{w}_p + \frac{d}{dx}\left(\frac{2\mu + \eta}{P_f} L_f \frac{dc_p}{dx}\right)$$

$$\frac{d}{dx}(\rho u c_p e_{Vp}) = \dot{w}_{Vp} + \frac{d}{dx}\left[\frac{2\mu + \eta}{P_f}\left(L_f e_{Vp}\frac{dc_p}{dx} + F_p c_p \frac{de_{Vp}}{dx}\right)\right] \quad (10.25)$$

with

$$P_f = \frac{2\mu + \eta}{\lambda_{TR}} C_{TR}^p, \quad L_f = \frac{\rho D C_{TR}^p}{\lambda_{TR}}, \quad L_{Vp} = \frac{\rho D C_{Vp}^p}{\lambda_{Vp}}, \quad \text{and}$$

$$F_p = \frac{C_{TR}^p \lambda_{Vp}}{C_{Vp} \lambda_{TR}} = \frac{L_f}{L_{Vp}}$$

The following 'shock slip' relations (generalized Rankine–Hugoniot relations) may then be written at $x = 0$, instead of the classical Rankine–Hugoniot relations (Eqn. (8.12)). Of course, frozen conditions are assumed for species concentrations and vibrational energies.

$$\rho_1 u_1 = \rho_2 u_2$$

$$p_1 + \rho_1 u_1^2 = p_2 + \rho_2 u_2^2 - \left[(2\mu + \eta)\frac{du}{dx}\right]_2$$

$$\rho_1 u_1 \left(h_1 + \frac{u_1^2}{2}\right) = \rho_2 u_2 \left(h_2 + \frac{u_2^2}{2}\right)$$

$$- \left\{\frac{2\mu + \eta}{P_f}\left[\frac{dh_0}{dx} + (P_f - 1)\frac{d}{dx}\left(\frac{u^2}{2}\right)\right]\right\}_2$$

$$- \left[\sum_p (L_f - 1)\frac{2\mu + \eta}{P_f} h \frac{dc_p}{dx}\right]_2$$

$$- \left[\sum_p (F_p - 1)\frac{2\mu + \eta}{P_f}\frac{de_{Vp}}{dx}\right]_2$$

$$\rho_1 u_1 c_{p1} = \rho_2 u_2 c_{p2} - \left(\frac{2\mu + \eta}{P_f} L_f \frac{dc_p}{dx}\right)_2$$

$$\rho_{p1} u_1 e_{Vp_1} = \rho_{p2} u_2 e_{Vp_2} - \left[\frac{2\mu + \eta}{P_f}\left(L_f e_{Vp}\frac{dc_p}{dx} + F_p c_p \frac{de_{Vp}}{dx}\right)\right]_2$$

(10.26)

Appendix 10.3 Unsteady boundary layers

When the boundary layer moves at constant velocity, the coordinate transformation presented in Appendix 8.4 may be applied. Otherwise, a somewhat different transformation must be used in order to preserve the advantages enumerated in Appendix 8.4.

As an example, we take the unsteady one-dimensional boundary layer behind a reflected shock wave (Section 10.3.3, Fig. 65). We can assume that the velocity is small and therefore that the viscosity effects are negligible ($E_e \sim 0$). For a pure gas in vibrational non-equilibrium, we therefore have

$$\frac{\partial \rho}{\partial t} + \frac{\partial \rho u}{\partial y} = 0 \quad \text{and} \quad \frac{\partial p}{\partial y} = 0, \quad \to p = p_e(t)$$

$$\frac{dh}{dt} = \frac{\partial}{\partial y}\left(\lambda_{TR}\frac{\partial T}{\partial y} + \lambda_V \frac{\partial T_V}{\partial y}\right) + \frac{\partial p_e}{\partial t} \quad \text{and}$$

$$\rho \frac{de_V}{dt} = \dot{w}_V + \frac{\partial}{\partial y}\left(\lambda_V \frac{\partial T_V}{\partial y}\right) \tag{10.27}$$

Similar considerations to those of Appendix 8.4 lead us to use the following transformation (von Mises):

$$\tilde{t} = \int_0^t \rho_e \mu_e dt \quad \text{and} \quad \psi = \frac{\rho_e}{(2\tilde{t})^{1/2}} \int_0^y \frac{\rho}{\rho_e} dy \tag{10.28}$$

Here, ψ (as η) represents a transverse coordinate in the boundary layer, and the integration domain is quasi-rectangular. After solving the transformed equations, we obtain the time evolution of the flow quantities. The case with chemical reactions does not present further difficulty, and in the continuum regime, the accommodation is quickly negligible (the time constant is examined in Chapter 12). Thus, in the absence of chemical reactions, the wall heat flux remains close to a conduction flux, that is $q_w = (\rho C^p \lambda)^{1/2} \frac{T_e - T_w}{\sqrt{\pi t}} \sim \frac{A}{\sqrt{t}}$, and therefore the increase in wall temperature ΔT_w is practically constant (Appendix 7.2).

If the Navier–Stokes system is not solved, the influence of the boundary layer on the inviscid flow may be taken into account.[132] The displacement thickness is negative and therefore provokes an attenuation of the reflected shock wave, but this effect is generally negligible in comparison with other effects such as the relaxation or the presence of side-wall boundary layers (Chapter 11).

Appendix 10.4 CO_2/N_2 gas-dynamic lasers

Vibrational model for the CO_2/N_2 system

The average temperatures in the nozzles and laser cavity are moderate (200–2000 K), so that we can use a simplified physical model that includes only a few vibrational levels (Figs 82 and 83); the population of higher levels is lower than 1%.

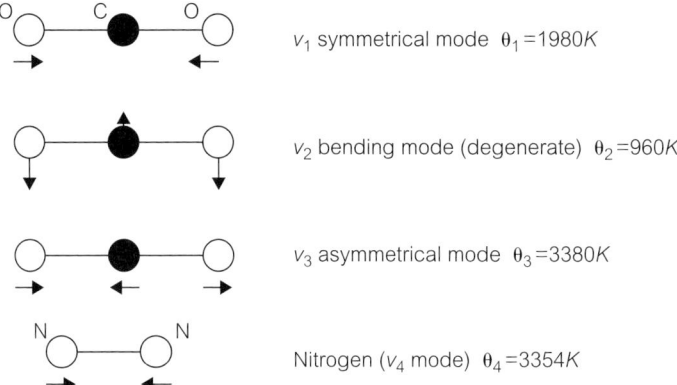

Figure 82. Vibrational modes of CO_2 and N_2.

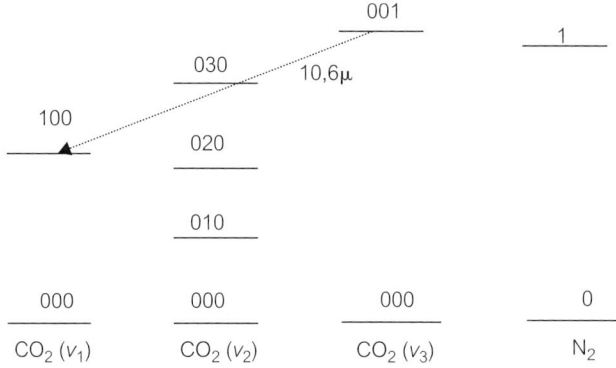

Figure 83. First vibrational levels of CO_2 and N_2.

If we take into account the diagram of Fig. 83, the dominant exchange processes are the following:

TV exchanges:

$$CO_2^*(\nu_2) + M \to CO_2 + M \quad (667 \text{ cm}^{-1})$$
$$N_2^* + M \to N_2 + M \quad (2331 \text{ cm}^{-1})$$

CO_2/CO_2 VV exchanges:

$$CO_2^*(\nu_3) + M \to CO_2^{***}(\nu_2) + M \quad (416 \text{ cm}^{-1})$$
$$CO_2^*(\nu_1) + M \to CO_2^{**}(\nu_2) + M \quad (102 \text{ cm}^{-1})$$

CO_2/N_2 VV exchanges:

$$CO_2^*(\nu_3) + N_2 \to CO_2 + N_2^* \quad (18 \text{ cm}^{-1})$$

The last, quasi-resonant, reaction is at the origin of the laser effect (10.6 μm).

Vibrational relaxation equations

Starting from the general relaxation equation (Eqn. (2.17)), written for each i_v level of each m mode ($m = 1, 2, 3$ for CO_2; $m = 4$ for N_2), we multiply each equation by $\varepsilon_{i_{vm}}$ and sum over the levels (represented in Fig. 83). Thus, we obtain four equations for the average vibrational energy of each mode e_{Vm}. If we adopt the harmonic oscillator model, these equations can be written in the following form, where q_m represents the number of vibrational quanta of the mode m per unit volume, that is, $q_m = n_{Vm} \frac{E_{Vm}}{h\nu_m}$:

$$\frac{dq_m}{dt} = \frac{\bar{q}_m - q_m}{\tau_{TV}} + \sum_n \frac{k_m}{g_m \tau_{VVmn}}$$

$$\times \left[q_n^{k_n} (q_m + 1)^{k_m} \exp\left(\frac{k_n \theta_{Vn} - k_m \theta_{Vm}}{T}\right) - q_m^{k_m} (q_n + 1)^{k_n} \right]$$

(10.29)

where $h\nu_m = k\theta_{Vm}$, and k_n is the number of quanta exchanged in a transition.

Small-signal gain coefficient

The relative intensity variation of radiation that has a wave number ω crossing a dz layer of a gaseous medium is equal to $dI/I = \alpha(\omega) dz$, where $\alpha(\omega)$ is the amplification coefficient of the medium if α is positive (Lambert–Beer law).

For a medium composed of molecules in levels i_r, i_v and for radiative transitions $\Delta i_v = \pm 1$ and $\Delta i_r = +1$ (R branch) or $\Delta i_r = -1$ (P branch), the coefficient $\alpha(\omega)$ is equal to

$$\alpha(\omega) = C(i_r, i_v) \, \rho N \phi \omega (i_r, i_v) \left(\frac{g_b}{g_h} n_{i_v+1, i_r \pm 1} - n_{i_v, i_r}\right) \quad (10.30)$$

Here, $C(i_r, i_v)$ is the absorption cross section per molecule and per unit solid angle, g_b and g_h are the eventual degeneracy of low and high states respectively, and ϕ is a shape parameter that takes into account the broadening of the line (Doppler and collisional effects). We generally have a Boltzmann distribution for the rotational population, that is:

$$\frac{n_{i_r}}{n} = \frac{g_{i_r}}{Q_R} \exp\left(-\frac{\varepsilon_{i_r}}{kT}\right)$$

Therefore, there is an amplification if we have

$$g_b n_{i_v+1, i_r \pm 1} > g_h n_{i_v, i_r} \quad (10.31)$$

This inequality involves a population inversion.

CHAPTER 10 REACTIVE FLOWS IN THE DISSIPATIVE REGIME

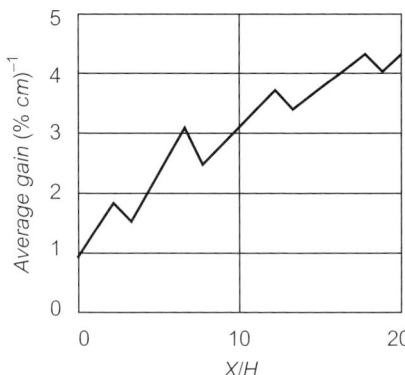

Figure 84. Variation of average small-signal gain along the cavity.

The emission line used here corresponds to the transition $001 \to 100$ (P branch, wavelength $10.6\,\mu m$). The average small-signal gain coefficient across the cavity is shown in Fig. 84, where we observe a strong increase along the cavity (1% at the inlet to 4% at the outlet) despite local decreases due to compression zones (Fig. 79).

Appendix 10.5 Transport terms in the non-equilibrium regime

Those transport terms that are the most sensitive to non-equilibrium conditions, that is relaxation pressure and vibrational conductivity, are analysed and developed in typical cases such as expansion flows and flows behind shock waves.

Relaxation pressure

As discussed in Chapter 5, this transport term appears only when some physical or chemical processes are in equilibrium and others are in non-equilibrium (at the zeroth order of the distribution function). Thus, for example, this term appears when the rotational mode is in equilibrium and the vibrational mode in non-equilibrium, or when all internal modes are in equilibrium but the chemistry is in non-equilibrium.

We have seen that, in the first case, p_r/p is of order τ_R/τ_V, and in the second case it is of order τ_V/τ_C. For typical cases, however, a more quantitative estimation of the relaxation pressure is proposed below.

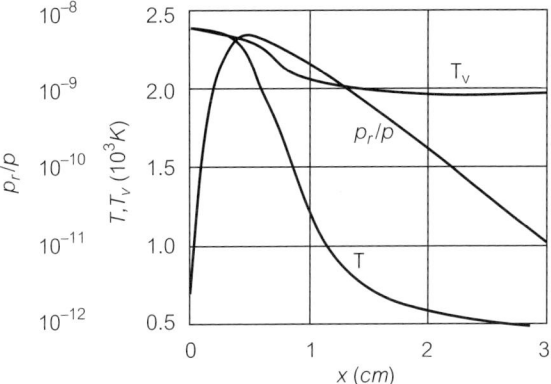

Figure 85. Relaxation pressure and temperatures along a supersonic nozzle (Conditions of Fig. 49(a)).

Vibrational non-equilibrium flow in a supersonic nozzle

We consider the nozzle flow corresponding to Fig. 49(a), Chapter 9 (nitrogen: $T_0 = 2400$ K and $p_0 = 10^7$ Pa). We apply Eqn. (5.27), giving an approximate expression for p_r/p, and the values found along the nozzle[14] are represented in Fig. 85: the corresponding curve presents a maximum close to the throat of the nozzle, since p_r/p is null in the reservoir and decreases strongly in the divergent part where the relaxation time rapidly increases. However, the value of p_r/p remains low and may be neglected in this type of flow.

Flow behind a straight shock wave

Considering a reactive mixture in vibrational equilibrium but in chemical non-equilibrium at zero order (Chapter 5), we calculate the relaxation pressure behind a straight shock wave[116] (Appendix 5.4). An example is given in Fig. 86 for an intense shock wave in air ($M_s = 25$). Thus, close to the shock wave, the ratio p_r/p is about 3% and decreases rapidly, becoming negligible 1–2 cm from the shock wave. This ratio is smaller still for lower Mach numbers.

Vibrational conductivity

In a vibrational non-equilibrium flow, vibrational conductivity λ_V is very sensitive to vibrational temperature T_V (Chapter 5).[28] An example of such a calculation is represented in Figs 87 and 88 for the vibrational conductivity of nitrogen λ_{VN_2} behind a shock wave in air and for two Mach numbers, 8 and

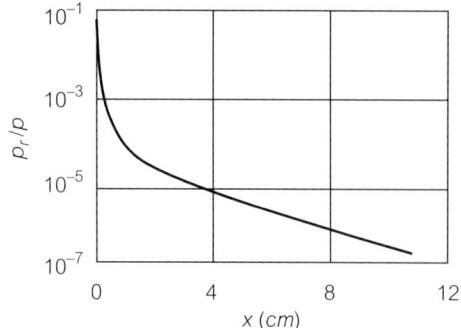

Figure 86. Relaxation pressure behind a shock wave in air (Conditions of Fig. 47(a)).

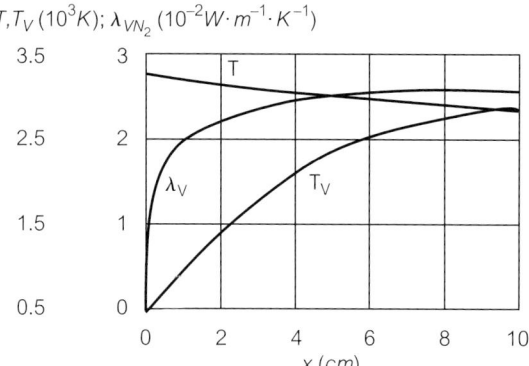

Figure 87. Vibrational conductivity of nitrogen behind a shock wave in air ($M_S = 8$, $T_1 = 271$ K, $p_1 = 102$ Pa).

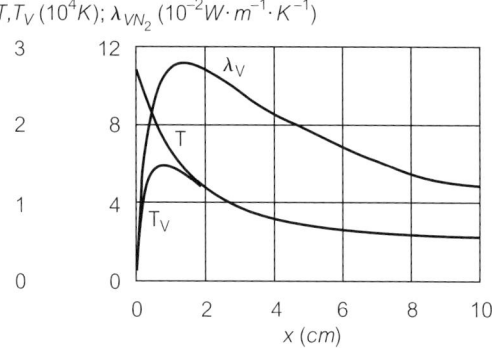

Figure 88. Vibrational conductivity of nitrogen behind a shock wave in air ($M_S = 25$, $T_1 = 205$ K, $p_1 = 8.5$ Pa).

25. Thus, in Fig. 87 ($M_s = 8$), the trend of λ_{VN_2} is similar to the evolution of T_{VN_2}, that is the conductivity increases in the relaxation zone while the temperature T decreases. For $M_s = 25$ (Fig. 88), because of the rapid relaxation, T_{VN_2} quickly becomes equal to T and then decreases with T, so that λ_{VN_2} presents the same behaviour. For λ_{VO_2}, the behaviour is similar, but the evolution is faster.

Appendix 10.6 Numerical method for solving the Navier–Stokes equations

Because the steady Navier–Stokes equations are of hyperbolic–elliptic type, they are relatively difficult to solve. However, if they are written in an unsteady form, they are of hyperbolic–parabolic type, and simpler solving methods can therefore be used.

Let us consider the general case of an axisymmetric hypersonic non-equilibrium airflow, which has been greatly analysed above.[112] In a cylindrical coordinate system (x, y), the unsteady governing equations of the flow are the Navier–Stokes equations, with mass conservation equations for each chemical species (1: N_2; 2: O_2; 3: NO; 4: N; 5: O) and two vibrational energy relaxation equations for N_2 and O_2 (NO can be assumed to be in vibrational equilibrium).

This system of ten equations is solved by a non-iterative implicit finite-difference scheme with a flux splitting technique in the implicit operator.[133,134] This method is chosen in order to overcome two main difficulties: the first is the stiffness of the problem due to the chemical and vibrational processes, and the second is the necessity of using fine meshes near the walls in order to take into account the boundary-layer effects. After discretization, the linear system with a block-pentadiagonal matrix is solved by a Gauss–Seidel line relaxation method.[135] This implicit approach also enables us to use a larger integration time step than that imposed by the CFL condition, and then to reduce CPU time. Details on the method are given below.

Thus, the ten governing equations may be expressed in the following vectorial form:

$$\frac{\partial U}{\partial t} + \frac{\partial F}{\partial x} + \frac{\partial G}{\partial y} + H = \Omega \tag{10.32}$$

where the conservative vector U includes the unknown quantities

$$U = [\rho_s(s = 1 \cdots 5), \rho u, \rho v, \rho e, \rho_i e_{Vi} \quad (i = 1, 2)]$$

Here, F and G are vectors that include the convective fluxes (F_C, G_C) and the diffusive fluxes (F_D, G_D) in each direction, x and y. The vector H includes the axisymmetric terms of the equations, and Ω includes the chemical and vibrational energy source terms.

The x, y physical plane is transformed in a rectangular (ξ, η) plane by a suitable transformation depending on the geometry of the flow, and the system is solved by an implicit finite-difference scheme: assuming that the flow parameters are known at time $n\Delta t$, for each node (i, j) of the grid including $i_M \times j_M$ points, at the time step $(n + 1)\Delta t$, the system may be written as follows:

$$\frac{\delta U^{n+1}}{\Delta t} + \frac{DF^{n+1}}{\Delta \xi} + \frac{DG^{n+1}}{\Delta \eta} + H^{n+1} = \Omega^{n+1} \tag{10.33}$$

where $\delta U^{n+1} = U^{n+1} - U^n$ and D/Δ represent finite-difference operators.

Each vector Φ of Eqn. (10.33) is linearized, so that

$$\Phi^{n+1} = \Phi^n + \left(\frac{\partial \Phi}{\partial U}\right)^n \delta U^n$$

where $\frac{\partial \Phi}{\partial U}$ represents the Jacobian matrix of the vector Φ relative to the vector U.

Splitting the vectors F_C and G_C into a positive and a negative part, the system (Eqn. (10.33)) is equivalent to

$$\left[1 + \Delta t \left(\frac{D_- A^n_+}{\Delta \xi} + \frac{D_+ A^n_-}{\Delta \xi} + \frac{D_- B^n_+}{\Delta \eta} + \frac{D_+ B^n_-}{\Delta \eta} + \frac{D^2 M^n_V}{\Delta \xi^2} + \frac{D^2 N^n_V}{\Delta \eta^2} + D^n - C^n\right)\right]$$
$$\times \delta U^{n+1} = \Delta U^n$$

where

$$\Delta U^n = -\Delta t \left(\frac{DF^n}{\Delta \xi} + \frac{DG^n}{\Delta \eta} + H^n - \Omega^n\right) \tag{10.34}$$

and D, D_+, D_- are the central, forward, and backward difference operators respectively, while $A_+, A_-, B_+, B_-, M_V, N_V, D,$ and C are the Jacobian matrices of $F_{C+}, F_{C-}, G_{C+}, G_{C-}, F_V, G_V, H,$ and Ω respectively.

After discretization, the system (Eqn. (10.34)) may be written as a pentadiagonal matrix linear system:

$$\widehat{A}_{i,j}\delta U^{n+1}_{i,j} + \widehat{B}_{i,j}\delta U^{n+1}_{i,j+1} + \widehat{C}_{i,j}\delta U^{n+1}_{i,j-1} + \widehat{D}_{i,j}\delta U^{n+1}_{i+1,j} + \widehat{E}_{i,j}\delta U^{n+1}_{i-1,j} = \Delta U^n_{i,j} \tag{10.35}$$

By solving this system for a given value of i, we can obtain the vector $\delta U^{n+1}_{i,j}$ for all j values. For this purpose, at each time step, a predictor–corrector scheme is used, and the system (Eqn. (10.35)) is solved by a Gauss–Seidel line relaxation method with alternating sweeps in backward and forward ξ directions.

With this algorithm, we must use suitable boundary conditions depending on geometrical and physical conditions. A few thousand grid points are generally required.

For example, for the nozzle analysed in Section 10.4.1, a mesh of 60×80 grid points is used, with $\Delta x_{min} = 3 \times 10^{-4}$ m and $\Delta y_{min} = 2 \times 10^{-8}$ m. Owing to the implicit method used here, during the iterative procedure the integration time step increases from 1 to $10^3 \Delta t_{CFL}$, and steady state is obtained after about 7000 iterations.

ELEVEN
Facilities and Experimental Methods

11.1 Introduction

The experimental study of non-equilibrium gas flows requires the design, creation, and operation of specific facilities and the development of particular diagnostic techniques.

Although they give interesting and timely results and are designed to test particular equipment or to explore specific situations (an example is given in Appendix 11.1),[136] we will not discuss here experiments concerning the actual flights of space probes and vehicles. It is probable that, in the future, new data will be provided by this type of experimentation, leading to further developments of space vehicles.

Most experiments are carried out in ground-based facilities which can generate non-equilibrium flows. These facilities generally require vital equipment and investment. The essential purpose is to create high-enthalpy gas flows undergoing more or less intense perturbations (shock wave, rapid expansion, and so on), so that, as analysed in the preceding chapters, physical and chemical processes evolve on a timescale equal to or longer than the characteristic flow timescale.

Two main types of facilities are used, depending on the type of phenomena or the processes analysed. Thus, if the analysis of physical and chemical processes is considered essential, simple facilities such as shock tubes generating one-dimensional, non-dissipative flows are used. In contrast, if simulating the conditions of real flight is required, facilities such as arc and shock tunnels generating hypersonic flow around various bodies must be used. We should also mention here plasma generators, which are an example of more complex high-enthalpy equipment.

11.2 The shock tube

The simplest model of a shock tube consists of creating in a tube of constant (circular or rectangular) cross section a moving shock wave generating a flow at high temperature and out of equilibrium. Ideally, this flow is one-dimensional and non-dissipative.[100,137–141]

11.2.1 Simple shock tube theory

Schematically, a tube initially containing the test gas (low-pressure chamber) is separated by a diaphragm from another chamber (high-pressure chamber or driver section) containing another gas (driver gas). After the rupture of this diaphragm, the driver gas, acting as a piston, expands into the low-pressure chamber and generates a shock wave which propagates in the test (driven) gas (Fig. 89(a)). The shock wave gives to the test gas a violent acceleration, which

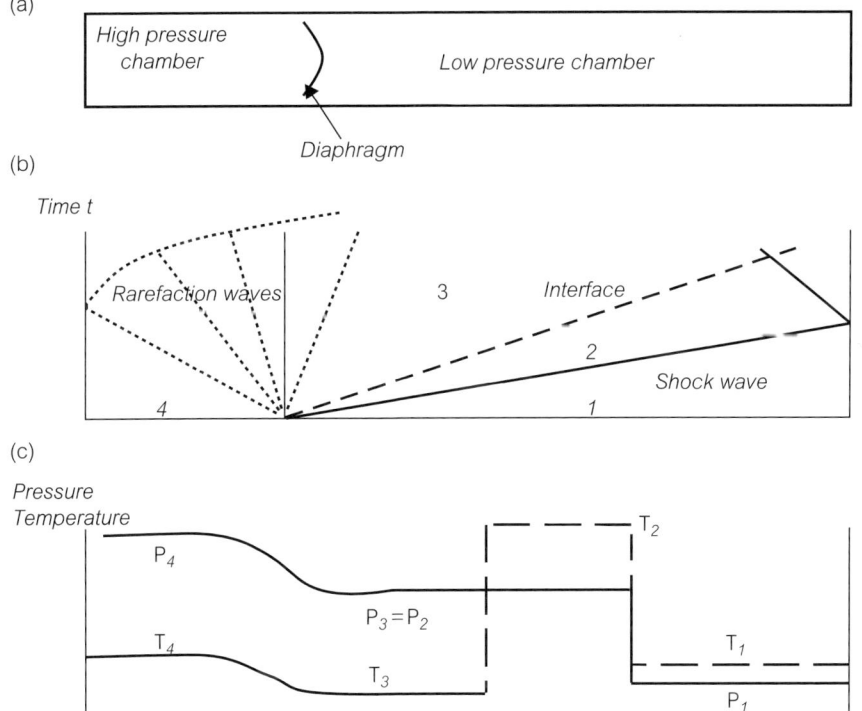

Figure 89. Shock tube: principle and operation. (a) Simple shock tube; (b) Wave system in a shock tube; (c) Pressure and temperature distribution at a given time t.

is accompanied by a jump in temperature, pressure, and density (Chapters 8 and 9). Physical and chemical processes can then start and possibly evolve to their equilibrium state.

The test-gas flow is limited by a contact surface (or interface) separating it from the driver gas (Fig. 89(b)), and in current installations a few metres long, this flow generally lasts a few hundred microseconds. In the assumed absence of dissipative phenomena, the shock wave preserves a constant speed, and therefore, in a reference frame fixed to this shock wave, the flow is one-dimensional and stationary. Moreover, if the rupture of the diaphragm is assumed to be instantaneous, a system of centred rarefaction waves develops in the expanding driver gas (Fig. 89 (b) and (c)). In addition, pressure and velocity are preserved through the interface, whereas temperature and density undergo a discontinuity (Chapter 8).

As also discussed in Chapter 8, the flow parameters of the test gas (region 2 of Fig. 89(b)) can be deduced from the initial quantities (region 1) and from the shock-wave velocity U_s (more precisely, the Mach number $M_s = U_s/a_1$). This is why measurement of the shock-wave velocity is fundamental to attaining further experimental data and must be carried out in the low-pressure chamber.

Principles of a simple shock tube

We can generally assume that the driver gas behaves as an ideal gas (γ is constant) and that the expansion is an isentropic process in the form of centred waves. This is justified because the driver gas is often a monatomic gas and the temperature is relatively low during the expansion; exceptions are mentioned below.

Thus (see Chapter 8), across the centred wave system (Fig. 89), the quantity $u + \frac{2a}{\gamma-1}$ remains constant, and we have

$$\frac{2a_4}{\gamma_4 - 1} = \frac{2a_3}{\gamma_4 - 1} + u_2 \tag{11.1}$$

because $\gamma_3 = \gamma_4$, $u_2 = u_3$ (interface), and $u_4 = 0$ (high-pressure chamber).

We also have

$$\frac{p_4}{p_2} = \left(\frac{a_4}{a_3}\right)^{2\gamma/\gamma_4 - 1} \quad \text{with } p_2 = p_3 \tag{11.2}$$

so that

$$\frac{p_4}{p_2} = \left(\frac{a_4}{a_4 - \frac{\gamma_4-1}{2} u_2}\right)^{2\gamma_4/\gamma_4 - 1} \tag{11.3}$$

Here, u_2 and p_2 are related to the initial conditions of the test gas (p_1, T_1) by the RH relations (Eqn. (8.12)). In the frozen case $(\gamma_1 = \gamma_2)$, we obtain from Eqns (8.13), (8.14), and (11.3):

$$\frac{p_4}{p_1} = \frac{\frac{\gamma_1-1}{\gamma_1+1}\left(\frac{2\gamma_1}{\gamma_1-1}M_s^2 - 1\right)}{\left[1 - \frac{\gamma_4-1}{\gamma_1-1}\frac{a_1}{a_4}\left(M_s - \frac{1}{M_s}\right)\right]^{2\gamma_4/\gamma_4-1}} \tag{11.4}$$

This expression gives the intensity of the shock wave (M_s) as a function of the initial conditions in both chambers, but under restrictive conditions (ideal gas). This is why it is preferable to determine the flow quantities 2 from the measured velocity of the shock wave, as indicated above. However, Eqn. (11.4) gives a qualitatively correct idea of the importance of the various parameters. Thus, in order to obtain the highest possible Mach number, the ratio of initial pressures p_4/p_1 must be as high as possible, which is intuitive, but the ratio a_4/a_1 must also be maximum. In particular, when $\frac{p_4}{p_1} \to \infty$, $M_s \to \frac{\gamma_1+1}{\gamma_4-1}\frac{a_4}{a_1}$. Thus, for a given test gas, we see the advantage of using a light and hot gas as a driver gas. We can also deduce from Eqn. (11.4) the maximum expected values for M_s in the case of a given gas pair.

These results are qualitative if chemical processes are significant behind the shock. Thus, the shock Mach number must be generally deduced from assumed equilibrium conditions behind this shock (Figs 39 and 40, for example), and of course, the maximum values for M_s are lower than those given by Eqn. (11.4).

Technological limitations and constraints

As already indicated, the test-gas flow between the shock wave and the interface has a very short duration and can be disturbed by the various wave systems which propagate in the tube because of its limited dimensions. Thus, the rarefaction waves moving up in the driver section after the rupture of the diaphragm are reflected at the end of this chamber and then move back down (while accelerating), until they potentially overtake the interface and the test gas (Fig. 118). Similarly, the incident shock wave may be reflected at the end of the tube and may interact with the incident test-gas flow: this last point is not always a disadvantage (Section 11.2.3). However, if we take into account these configurations, it is of course possible (see calculations in Appendix 11.2) to optimize, for example, the duration of the test-gas flow at a given abscissa along the tube (such as at the test section) independently of the disturbing phenomena described below.

11.2.2 Disturbing effects

Obviously, this ideal scheme of operation corresponds only roughly to reality, and various phenomena contribute to somewhat modify this scheme and have an influence on the analysed non-equilibrium phenomena. The most significant effects concern the perturbations related to the presence of the wall boundary layer and, to a lesser extent, those arising from the non-instantaneous rupture of the diaphragm.

Wall boundary layer

The boundary layer which develops along the walls of the shock tube between the incident shock and the interface acts like a well for the non-dissipative part of the test gas, and a loss of this gas occurs through the interface in the boundary layer (Fig. 90). This is because the boundary layer of the driver gas has a negligible thickness since the value of the Reynolds number is much higher than the value of the test-gas flow (low temperature, high density). This leads to a deceleration of the shock wave, an acceleration of the interface, and thus a non-constant value of the flow quantities. This unsteady regime tends to a stationary limiting regime, theoretically obtained when the total mass flux through the shock wave is equal to that lost through the interface inside the boundary layer. This last mass flux increases because the separation distance between the shock wave and the interface initially increases (Fig. 89). The shock wave and the interface in the limiting regime have the same (constant) velocity, but the flow quantities, while being stationary, vary between the shock and the interface.

An example of a calculation of the trajectories of the shock wave and contact surface is shown in Fig. 91 in the case of a low initial pressure for which the boundary layer is laminar.[142–144] For higher values of the initial pressure, the boundary layer is turbulent, but approximate models are available.[145,146]

These effects are all the more significant as the initial pressure and the cross section of the tube are lower (low value for Re). Moreover, the hot gas loss across

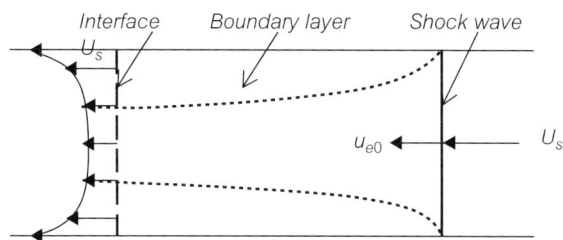

Figure 90. Scheme of flow in a shock tube (Coordinate system fixed to the shock wave).

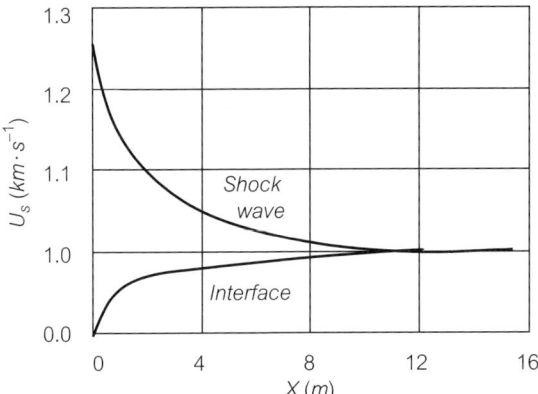

Figure 91. Spatial variation of the shock wave and the interface (Driver gas: Helium; Test gas: Argon: p_1 = 132 Pa, T_1 = 293 K, $M_{s(id)}$ = 4).

the interface tends to create a pressure gradient normal to the wall and therefore tends to give to the interface an increasingly convex form (Fig. 94).

Non-instantaneous opening of the diaphragm

In a 'real' shock tube, the shock wave is not instantaneously created but is formed by coalescence of the compression waves that arise during the progressive opening of the diaphragm, and thus the shock wave accelerates little by little (Appendix 8.1) a long time after the diaphragm has completely opened. Several metres of tube are often required to obtain a shock at constant speed.

Various models of this acceleration phase exist, which simultaneously take into account the mechanical opening process, the presumably isentropic and stationary flow through the aperture, the recompression stationary shock, and the successive compression waves that propagate downstream and progressively accelerate the shock wave.[147] Thus, knowing the total duration of the diaphragm opening, the initial pressure ratios of the driver gas and test gas, and their composition,[148,149] it is possible to describe the acceleration phase of the shock wave and the related properties of the flow. The shock wave is thus strongly accelerated close to the diaphragm until reaching a maximum speed; it then slows down slowly up to the ideal value[150] given by Eqn. (11.4) (Fig. 92). The acceleration phase is all the shorter as the ratio of the initial pressures is higher, the driver gas is lighter, and the total opening time t_{open} is shorter.

Combined effects of boundary layer and diaphragm opening

The result of the simultaneous action of the preceding effects is an initial acceleration of the shock wave followed by a continuous deceleration. The predominance

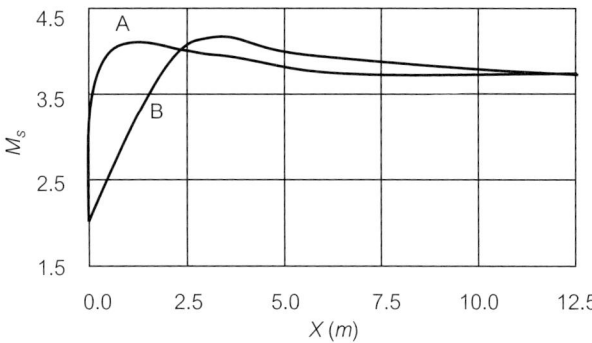

Figure 92. Influence of the opening time of the diaphragm on the evolution of the shock wave (air/air, $p_4/p_1 = 17\,700$, A: $t_{open} = 300$ ms, B: $t_{open} = 600\,\mu s$).

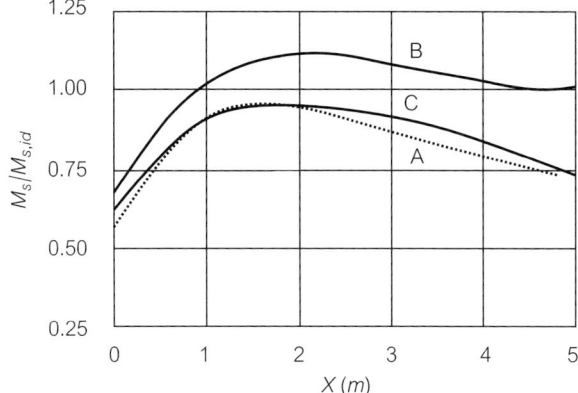

Figure 93. Example of a shock wave profile (air/air, $p_4/p_1 = 2134$, $p_1 = 526$ Pa, $t_{open} = 618$ ms). A: Experimental,[152] B: Computation without boundary layer,[150] C: Computation with boundary layer.[151]

of one or the other effect depends on the experimental conditions. An example of a computational result of spatial variation of the shock wave[151] is shown in Fig. 93 and compared with an experimental evolution.[152] We can observe a drastic variation due to low initial pressure and the use of air as a driver gas. This variation is naturally less marked for more 'usual' conditions (higher initial pressure, light driver gas, and so on).

Particular experimental aspects

Boundary layer and two-dimensional effects

Heat flux measurements by hot wire afford the possibility of distinguishing the passage of the shock wave, the interface, and more qualitatively, the boundary layer.[153] Thus, in Fig. 94, a sequence of the flow structure at three successive abscissas along a shock tube is presented: the convexity of the interface is clearly visible as well as its quasi-final shape in the limiting regime practically reached at the last abscissa. Moreover, the contact surface is not really a discontinuity but rather a turbulent mixing zone.

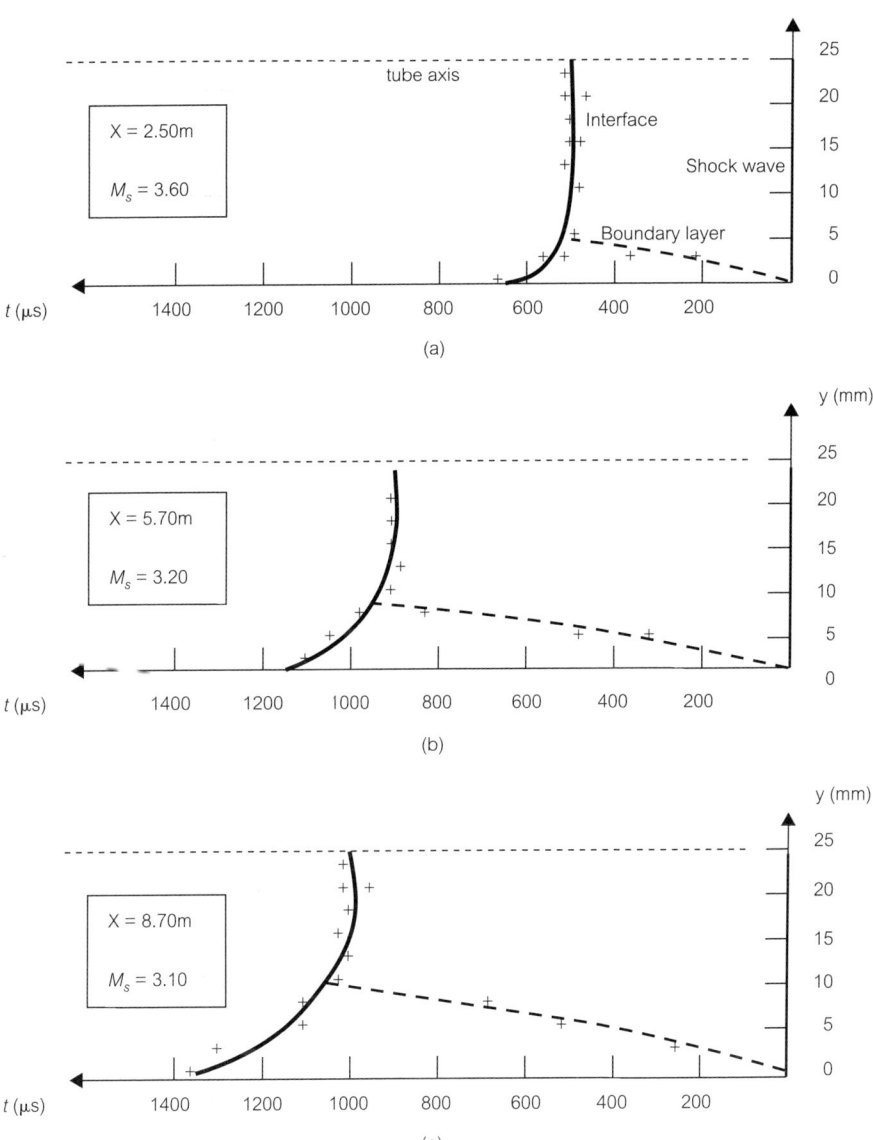

Figure 94. Evolution of the structure of a shock tube flow. Time origin: Passage of the shock wave (air/air; ideal $M_S = 3.80$; $p_1 = 130$ Pa).

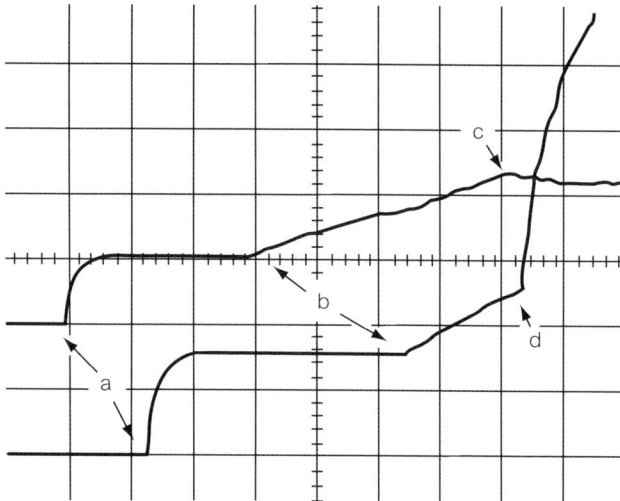

Figure 95. Oscillogram of wall temperature in a shock tube (200 μs/div.; a: Incident shock wave, b: Transition, c: Interface, d: Reflected shock).

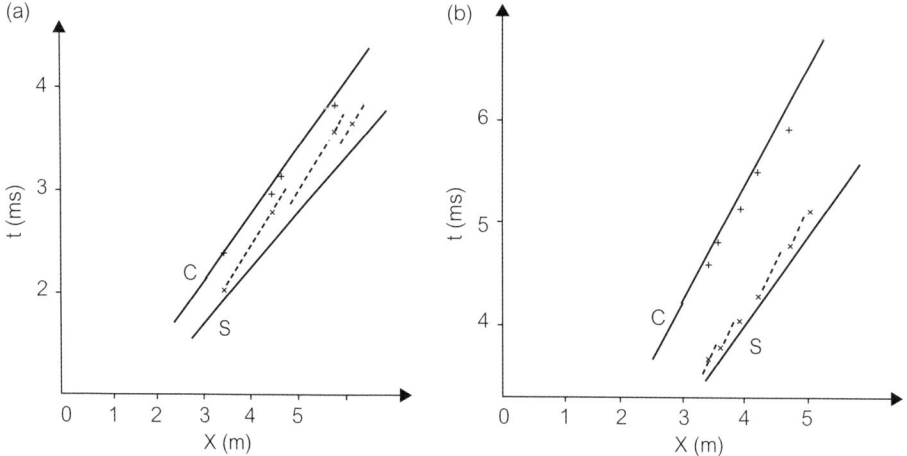

Figure 96. Experimental evolution of the transition in a shock tube (S: Shock wave, C: Contact surface (+), x: Transition). (a) $M_s = 5.7$; $p_1 = 921$ Pa; $Re_l = 64 \times 10^4$ m^{-1}; (b) $M_s = 3.8$; $p_1 = 6934$ Pa; $Re_l = 406 \times 10^4$ m^{-1}.

Transition in the boundary layer

The laminar–turbulent transition may be determined with thin platinum heat flux gauges[154,155] placed flush with the wall and therefore sensitive to the boundary layer regime (Fig. 95).

With several gauges placed along the shock tube, it is possible to follow the evolution of the transition point.[156] We observe that it strongly depends on the

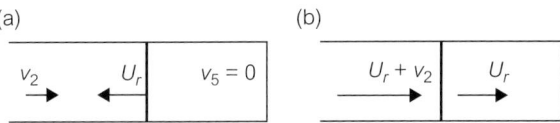

Figure 97. Reflected shock wave (a) Stationary coordinates, (b) Shock-fixed coordinates.

Reynolds number per unit length Re_l: thus, for low values of Re_l, the transition appears in the form of large structures called 'turbulent spots', which are regularly created along the tube and regress towards the contact surface (Fig. 96(a)). For higher values of Re_l, the size of the disturbances decreases, and their frequency increases, so that a compact transition front appears little by little and moves at the same speed as the shock (Fig. 96(b)).

A precise and general stability criterion is difficult to define, but regimes of global stability can, for each type of installation, be experimentally defined,[156] apparently independently from the shock wave Mach number.

11.2.3 Reflected shock waves

At the end of a closed tube, the shock wave is reflected and comes back into the gas already compressed and heated by the incident shock wave. Thus, there is a further increase in the temperature, pressure, and density of the test gas, which, in principle, affords more favourable conditions in which to start chemical processes. Moreover, in theory, the gas is without velocity behind the reflected shock waves (region 5, Fig. 97(a)).

Generalities

Of course, the gas parameters of region 5 of Fig. 97(a) (frozen or in equilibrium) may be computed using the usual RH relations (Eqn. (8.12)) across the reflected shock wave in a coordinate system fixed to this wave (Fig. 97). If the ideal gas model is used, we obtain the following analytical relations for the pressure and temperature ratios as functions of p_2/p_1. This pressure ratio itself is simply related to M_s (Eqn. (8.13)):

$$\frac{p_5}{p_2} = \frac{\gamma + 1 + (2 - p_1/p_2)(\gamma - 1)}{\gamma - 1 + (\gamma + 1)(p_1/p_2)} \quad \text{and} \quad \frac{T_5}{T_2} = \frac{p_5}{p_2}\left[\frac{\gamma + 1 + p_5/p_2(\gamma - 1)}{\gamma - 1 + (\gamma + 1)p_5/p_2}\right] \tag{11.5}$$

The velocity of the reflected shock wave U_r is given by the following relation:

$$U_r = 2U_s \frac{\gamma - 1 + p_1/p_2}{\gamma + 1 - (\gamma - 1)p_1/p_2} \tag{11.6}$$

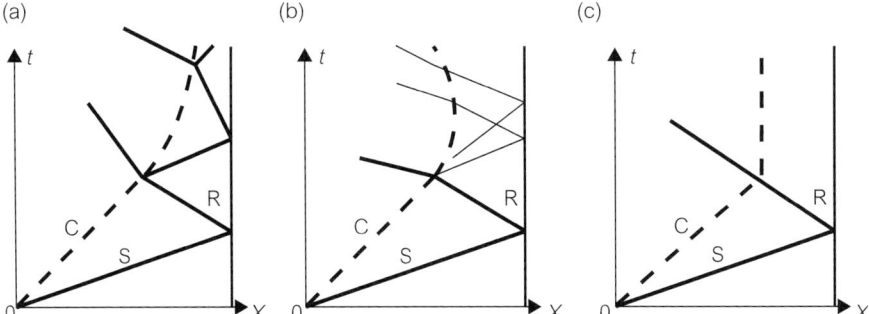

Figure 98. Wave systems generated by the interaction of the reflected shock and the interface. (a) Over-tailored case, (b) Under-tailored case, (c) Tailored case.

For a 'real' gas (in equilibrium), the values obtained for p_5 and T_5 are of course lower.

Disturbing effects

As in the case of the incident shock, various aerodynamic processes can disturb the test gas downstream from the reflected shock.

One of the disturbing effects relates to the interaction of the reflected shock and the contact surface: this interaction is summarized in Fig. 98.

Three interaction cases are possible: either the reflected shock is partially reflected on the interface in the form of shock (case a) or in the form of rarefaction waves (case b), or it crosses the interface without reflection (intermediate case c). In all three cases, a shock wave propagates into the driver gas. In the first two cases, which are the most common, the properties of the test gas downstream from the reflected shock are modified, and the useful test time can be strongly reduced, whereas the test time is theoretically very large in the third case, in which the interface comes to rest ('tailored case'). This occurs, however, only for quite precise initial conditions (for example, for $M_s = 6$ in the case of a gas pair He/N$_2$).[157] Nevertheless, if chemical processes are to be analysed behind the reflected shock, the tailored case represents the best experimental condition. Complete calculations corresponding to the above three cases can be carried out by using the methods presented in Chapter 8.

This scheme itself is disturbed by the presence of the boundary layer developing along the side walls downstream from the incident shock.[158] Under the action of the reflected shock, this boundary layer tends to separate from the wall (too-low stagnation pressure), and a gas 'bulb' is created on which the reflected shock adopts a structure in λ (Mach reflection, Chapter 8). This phenomenon (Fig. 99) is all the more accentuated as the gas atomicity is high (small γ value).

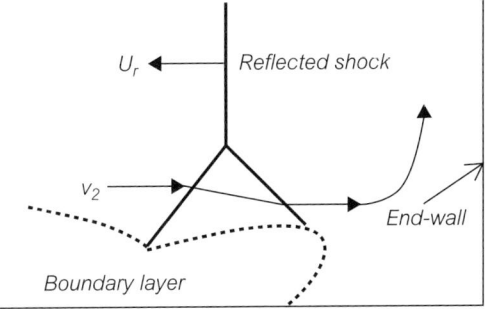

Figure 99. Scheme of interaction between the reflected shock and the boundary layer.

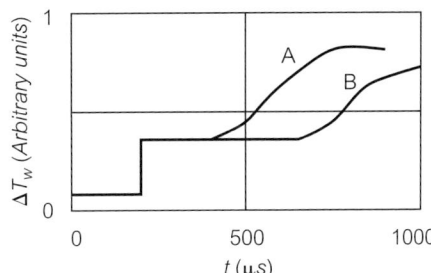

Figure 100. Evolution of wall temperatures at the end wall of a shock tube. A: Non-centred gauge, B: Centred gauge.

Then, the propagation of the reflected shock is obviously affected, and across the 'feet' of the λ shock, there remains a gas velocity component directed towards the end of the tube. This disturbs (primarily cools) the test gas. Moreover, as experimentally confirmed, when the reflected shock encounters the interface, the driver gas itself flows along the side walls, preceding the central part[159] (Fig. 100).

11.2.4 General techniques: configurations and operation

It is not the objective here to describe the technologies used in the construction of shock tubes, or to examine in detail the operation conditions. Instead, the intention is simply to present an outline of the existing or possible configurations, techniques, and devices used to obtain intense shock waves.

General design features

As described above, the 'simple' shock tube composed of two chambers represents the majority of existing installations. However, these installations differ according to the type of studies planned. Thus, for example, tubes with circular cross sections, which are easier to construct, lend themselves less easily to visualizations than those with square or rectangular cross sections. Moreover, these require a transition section between the high pressure (HP) chamber (of circular cross section for safety reasons) and the low pressure (LP) chamber.

As already discussed, the length of both chambers is important when optimizing the flow test time because of the various wave systems and end-wall reflections. Moreover, a third chamber, of large size and placed downstream from the low pressure chamber, is often used when experiments are limited to the flow downstream from the incident shock. This chamber (dump tank), separated by a second diaphragm from the driven section and in which high vacuum conditions prevail, makes it possible to obtain after the conclusion of experiments a low residual pressure, useful in the case of high pressure and/or combustible driver gases (H_2) or in the case of toxic test gases (CO, CN, and so on). The residual initial pressure, obtained after pumping, especially in the test section, must be sufficiently low (10^{-2}–10^{-4} Pa) so as to have no influence on the purity of the test gas, particularly for spectroscopic studies.

The diaphragms separating the HP and LP chambers are generally metallic: aluminium or copper for moderate pressures in the HP chamber (lower than 10^7 Pa), and steel for higher pressures. They can be composed of plastic for lower pressures. The metallic diaphragms are scored (cross-shaped scores with variable depths of 1/2 to 2/3 of their thickness) and calibrated to open at well-defined pressures. We thus obtain a dispersion of the incident shock Mach number which does not exceed 1%. Generally, for moderate pressures, the diaphragms break themselves by increasing the pressure of the HP chamber. For more precision, and especially in the case of very high pressures, a double-diaphragm system is used: it consists of a small chamber inserted between the HP and LP sections and in which the pressure is the intermediate of these two chambers. The sudden pumping of this chamber produces a precise and reproducible bursting of the diaphragms for given conditions.

Configurations: performances

Various possibilities exist to improve the performances of the simple shock tube, i.e. to increase the incident shock Mach number: these possibilities are briefly described below. Most of them are put into practice.

Area reduction close to the diaphragm

The HP chamber has a section larger than that of the LP chamber: there is a quasi-stationary expansion of the driver gas in the area transition zone, which increases the efficiency of the thrust. From a classical calculation,[141] we define a parameter g equal to

$$g = \frac{(p_4/p_1)_{A_4/A_1=1}}{(p_4/p_1)_{A_4/A_1}} \tag{11.7}$$

so that the tube with varying cross section is equivalent to a tube of constant cross section working with an initial pressure ratio equal to $g \cdot p_4/p_1$ and an initial sound velocity ratio equal to

$$a_4/a_1 \cdot g^{(\gamma_4-1)/2\gamma_4} \tag{11.8}$$

It should be noted that the increase in the Mach number is significant.

Double-diaphragm shock tube

A third section is added to the LP chamber and is used as the test section. It is separated from the intermediate chamber by a diaphragm on which the shock wave is reflected before breaking it, thus creating conditions of high pressure and temperature in the gas of this chamber. This gas is used as a driver gas for the test gas of the third section. In this case, the Mach number is significantly higher, but the test time is greatly reduced (Fig. 101).

It is also possible to use the expanded gas of the intermediate chamber as the test gas in the supersonic or hypersonic regime.[160]

Combustion shock tube

The increase of the sound speed in the driver gas can be obtained not only by using a light gas but also by raising its temperature. One effective way is to create combustion in the HP chamber.

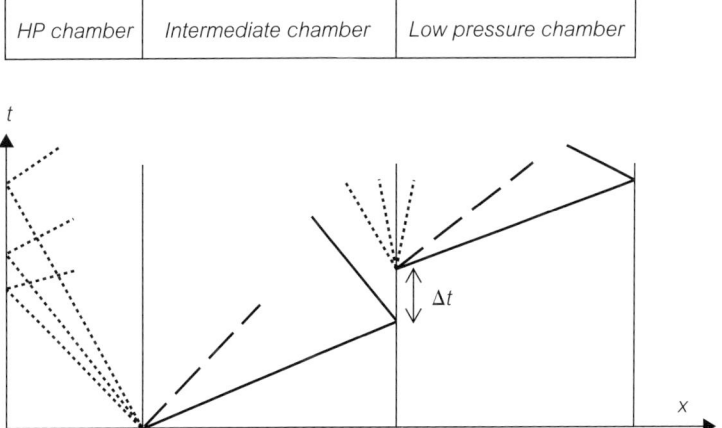

Figure 101. Principle of double-diaphragm shock tube.

This combustion is generally achieved by using a stoichiometric mixture of hydrogen and oxygen diluted in helium (approximately 70%). The main difficulty is in obtaining a uniform combustion without detonation; this is generally realized with a significant number of spark plugs arranged in spiral along the HP chamber. The gain in Mach number, however, is partially compensated by a stronger deceleration of the shock wave, caused by the sharp pressure fall after the combustion and also sometimes by a rebound of the 'petals' of the diaphragm on the side walls.

An alternative solution is the precise creation close to the diaphragm of a detonation wave that propagates upstream in the HP chamber: this results in a better uniformity for the pressure and temperature in the driver gas after the diaphragm ruptures.[161]

Free-piston shock tube

The fast compression of a light gas is also a means for increasing the pressure and temperature of this gas, used as a driver gas. This compression is carried out by a piston launched at high speed in a tube serving as a compression chamber: the compressed hot gas ensures that the diaphragm ruptures. This method is undoubtedly the most efficient process of creating a shock wave of high intensity.[162]

A diagram of this device is shown in Fig. 102. In a first chamber (tank R), a gas (generally air) is compressed up to several hundred atmospheres and, owing to a double-diaphragm system D1–D2, pushes a piston P (10 to 500 kg), which compresses the driver gas (generally helium or a helium–argon mixture) of the HP chamber. The diaphragm D3, which must be initially calibrated and is located at the end of this chamber, is then ruptured, creating a shock wave in the LP chamber. After the rupture, the piston continues to move and maintains a pressure sufficiently high to delay the propagation of rarefaction waves towards the LP chamber.

Of course, the piston must be rapidly stopped for safety reasons and also to avoid a rebound.[163–165]

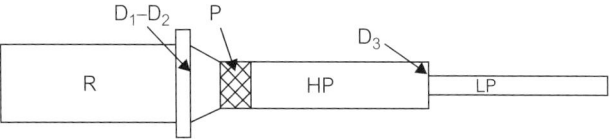

Figure 102. Diagram of a free-piston shock tube.

In addition, special configurations of the piston are used to attain a continuous rise in pressure at the end of the HP chamber and thus to obtain a reproducible rupture of the D3 diaphragm. An example of such an operation, in the form of an x, t diagram (trajectory of the piston, wave systems),[166] is presented in Appendix 11.2.

Important shock Mach numbers (10–25) are thus generated in gases or gas mixtures representative of various planetary atmospheres (Chapter 12).

11.2.5 General methods of measurement

Only those methods of measuring usual flow parameters such as pressure, density, heat flux, and traditional visualization techniques are briefly discussed here, and they are illustrated by a few examples. Methods specific to non-equilibrium flows are described in Chapter 12.

Pressure measurements

The range of the pressures to be measured is very important, from a few hundred pascals downstream from relatively weak shocks that propagate in gases with low initial density, up to 10^7–10^8 Pa downstream from reflected shocks.

Piezoelectric-type gauges with response times often shorter than 1 μs are generally used, either directly flush with the wall (for the lowest pressures and for short-duration flows, or inside a cavity connected to the wall by a small pipe (for the highest pressures or in corrosive or fluctuating atmospheres). Two extreme examples are given in Fig. 103 (a) and (b).[167,168]

Heat flux measurements

Measurements are generally carried out with thin metallic films (platinum), deposited on an insulating support whose resistance varies with temperature. Thus, after calibration, the surface temperature of the support may be obtained (Appendix 7.2). Mounted flush with the wall of a shock tube, if the boundary layer is laminar and non-reactive, the temperature rise is quasi-constant downstream from the incident shock (Appendix 10.3) and downstream from the reflected shock on the end wall of the tube, before the arrival of eventual disturbances (Fig. 100). On a body placed in the steady flow, where the heat flux is constant (during the useful test time), the temperature rise is parabolic (Fig. 104).

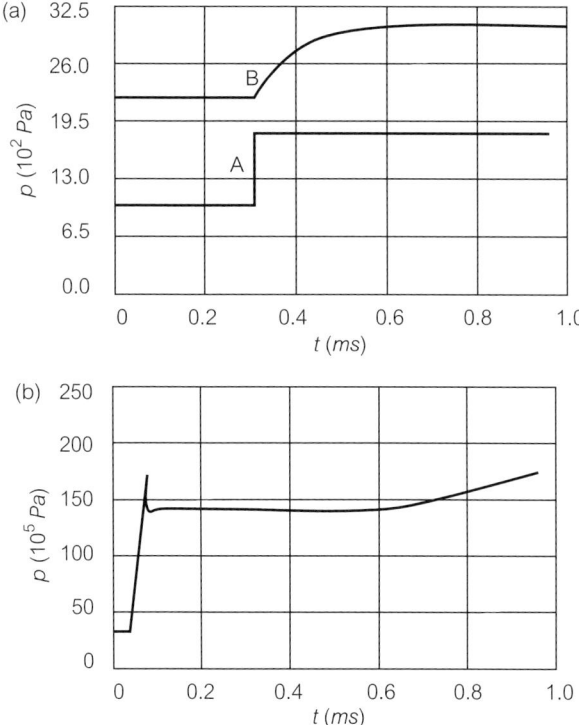

Figure 103. (a) Pressure records on shock tube wall (with filter). A: Gauge flush with the wall, B: Gauge in cavity. (b) Pressure record at the end wall of a shock tube (with filter).

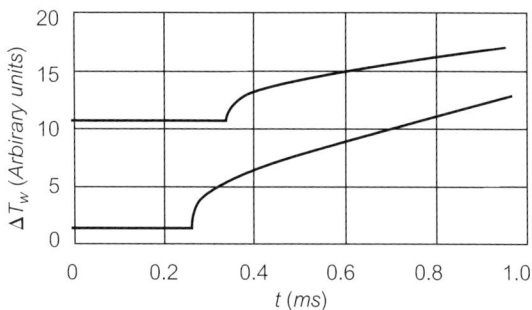

Figure 104. Temperature records along a cone in a shock tube flow.

For very important heat fluxes, special thermocouples are used, which give similar temperature signals. Other methods are also used (Appendix 11.3).

Density measurements

The density change across the shock ($\rho_2 - \rho_1$) and its evolution, for example in a relaxation zone, is measured with a fringe system created by the interference

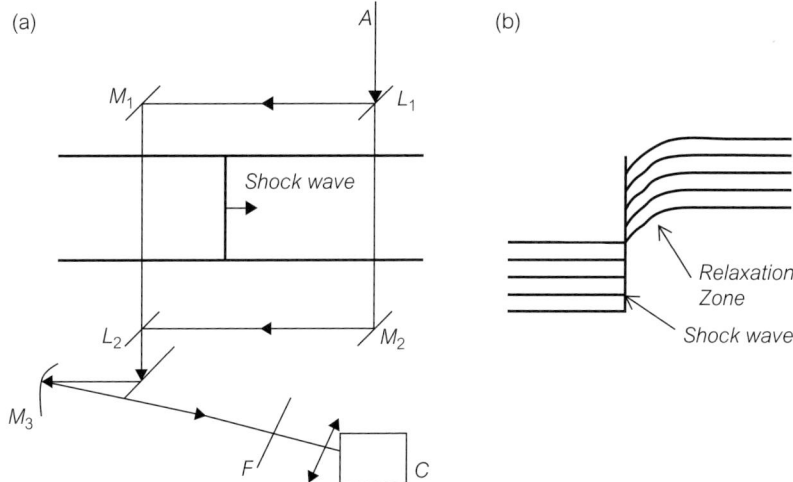

Figure 105. (a) Scheme of a Mach–Zehnder interferometer. A: Light source, L_1, L_2: Semi-silvered mirrors, M_1, M_2, M_3: Mirrors, F: Slit, C: Camera; (b) Fringe system.

of light beams crossing the tube (Fig. 105(a)). If this change corresponds to a displacement of k fringes, we have for a tube of width L and wavelength λ:

$$k\lambda = (n_2 - n_1) L \tag{11.9}$$

The index of refraction n is proportional to the density ρ, that is:

$$n - 1 = K\rho \quad \text{(Gladstone–Dale relation)} \tag{11.10}$$

where K depends only on the nature and temperature of the gas and on the wavelength.

Then we have

$$\rho_2 - \rho_1 = \frac{k\lambda}{KL} \tag{11.11}$$

The interferometer that is generally used is of Mach–Zehnder type (Fig. 105(a)), and the interfering beams cross the tube, one in the test gas flow, the other in the undisturbed zone. The quantity λ/KL may be determined from a preliminary calibration. Interferograms are obtained either by using a monochromatic light source of short duration (0.1–1 µs) or by using a continuous source and a high-speed sweeping camera with a slit perpendicular to the flow in order to obtain y, t diagrams (Fig. 105(b)).

Visualizations

Visualizations can be carried out by interferometry, as described above. They are then quantitative (density measurements). More qualitatively, the significant and different deviation of light rays crossing the shock tube by flow regions of different densities affords the possibility of visualizing those flow areas presenting density gradients, in particular the shock waves. This is the schlieren technique, also associated with fast photography or cinematography (x, t or y, t diagrams). Examples of shock tube flows[169] are thus represented in Figs 106–109.

In Fig. 106, the flow around a wedge is shown at four successive instants after the passage of the incident shock wave: (a) during the phase of flow establishment, (b) during the stationary phase of the 'hot' test gas flow, (c) on arrival of the mixing zone (interface), and (d) in the (turbulent) expansion of the driver gas, where we observe that the local Mach number is much higher than during the test-gas flow.

Two other examples are shown in Fig. 107, where the incident shock wave is propagating along a striate dihedron, and in Fig. 108, which is a schlieren photograph of the detached shock wave (and relaxation zone) in front of a cylinder.

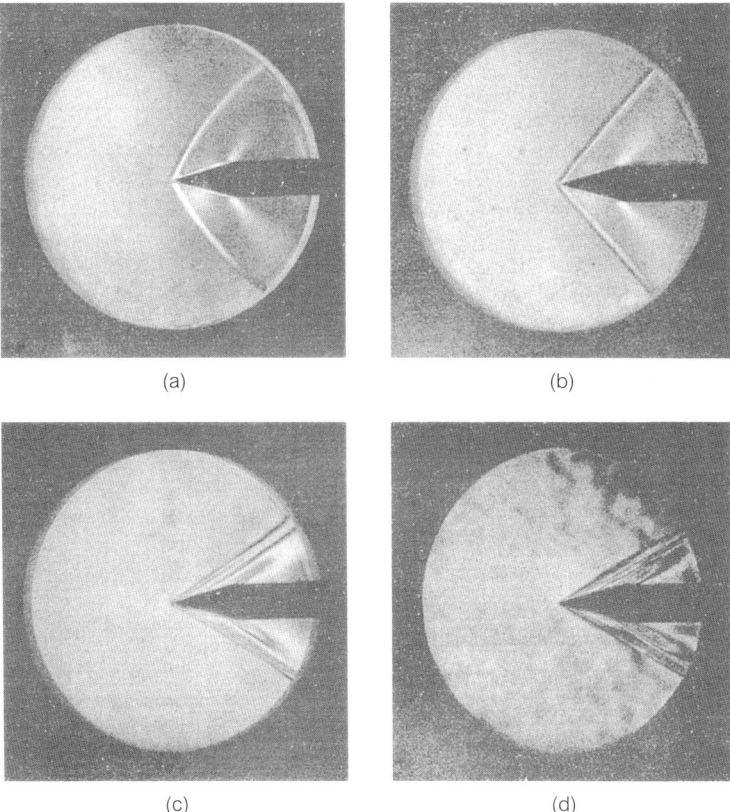

Figure 106. Flow around a wedge (semiangle 15°) (air/air, $M_s = 3.86$, $p_1 = 130$ Pa) (θ: Instant after the passage of the shock wave). (a): $\theta = 23$ μs, (b): $\theta = 100$ μs, (c): $\theta = 700$ μs, (d): $\theta = 1300$ μs.

Figure 109 illustrates an experimental x, t diagram of the shock wave reflection at the end of a tube: regular reflection for a shock wave in argon (Fig. 109(a)) and Mach reflection for a shock wave in carbon dioxide (Fig. 109(b)). This last type of reflection, as discussed above, results from the separation of the boundary layer.

A simpler method consists of using the self-luminosity of the high-temperature flows: an example is provided in Fig. 110 (a) and (b), representing x, t diagrams of the reflection of a shock wave and its interaction with the contact surface (over-tailored case).

Figure 107. Incident shock wave propagating along a striate wedge (Carbon dioxide, $M_s = 4.0$, $p_1 = 130$ Pa, $M_2 = 2.35$).

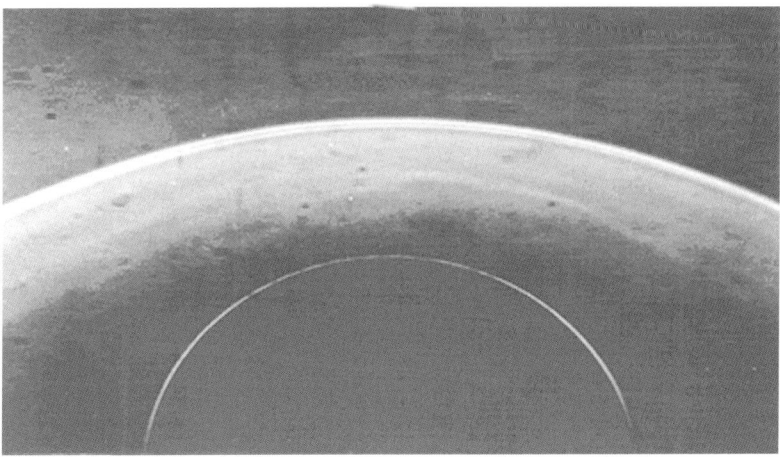

Figure 108. Carbon dioxide flow around a cylinder (Conditions of Fig. 107).

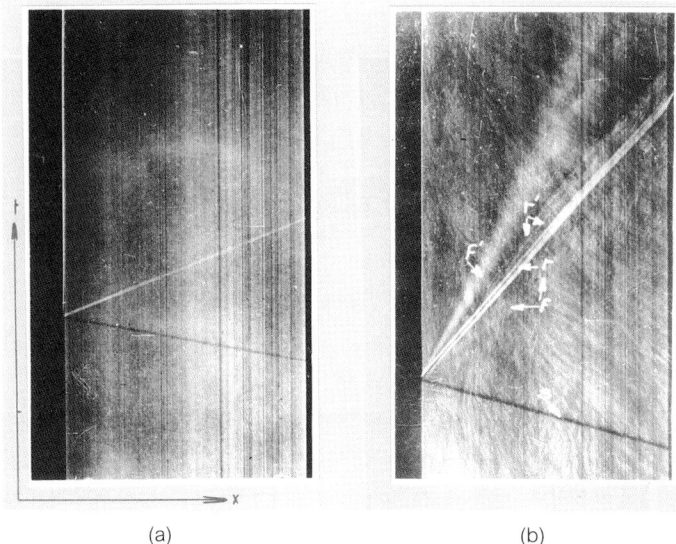

(a)　　　　　　　　　　　　(b)

Figure 109. Reflection of a shock wave at the end wall of a shock tube (x, t diagram). (a): Argon, $M_S = 3.30$, $p_1 = 790$ Pa; (b): Carbon dioxide, $M_S = 3.40$, $p_1 = 1315$ Pa. s: Incident shock, r: Reflected shock, r': Secondary shocks, f: Front of the bifurcated shock, f': Rear of the bifurcated shock.

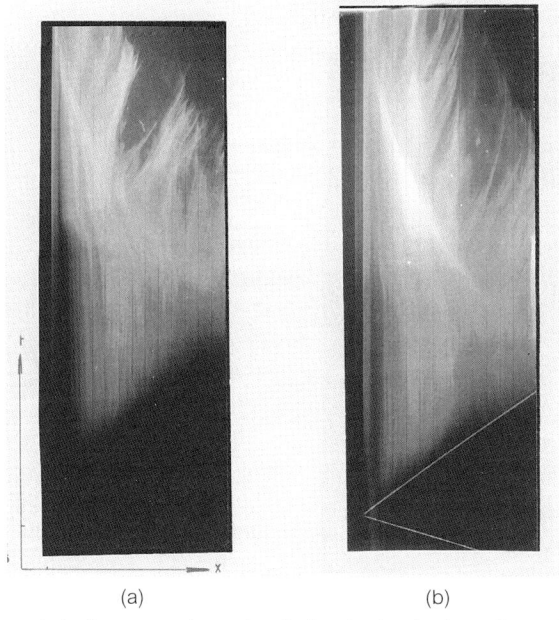

(a)　　　　　　　　　　　　(b)

Figure 110. Reflected shock wave at the end wall of a shock tube (x, t diagram). (a): Argon, $M_S = 5.20$, $p_1 = 395$ Pa; (b): Argon, $M_S = 6.10$, $p_1 = 395$ Pa] (Incident wave: right→left; x: 4 cm/div., t: 100μ s/div.).

11.3 The hypersonic tunnel

11.3.1 Generalities

As specified in the introduction of this chapter, non-equilibrium flows can be generated with a high-pressure hot gas, assumed to be in equilibrium, then quickly expanded in a nozzle in order to obtain a hypersonic flow at low TR temperature but in chemical and/or vibrational non-equilibrium.

The processes used to obtain a gas at high temperature are generally of two types of heating: by a turning arc in a pressurized atmosphere[170] or by a shock wave in a shock tube.[157,171] Installations with a continuous arc also exist.

In the arc tunnels, it is possible to obtain very important reservoir temperatures and very high pressures in an important gas mass. The gas is then expanded in a nozzle until reaching hypersonic Mach numbers and also important Reynolds numbers with a rather long test time (about several milliseconds): the flow can thus more or less satisfy similarity criteria for studies of models, probes, vehicles, and so on. The main disadvantage is the relatively high rate of pollution of the flow—and the significant cost of the installation.

11.3.2 The hypersonic shock tunnel

Principle

In the latter part of this chapter, we describe in more detail the principle of the shock tunnel derived from the shock tube concept. Thus, the gas downstream from the incident shock wave can be directly expanded in a nozzle placed at the end of the shock tube. The flow can then become hypersonic on exiting the divergent part of the nozzle. However, this process is seldom used for various reasons, primarily because of the very short test time. It is better to use the gas downstream from the reflected shock as a reservoir gas. In theory, it is at rest, in equilibrium, and at high temperature and pressure during a relatively long time in the 'tailored' conditions (Section 11.2.3). Thus, at the exit of the nozzle, we can expect test times of about one millisecond. The main drawbacks arise from the possible pollution of the test gas by the premature arrival of the driver gas (Section 11.2.3) and from the 'starting process' of the nozzle lasting a non-negligible time. Thus, the duration of the 'useful' (stationary) test time is only a few hundred microseconds, even for relatively large installations. A diagram of a free piston shock tunnel is given in Fig. 111.

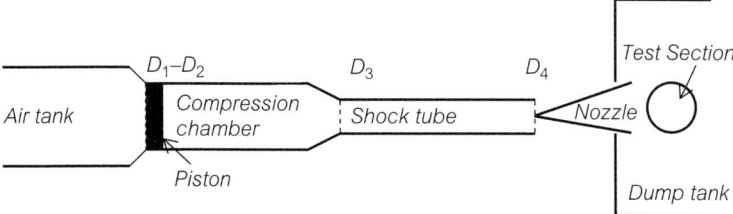

Figure 111. Scheme of a free-piston shock tunnel.

Operation

At the rupture of the diaphragm D4, a shock wave propagates in the nozzle, compressing the residual gas of the expanding chamber (dump tank), initially at very low pressure. Another shock wave, known as a secondary shock, is formed downstream to ensure the continuity of pressure, with the test gas expanding into the nozzle. This shock interacts with the very thick boundary layer of the residual gas (Mach reflection) but is pulled downstream by the test-gas flow, the stationary expansion phase of which constitutes the useful part of the flow, analysed for example in Chapters 9 and 10.

Traditional nozzles, known as 'contoured' or 'adapted' nozzles, used in supersonic ideal gas flows are difficult to use in hypersonic non-equilibrium flows because they are in fact 'adapted' only for conditions close to one single point of operation. Thus, they are generally replaced by nozzles that include a conical divergent part. These nozzles require a suitable convergent–divergent connection but are simpler to build and more easily calculable. However, they also suffer from some defects (Appendix 11.6).

Calibration

Measurements of stagnation pressure (Pitot pressure) with small probes placed normally to the flow are used to validate flow computations and to determine the non-dissipative part of the test section, in particular at the exit of the nozzle, where the models are placed. An example[128] of the distribution of Pitot pressure is shown in Fig. 112. This pressure is practically insensitive to non-equilibrium and physical models. These measurements may also give an idea of the effective duration of the useful test time.

Measurements of static pressure and wall heat flux generally complete our knowledge of nozzle flow. Thus, an example of the distribution of wall pressure for an airflow along a hypersonic nozzle is shown in Fig. 113[172] and is compared with the pressure calculated by taking into account non-equilibrium and by assuming that the boundary layer is laminar (Chapter 10). The agreement

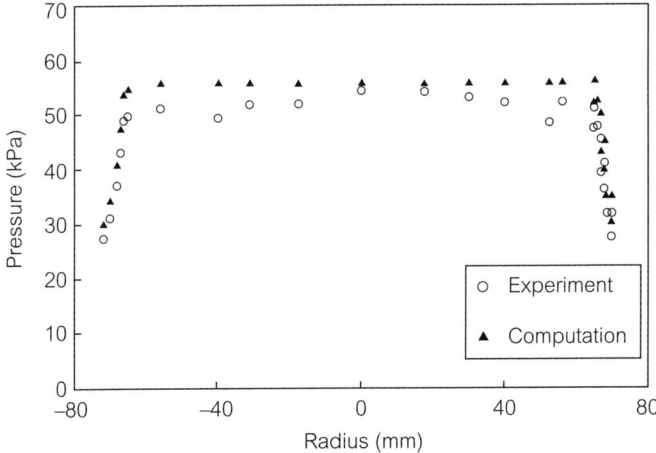

Figure 112. Pitot pressure distribution at the exit of a nozzle ($p_0 = 18$ MPa, $T_0 = 6000$ K, $M_\infty = 7.2$).

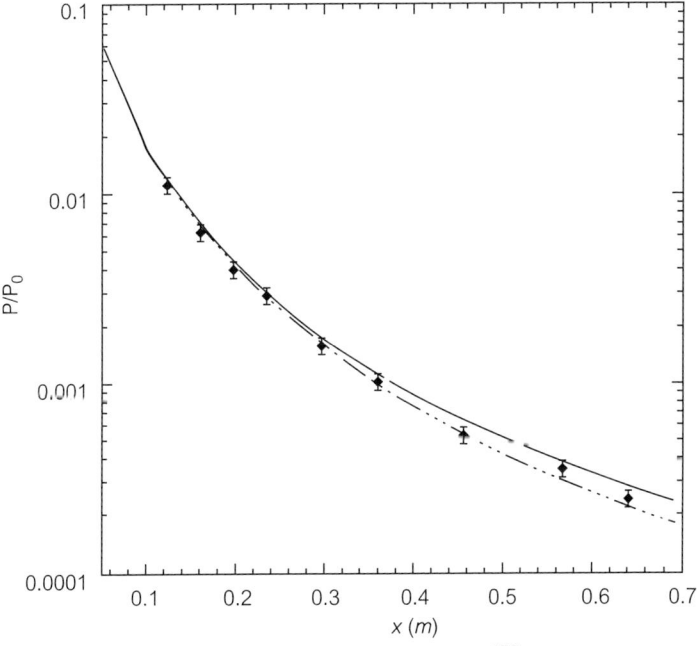

Figure 113. Static pressure distribution along a hypersonic nozzle[172] (Conical nozzle, length: 70 cm, semiangle: 8°, $A/A_c = 164$) (Air, $M_s = 8.75$, $p_0 = 45 \times 10^6$ Pa, $T_0 = 6500$ K). ●: Experimental points, ———: Computation with vibrational coupling, —··—: Computation without coupling.

appears reasonable, but it does not allow us to solve the problem of the influence of vibrational coupling, which, in any case, appears to be barely significant (Chapter 5).

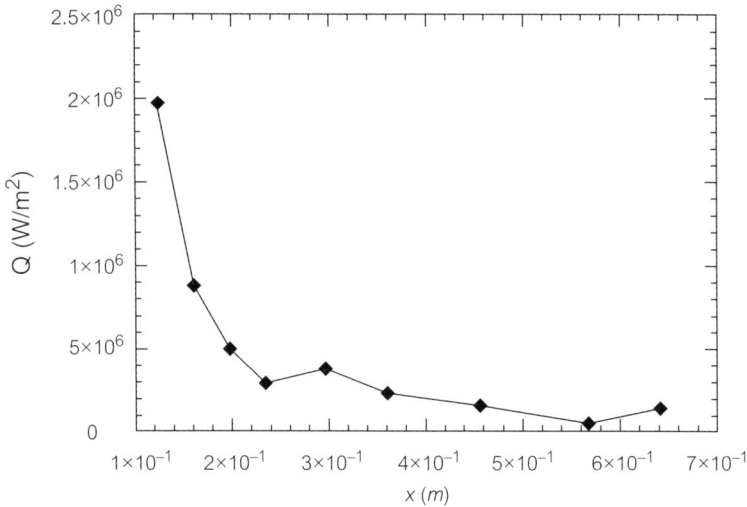

Figure 114. Wall heat flux distribution along a hypersonic nozzle[172] (Nozzle of Fig. 113; air; $M_s = 5.76$, $p_0 = 23 \times 10^6$ Pa, $T_0 = 3530$ K). •: Experimental points.

Similarly, an example of the wall heat flux distribution along the same nozzle is shown in Fig. 114. The flux decrease observed along the nozzle approximately corresponds to that computed with a laminar boundary layer. However, in the distribution, we observe local increases similar to those observed in a shock tube and which correspond to the appearance of turbulent spots that seem to quickly spread: this would also correspond to the longitudinal variation of the Reynolds number, which, initially growing (because of the increasing distance), passes by a maximum and then decreases very quickly, mainly because of the decreasing density, so that it is difficult to find here a definite transition.

Appendix 11.1 Experiments in real flight

Until now, most experiments in real flight (shuttles) have mainly consisted of measuring pressure and heat flux so as to validate the calculation methods presented in Chapter 10 and to test the catalytic properties of materials for thermal coating.

As an example, Fig. 115 shows the characteristics of the STS2 shuttle flight (altitude, speed, and so on) with time during the re-entry phase into Earth's atmosphere,[136] where $t = 0$ corresponds arbitrarily to the altitude of 120 km.

Surface temperature measurements along the symmetry line of the windward side of the shuttle were carried out during this re-entry. Heat fluxes were deduced

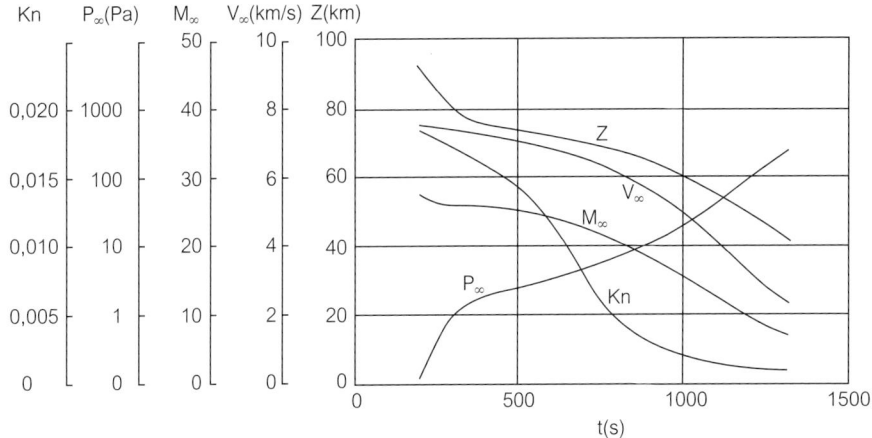

Figure 115. Re-entry trajectory: Shuttle flight STS2 (Kn: Knudsen number, Z: Altitude).

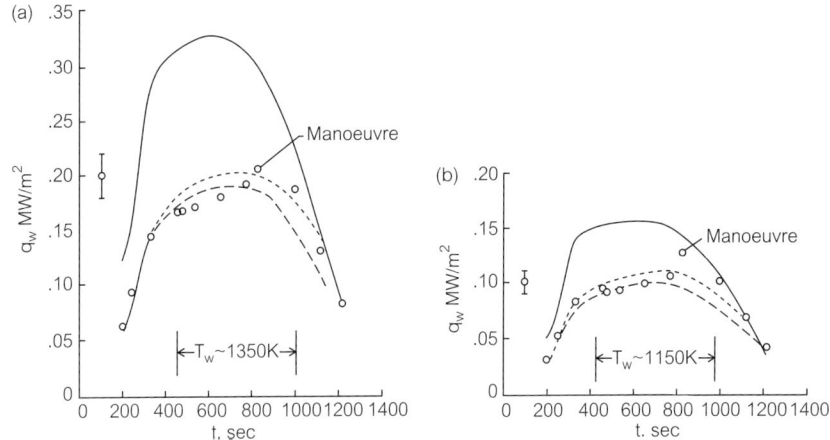

Figure 116. Time evolution of the heat flux along the symmetry line of the windward side of the shuttle (STS2) during the re-entry phase. (a) x/L = 0.025; (b) x/L = 0.10; ——— Equilibrium (computed), ··· Non-equilibrium (k_{Aw} = 100 cm/s), --- Non-equilibrium (k_{Aw} = 200 cm/s), o: Experimental points.

from these measurements and compared with calculations similar to those of Chapter 10 and made with different values for the recombination-rate coefficient of oxygen atoms at the wall, k_{Aw}. An example of the time evolution of these fluxes at two different x coordinates is shown in Fig. 116. The experimental values are in agreement with those calculated with k_{Aw} = 100 cm/s (value before correction of trajectory) and confirms the good behaviour of the protection material (reaction-cured glass (RCG) coating) used. However, a phenomenon of ageing was observed during the following flights.

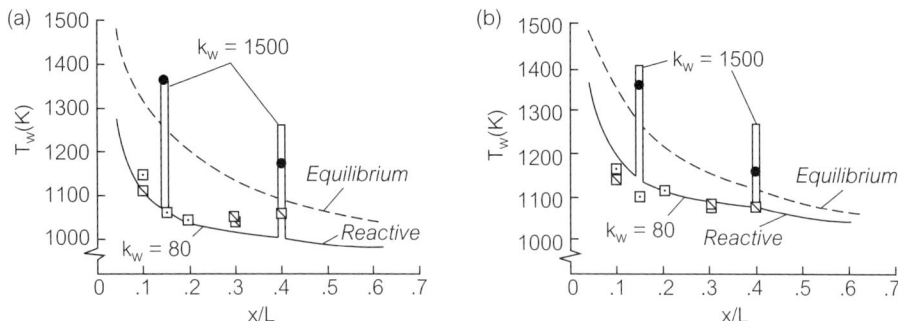

Figure 117. Wall temperature distribution along the symmetry line of the windward side of the shuttle. (a): $t = 450$ s; (b): $t = 650$ s. □: Experimental points (RCG coating); •: Experimental points (catalytic coating).

We can also notice the specific evolution of heat fluxes showing the competition between the altitude (or density) and the velocity, and highlighting a critical maximum value at an altitude of 60–65 km ($M_\infty \sim 20$–25).

Another experiment consisted of replacing the preceding coating of two 'tiles' by a fully catalytic coating placed at the reduced x coordinates 0.15 and 0.40. The distribution of the wall temperature along the intrados, measured and calculated, is shown in Fig. 117 at two instants of the re-entry. A very strong increase in the temperature at these two x coordinates may be observed that is even higher than that given by an equilibrium calculation, because of the 'excess' of non-recombined atoms upstream from these tiles.

Appendix 11.2 Optimum flow duration in a shock tube

Within the framework of the ideal shock tube theory, it is possible to determine the maximum duration of a test-gas flow, taking into account the constraints imposed by the length of the HP and LP chambers, the nature of the driver gas and of the test gas, and so on, or conversely, to determine these quantities by taking into account the desired time of measurement (test time). This is obviously independent of the other 'disturbing' phenomena (boundary layer, diaphragm, transition, and so on).

Figure 118 is a diagram of the interaction of various wave systems in a shock tube, where τ_m represents the maximum duration of the test-gas flow at an abscissa X_m, independently of the possible arrival of the reflected shock.

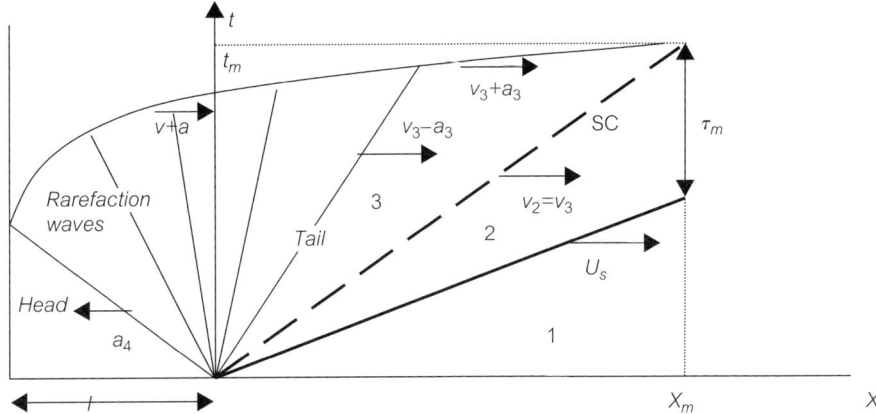

Figure 118. Wave diagram for determining the maximum test-gas flow duration.

With the method of characteristics (Chapter 8), we can show[100] that the maximum useful test-gas flow ends at a time t_m equal to

$$t_m = \frac{2l}{a_4}\left[1 - \frac{\gamma_4 - 1}{\gamma_1 + 1}\frac{a_1}{a_4}\left(\frac{M_s^2 - 1}{M_s^2}\right)\right]^{-\frac{\gamma_4+1}{2(\gamma_4-1)}} \quad (11.12)$$

The corresponding abscissa X_m and the maximum flow duration τ_m are then respectively equal to

$$X_m = t_m \frac{2a_1(M_s^2-1)}{(\gamma_1+1)M_s^2}$$
$$\tau_m = t_m\left[1 - \frac{2(M_s^2-1)}{(\gamma_1+1)M_s^2}\right] \quad (11.13)$$

Thus, for given conditions, a length of the LP chamber longer than X_m is useless (without reflected shock) and even prejudicial.

Appendix 11.3 Heat flux measurements in a shock tube

A commonly used technique of heat flux measurements is the thin film heat gauge method (Section 11.2.6), combining good sensitivity and short response time. However, in the case of constant heat flux (for example in the unperturbed test-gas flow), other methods, presenting specific advantages, may be used. This is the case for the techniques of 'hot wire' and luminescent sulphides briefly presented below.

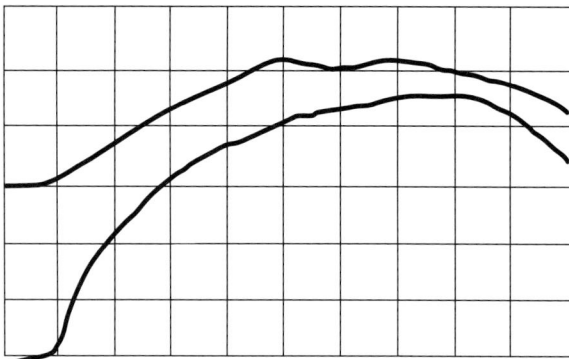

Figure 119. Example of hot wire records (200 μs/div.) (driver gas: nitrogen; test gas: argon) (p_1 = 130 Pa, T_1 = 293 K, M_S = 3.30, X = 5.70 m). Upper signal: Hot wire at the tube centre (y = 2.50 cm), Lower signal: Hot wire near the tube side-wall (y = 0.3 cm).

When a thin wire, crossed by a constant (small) electric current, is placed in the hot flow of a shock tube, its temperature increases, and the thermal balance of the wire, assumed to behave as a calorimeter, gives an exponential increase in temperature until an 'equilibrium' value is obtained. As the corresponding time constant is generally much longer than the test-gas flow duration, the voltage signal (proportional to the temperature) is quasi-linear (Fig. 119), and the heat flux may be deduced.[153] Another advantage of this technique is the possibility of testing the local structure of the flow. Thus, it is possible to detect the arrival of a shock wave, a contact surface, or even (qualitatively) a boundary layer. The flow structure presented in Fig. 94 derives from hot wire measurements.

Another less conventional method consists of using the properties of luminescent materials such as zinc sulphides, for which the luminescent signal depends on their temperature.[173] This property may be evidenced by depositing a thin layer of sulphide upon a platinum thin film which is used as a heat source and a thermometer. Excited by a continuous UV light source, the sulphide emits a luminescent line (3500 Å for SZn (Cu)). The platinum film is crossed by a constant current for 1 ms, and the sulphide layer gives an emission line, the intensity of which is shown in Fig. 120 and compared to the film temperature signal. Thus, a calibration may be operated.

An application to the measurement of heat fluxes along a hemisphere-cylinder is made in a shock tube. Seven measurement points are placed along the model from the stagnation point ($\theta = 0°$) up to $\theta = 90°$, and the (relative) heat flux values are shown in Fig. 121.

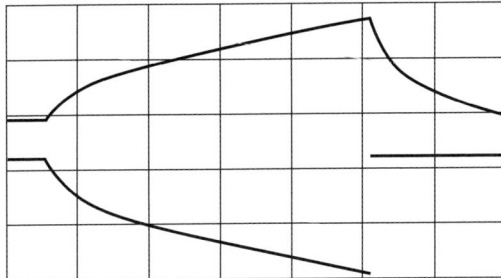

Figure 120. Luminescent signal of ZnS(Cu) (upper signal) and temperature signal of the film (lower signal) (Wall heat flux: 100 W / cm^2, 2 ms/div).

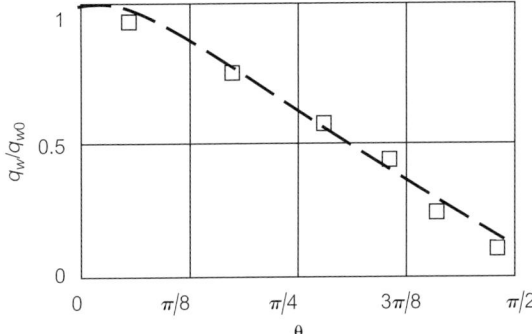

Figure 121. Distribution of heat transfer along a hemisphere-cylinder (Driver and test gas: Air, $p_1 = 260$ Pa, $M_s = 3.02$, $R = 2$ cm).

Appendix 11.4 Shock–interface interactions

As already discussed, in reality a contact surface is a turbulent mixing zone ('contact zone') between driver gas 3 and driven gas 2 (Fig. 89). The width and heterogeneity of this zone depends on the difference of the densities of both gases (in fact on the Atwood number $(\rho_3 - \rho_2)/(\rho_3 + \rho_2)$. Independently of the contact-surface–boundary-layer interaction (Section 11.2.2), the shape of the contact surface front changes along the tube because of the growth of turbulent instabilities within the mixing region,[174,175] starting from initial perturbations due, for example, to diaphragm or boundary-layer effects.

When perturbed by the reflected shock wave, the contact zone undergoes a deceleration (Section 11.2.3) and becomes more unstable depending on the importance of the deceleration, so that after interaction the mixing zone thickens.[176]

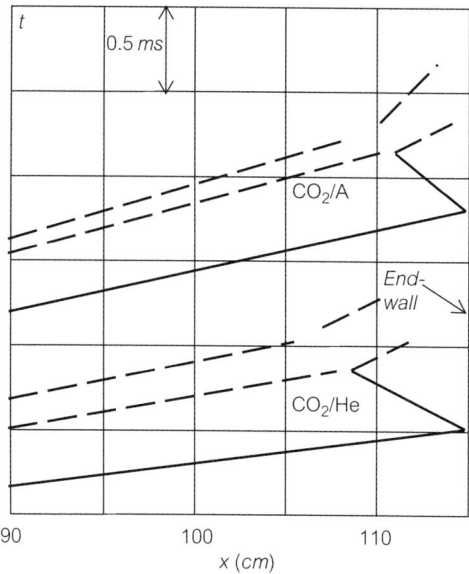

Figure 122. Experimental (x,t) diagram of shock–interface interaction (p_{4CO_2} = 1520 Pa, p_{1A} = p_{1He} = 925 Pa, M_{sAr} = 5.83, M_{sHe} = 3.20). ———: Exp. shock wave, – – –: Exp. interface.

These processes are observed in experiments carried out in a double-diaphragm shock tube (Section 11.2.4) with the gas combinations $H_2/CO_2/He$ and $H_2/CO_2/Ar$ and identical initial conditions.[177] Growth of the contact zone is determined by measuring the intensity of the IR emission (centred on the 4.3 μm wavelength) arising from the asymmetric vibrational mode of CO_2, experimental results of which are presented in Fig. 122.

Thus, as expected, we observe that the contact zone behind the incident shock wave is thicker, and more turbulent and heterogeneous, in the CO_2/He case compared to the CO_2/Ar case. After the interaction with the reflected shock, the contact zone thickens more in the CO_2/Ar case than in the CO_2/He case, because the deceleration is stronger (we are close to the 'tailored' conditions); however, the CO_2/He contact zone remains as unstable as before the interaction.

Note that processes related to interface instability may be important in problems of laser-induced nuclear fusion.

Appendix 11.5 Operation of a free-piston shock tunnel

Analysis of the operation of a hypersonic free-piston shock tunnel can be carried out by starting from the one-dimensional and unsteady Euler equations applied

APPENDIX 119.5 OPERATION OF A FREE-PISTON SHOCK TUNNEL

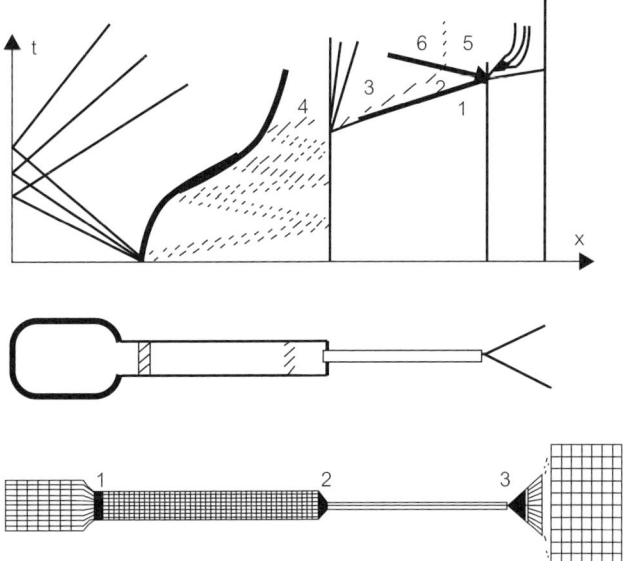

Figure 123. Scheme of operation and computation of a free-piston shock tunnel.

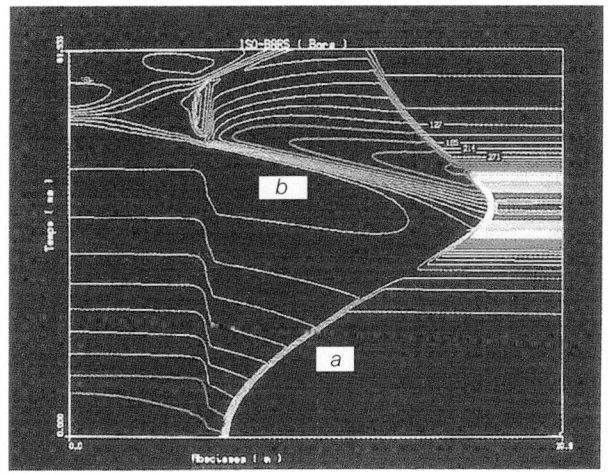

Figure 124. (x, t) Pressure diagram in a closed compression chamber (a: Piston trajectory, b: Rear shock wave).

to each element of the tunnel. It is indeed misleading and expensive to use the Navier–Stokes equations to gain an overview of the operation of such a tunnel.

An example of the result of such a calculation[166] is presented below, corresponding to the diagram of Fig. 123. We consider here the successive phases of the running process and analyse each in turn.

358 CHAPTER 11 FACILITIES AND EXPERIMENTAL METHODS

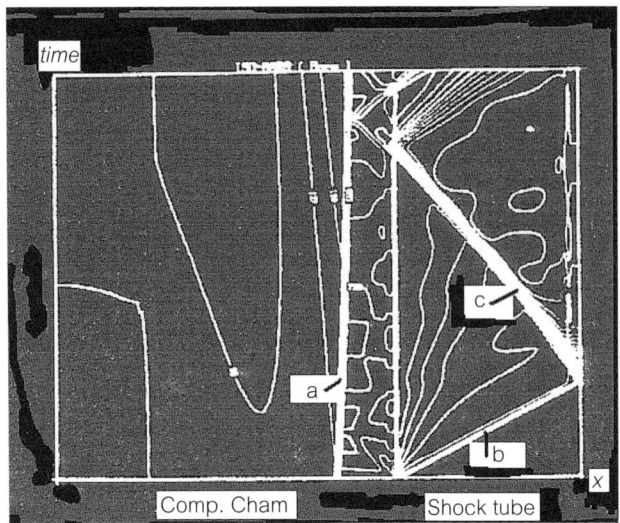

Figure 125. (x, t) Velocity diagram in a compression chamber and a shock tube (a: Piston, b: Incident shock wave, c: Reflected shock wave).

Thus, considering first the phase of compression by the piston, without any rupture of diaphragm 2 of Fig. 123, an x, t diagram of the isobars in the compression chamber can be obtained (Fig. 124). This diagram shows the trajectory of the piston, including its rebound close to the diaphragm, the strong compression in front of this piston, and the shock wave being formed at the rear during the rebound phase.

Similarly, if the rupture of diaphragm 2 of Fig. 123 is taken into account, we obtain an x, t diagram of isovelocity lines as presented in Fig. 125, which shows the piston slowing down (without rebound) after the rupture of diaphragm 2, with the shock wave propagating in the shock tube and reflecting on diaphragm 3.

Appendix 11.6 Source flow in hypersonic nozzles

The use of nozzles with a divergent conical part is justified by a straightforward calculation of the corresponding hypersonic flow (Chapter 10). However, they are not free from defects, the main one leading to a 'conical' (or 'spherical') flow. This defect is called the 'source effect', as though the flow in the divergent part of the nozzle is from a source point. Thus, the flow quantities vary not only along the nozzle but also in the test section and therefore along a body that could be placed there.

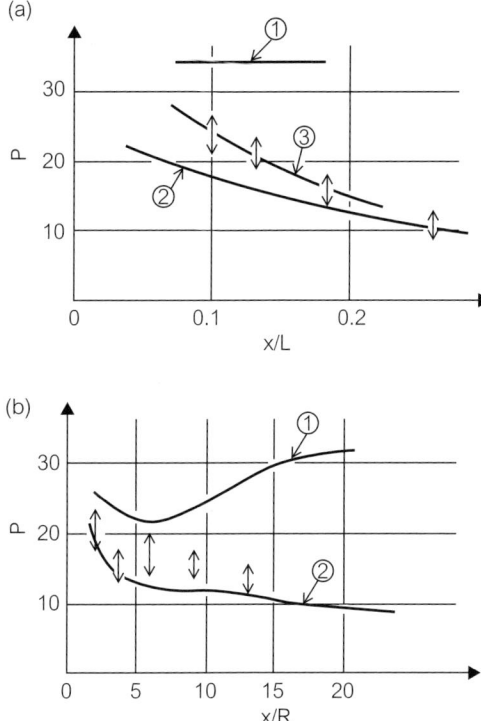

Figure 126. Pressure distribution (in torr) along a cone of 15° semiangle in a hypersonic flow (nitrogen, $\theta = 12.5°, p_0 = 2 \times 10^7$ Pa, $T_0 = 4000$ K, $M_\infty = 9$). (a) Sharp cone; (b) Blunted cone (R = 0.7cm). ↕: Experimental points, 1: Uniform flow, 2: Conical flow (Euler), 3: Conical flow (Navier–Stokes).

It is of course possible to take the source effect into account by interpreting the experimental results. Thus, the results of pressure measurements along a sharp cone and along a blunted cone placed without incidence at the exit of a conical nozzle of semiangle $\theta = 12.5°$ are illustrated in Fig. 126 so as to highlight clearly the source effect.[178] These results are compared with those deduced from an approximate calculation assuming a uniform and conical flow.

TWELVE

Relaxation and Kinetics in Shock Tubes and Shock Tunnels

12.1 Introduction

The aim of this chapter is to provide the reader with characteristic and representative data of relaxation processes and chemical kinetics in shock tubes and shock tunnels. Thus, experimental data on relaxation times, reaction-rate constants, and catalytic properties are presented and compared with the results of various theoretical models. Their effects on non-equilibrium flows are discussed, as well as the limit of their reliability. Of course, more complete or detailed results can be found in the specialized literature.

First, various results concerning vibrational relaxation are presented. A general outline is given regarding global relaxation times, as they can be deduced from macroscopic measurements of density and total vibrational energy and interpreted by using relaxation equations of Landau–Teller type. We then discuss results concerning the determination of state-to-state transition probabilities obtained with spectroscopic methods such as spontaneous Raman diffusion or absorption of lines characteristic of selected vibrational transitions. A few results of the measurement of vibrational populations in hypersonic nozzle flows are then presented. These results seem to put an end to a polemic on the relaxation times which could depend on the type of non-equilibrium flow, i.e. compression or expansion flows. Measurements of vibrational catalycity are also presented, showing the relative inefficiency of the exchanges between the vibrational mode and the wall.

For chemical kinetics, particular examples presented include the kinetics of complex reactions downstream from intense shock waves propagating in mixtures representative of planetary atmospheres. The methods of measurement used in these cases (essentially time-resolved emission spectroscopy) are also presented, as well as a comparison of experimental results with theoretical models.

12.2 Vibrational relaxation

The purpose of most measurements is to determine global relaxation times defined by using the Landau–Teller model, since this model is widely used in the computer codes of hypersonic flows. These include TV relaxation times for pure gases, and TV and VV relaxation times for mixtures (Chapter 2). More precise measurements aim to determine transition probabilities and the evolution of vibrational populations in the relaxation zone downstream from a shock wave (incident or reflected), thus providing a comparison with theoretical models.

12.2.1 Relaxation times: general methods

Pure gases

For the determination of global relaxation times, the system of RH equations (Eqn. (8.12)) is used, coupled with the Landau–Teller equation (Eqn. (9.38)) in which the relaxation time τ_V is regarded as unknown. The measurement of one macroscopic parameter is thus in theory sufficient to determine $\tau_V(p, T)$ at each point of the zone of relaxation: the most sensitive quantities to the non-equilibrium are mainly the density ρ and the total vibrational energy E_V.

Experimental techniques

Density measurements are made by using the interferometric method briefly described in Chapter 11. An example of an interferogram obtained downstream from the incident shock wave is shown in Fig. 105(b): the density profiles in the non-equilibrium zone are then deduced from the fringe system.[104]

The vibrational energy is deduced from the intensity of infrared (IR) emission due to the rotation–vibration lines of polyatomic molecules that have a permanent dipole such as CO. Thus, if we suppose that the vibrational distribution remains Boltzmann-like in the non-equilibrium zone, the intensity emitted by all the monoquantum transitions (fundamental) is proportional to the global average vibrational energy and the biquantum transitions (overtone) give a signal proportional to the square of this energy (Appendix 12.1). An example[179] of the time evolution of the intensity of the bands centred on 4.6 µm and 2.3 µm, corresponding respectively to the fundamental and the overtone coming from the flow downstream from a shock wave in CO, is represented in Fig. 127. A

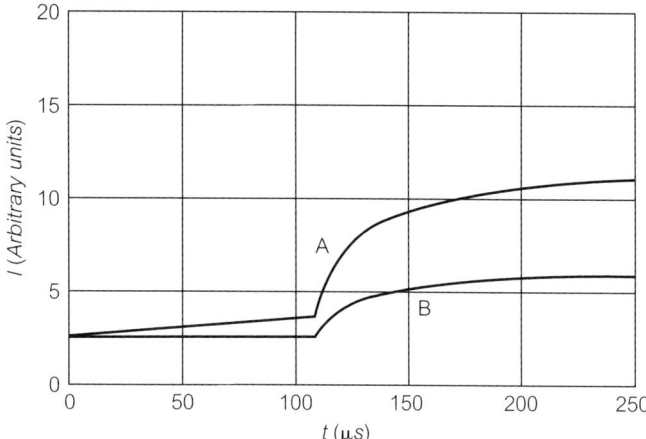

Figure 127. Infrared emission signals in CO ($M_s = 5.63$, $p_1 = 2500$ Pa)[142]. A: Fundamental; B: Overtone.

calibration of the equipment is necessary because the IR detector is sensitive to the emitting region before the arrival of the shock wave in the test section as it is visible in Fig. 127 (curve A). The time evolution of E_V is deduced from these recordings.

An alternative method also valid for homopolar molecules involves the measurement of the intensity of a Raman line of a gas subjected to laser radiation. It is thus possible to know the evolution of the vibrational energy but also the populations of the levels concerned (Section 12.2.2).

Results

It is possible to determine τ_V by solving the RH system and an equation of LT type, and with an experimental curve $\rho(t)$ or $E_V(t)$. Many results are thus obtained in the temperature range 500–5000 K and for a range of τ_V of 0.2 µs –1 ms.[11,180] However, these results are of two types:

a) Those for which only one (average) value for τ_V is retained in each experiment, in spite of the temperature variation during the experiment. These results are the oldest but also the most numerous results (method a).

b) Those for which the local variation of τ_V during the experiment is taken into account. These results are in theory more accurate, but some care in the interpretation of data has to be taken (method b).[181]

The representation of the results is obviously based on the LT–SSH model (Chapter 5), which leads to the following approximate dependence of $\tau_V p$ on T.

$$\tau_V p \sim F(T) \exp\left(T^{-1/3}\right) \tag{12.1}$$

where $F(T)$ depends weakly on T, so that we can write

$$Log\left(\tau_V p\right) = AT^{-1/3} + B$$

Another type of representation of the results is provided by the use of the quantity P_1^0, the average probability of de-excitation for the molecules in the first level,[180] or of the quantity $Z_1^0 = \left(P_1^0\right)^{-1}$, the probable number of TV collisions necessary for this de-excitation. From Section 2.4, we have

$$Z_1^0 \sim g(T) \exp\left(T^{-1/3}\right) \tag{12.2}$$

with $\tau_V = Z_1^0/Z$, where Z is the collision frequency.

Method a:

As an example, the average values of the relaxation times obtained for N_2 and O_2 are shown in Fig. 128. The following points may be emphasized.

For temperatures higher than 500 K, the dominant term of Eqn. (12.2) is the exponential term. The linear dependence of $Log\ Z_1^0$ with $T^{-1/3}$ is clearly verified for N_2 and O_2 and for many other gases.

For lower temperatures, the significant curvature of the graph, which is due to the attractive part of the interaction potential, is underestimated by the SSH theory.

For temperatures higher than 5000 K (not shown), few experimental results are available, and the extrapolation of SSH curves leads to values of τ_V which may become lower than those of elastic collision times! Empirical corrections are obviously possible[182] (Appendix 12.2), but it is quite unlikely that the LT–SSH correlation is valid for these temperatures, taking into account the possible multiquantum transitions, the effects of dissociation, and so on.

Method b:

If we exclude the values of τ_V obtained at the end of the relaxation zone where $E_V \to \overline{E}_V$ and $\frac{dE_V}{dt} \to 0$, there is agreement between the LT–SSH representation and the experimental results. Despite experimental uncertainties, these results validate this type of representation (sometimes questioned) and can contribute to give a more precise value of the slope of the LT–SSH straight line (see the example for CO in Fig. 129).[179]

Mixtures

Many binary mixtures have been studied, and the corresponding relaxation times τ_{TVpq} have been determined with similar experimental methods and represented

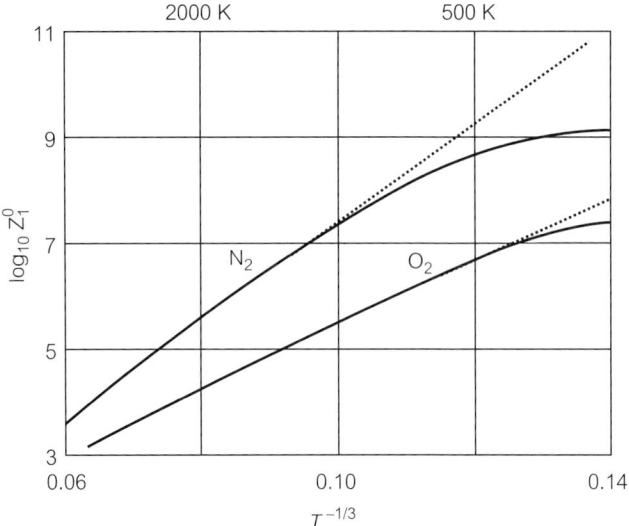

Figure 128. LT–SSH representation for N_2 and O_2. —: Experiments, ···: SSH.

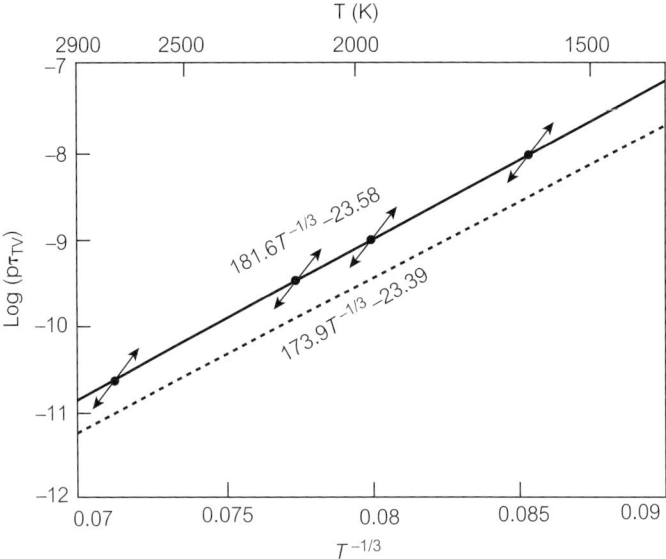

Figure 129. LT–SSH representation for CO. ···: Average values, —: Local values.

with LT–SSH diagrams (Chapters 3 and 9; Appendix 12.2). From these results, we can use in the case of more complex mixtures the barycentric formula (Eqn. (9.37)) for the relaxation times τ_{TV}. For the relaxation times τ_{VV}, various models, including the SSH model, are used (Chapter 9).

As an example of the experimental determination of relaxation times at moderate temperature without dissociation, we consider the quasi-resonant mixture $CO-N_2$; knowing the relaxation times $\tau_{TVCO-CO}$ and $\tau_{TVN_2-N_2}$, and taking into account Eqn. (2.70), we must determine τ_{TVCO-N_2} and τ_{VVCO-N_2}. If we assume a Boltzmann distribution for each gas at two temperatures T_{VN_2} and T_{VCO}, two different types of measurement are necessary. However, if we use the standard SSH formula (Eqn. (12.1)), that is, $Log(\tau_V p) = A_{CO-N_2} T^{-1/3} + B_{CO-N_2}$, two experiments of the same type with different conditions are enough to determine A and B. Thus, taking into account experimental uncertainties, we find

$$A_{CO-N_2} \simeq A_{CO-CO} \text{ and } B_{CO-N_2} \simeq B_{CO-CO}$$

so that $\tau_{TVCO-N_2} \simeq \tau_{TVCO-CO}$.

The same is true for other TV relaxation times between gases of similar molecular masses $(\tau_{TVN_2-O_2} \simeq \tau_{TVN_2-N_2}$, and so on$)$.

From this result, experimental interpretations give results for τ_{VVCO-N_2}, shown in Fig. 130, where we can see that they deviate significantly from the SSH curve.[183,184]

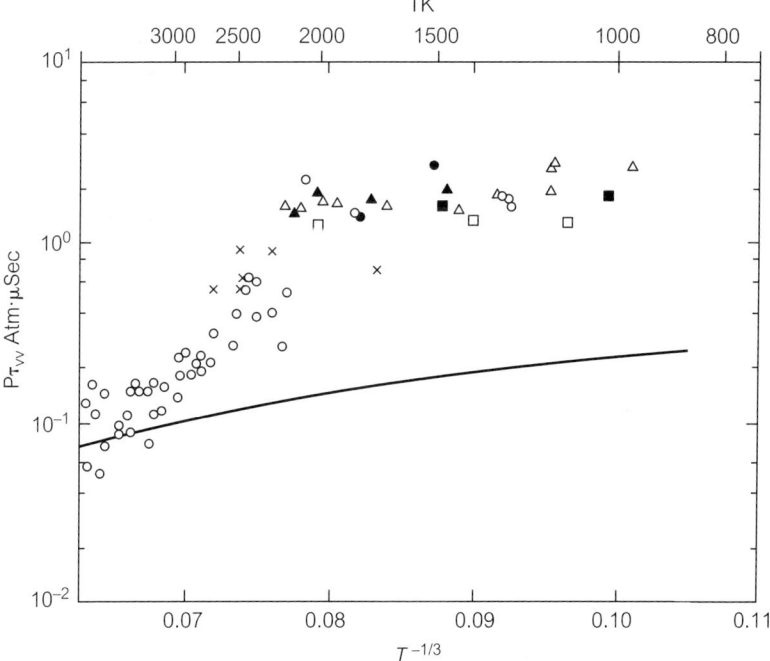

Figure 130. VV relaxation times for CO/N_2. (Experimental points), ———: SSH.

12.2.2 Vibrational populations

Experiments in a shock tube

Rigorously, as already pointed out, the evolution of vibrational populations does not generally take place according to a Boltzmann distribution, and consequently, the concept of vibrational temperature does not have a physical meaning. Thus, it is preferable to determine these populations directly by experiment. The measurements are essentially made in two ways: either by Raman diffusion[104] (spontaneous or stimulated, though this last technique has not been applied in shock tubes), or by infrared absorption.[185]

Spontaneous Raman diffusion

When molecules initially in levels i_v and i_{v+1} are excited by a line of frequency v and intensity I (Fig. 131), they re-emit diffusion lines of frequency $v - \Delta v$ (Stokes lines) and $v + \Delta v$ (anti-Stokes lines), with respective intensities I_S and I_{AS}, such that

$$\frac{I_S}{I} \sim n_{i_v} \quad \text{and} \quad \frac{I_{AS}}{I} \sim n_{i_v+1}$$

The recording of these quantities in the relaxation zone affords the possibility of following the evolution of the populations n_{i_v} and n_{i_v+1}, with a calibration operated in the equilibrium region. A powerful light source is necessary, and generally only the lowest levels (usually the most populated) are accessible.

A scheme of the experimental assembly is shown in Fig. 132. The intensities of the exciting line and anti-Stokes line corresponding to the transition $1 \to 0$ are recorded, and taking into account the relaxation equation of the population n_1 (Eqn. (2.17)), we can obtain the collision rate a_1^0 (and thus Z_1^0). Such a measurement realized in nitrogen gives values of Z_1^0, which confirms the global measurements (Fig. 128). Similar measurements concerning upper levels will of course be necessary.

Infrared absorption

The absorption coefficient α of a medium made up of heteropolar molecules and crossed by a spectral line corresponding to a transition $i_v, i_r \to i_v+1, i_r-1$ (P branch) is given by an expression of the type of Eqn. (10.30) as a function of the population of the levels concerned; if we assume a Boltzmann distribution for the rotational levels, this expression can be written as

$$\alpha_{i_r,i_v} = A_1^0 \left(8\pi \omega_{i_r,i_v}^2\right)^{-1} \left(\frac{v_{i_v+1,i_v}}{v_{1,0}}\right)^3 \left(\frac{R_{i_v+1,i_v}}{R_{1,0}}\right)^2 \phi \left(n_{i_v+1}\frac{\overline{n}_{i_r-1}}{n} - n_{i_v}\frac{\overline{n}_{i_r}}{n}\right)$$

(12.3)

For the same vibrational transition, there is a similar equation for the absorption coefficient α_{j_r,i_v} corresponding to another rotational transition $j_r \to j_r - 1$.

From the measurement of the two coefficients α_{i_r,i_v} and α_{j_r,i_v}, it is thus possible to deduce the populations of the levels i_v and $i_v + 1$, since we have

$$\frac{n_{i_v}}{n} = A_1 \alpha_{i_r,i_v} + B_1 \alpha_{j_r,i_v}$$

$$\frac{n_{i_v+1}}{n} = A_2 \alpha_{i_r,i_v} + B_2 \alpha_{j_r,i_v} \quad (12.4)$$

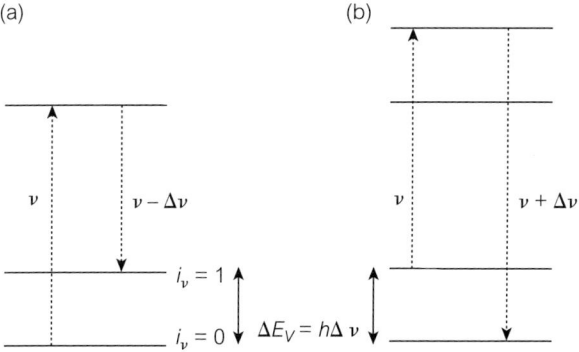

Figure 131. Principle of spontaneous Raman diffusion. (a) Stokes line; (b) Anti-Stokes line.

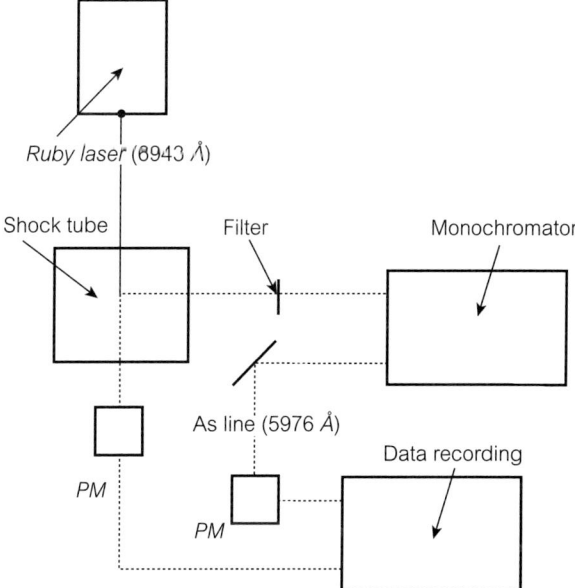

Figure 132. Scheme of Raman set-up (nitrogen).

In this system, the four coefficients A_1, A_2, B_1, and B_2 depend on spectral parameters, generally well-known, and on macroscopic flow parameters $(T, \rho, \ldots)$. The flow parameters depend very little on the vibrational distribution model (Chapter 9), so they are calculated before solving Eqn. (12.4), assuming for example a LT–SSH model. Thus, the coefficients α_{i_r, i_v} and α_{j_r, i_v} may be determined as well as the evolution of the vibrational populations, which may be compared with global distribution models (Boltzmann, Treanor, and so on).

As an example, measurements of the absorption coefficient α are carried out downstream from incident shock waves in carbon monoxide, CO, for various vibrational transitions, each including two rotational transitions.[186] The scheme of the experimental set-up is presented in Fig. 133: it includes a CO monomode laser (centred at 3.6 μm), capable of separating the rotation–vibration lines. The wavelengths are measured by a spectrometer, and for each line of the P branch corresponding to the vibrational transitions from $3 \rightarrow 2$ up to $7 \rightarrow 6$, the intensity I crossing the shock tube and the reference intensity I_0 are measured by IR detectors (InSb). The Mach number range lies between 4.5 and 7, which corresponds to an equilibrium temperature range 1500–3000 K.

Figure 134 shows an example of the recording of the difference of relative intensity $\Delta I = I - I_0$ of two P lines corresponding to the vibrational transition $4 \rightarrow 3$ in the relaxation zone. The evolution of the coefficient α is deduced from this recording and finally also the relative population of level 3, which is shown in Fig. 135. Also in Fig. 135, the evolution of the relative population of level 6 is represented. From these experiments, the following comments may be made.

The population of the lowest levels cannot be accurately determined from these measurements, because the schlieren effect, owing to the gradient of refraction index, important just downstream from the shock wave, primarily affects the first levels. The significant results thus concern only those levels equal to or higher than the third level. Because of the very small initial population of

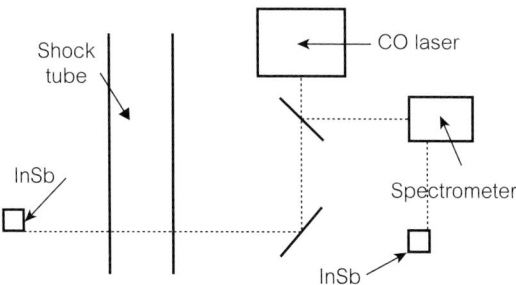

Figure 133. Scheme of IR absorption set-up.

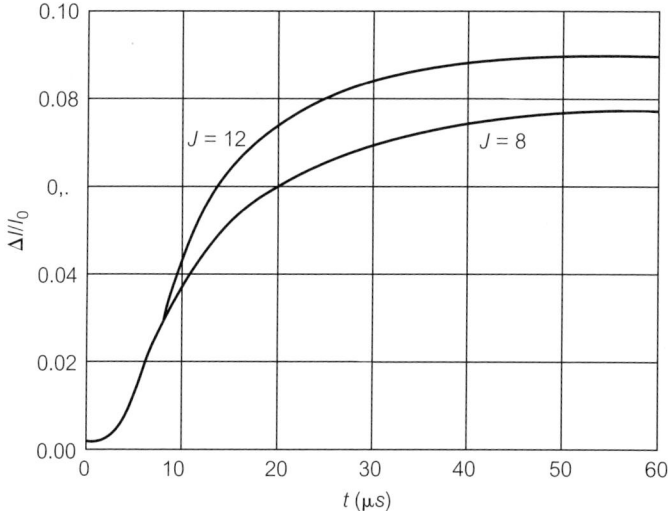

Figure 134. Evolution of the relative absorbed intensity for two rotational lines of the vibrational transition $4 \to 3$ (CO: $M_S = 5.57$, $T_1 = 291$ K, $p_1 = 196$ Pa).

these levels, this population starts to grow significantly one 'incubation time' after the passage of the shock wave, which, in these experiments, is about a few microseconds.

The experimental evolution of the population of these levels is relatively close to an evolution of Treanor type (higher in the relaxation zone and lower close to equilibrium). It remains, however, much higher than a Boltzmann distribution, the difference increasing with the level. Thus, we have experimental confirmation of the theoretical results presented in Chapter 9 (Appendix 9.1).

The characteristic time for reaching equilibrium for those levels equal to or higher than 3, including the incubation time, remains close to the total relaxation time, and this is essentially due to TV collisions, as discussed above.

Experiments in a shock tunnel

For global measurements, we can 'historically' distinguish the measurements carried out in shock tubes (compression) and those carried out in the nozzle of shock tunnels (expansion). Contradictory results have been obtained in the two types, though a synthesis of the results now seems to be generally accepted.

One of the main difficulties of conducting measurements at the nozzle exit of shock tunnels arises from the low density of the flow, which is essentially the price to be paid for the generation of a hypersonic flow. Thus, the techniques used in shock tubes (interferometry, IR emission or absorption) are not efficient, except in front of an obstacle—but here the gradients are important.

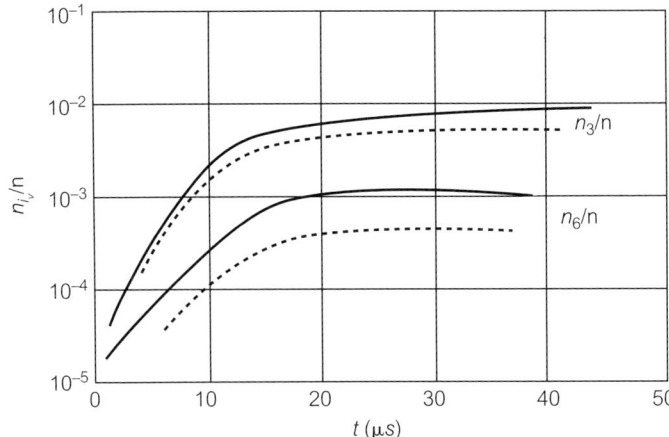

Figure 135. Evolution of the relative population of the third and the sixth vibrational level (CO: $M_S = 5.60$, $T_1 = 293$ K, $p_1 = 196$ Pa). ———: Experimental, Treanor, ----: Boltzmann.

This is why methods used in the past to determine global relaxation times were indirect measurements (sodium or chromium line reversal, inclusion of tracers, Rayleigh diffusion, and so on). The general conclusions of these studies were that the vibrational relaxation times deduced from these measurements were much lower than those deduced from measurements in shock tubes and, thus, that the flow at the nozzle exit was much closer to equilibrium than that calculated using the values of relaxation times determined in shock tubes. A large dispersion of values of τ_V also resulted from these data.[187–189]

As no plausible explanation could justify these results, direct measurements of vibrational populations in nitrogen were realized by spontaneous Raman diffusion initially in a nozzle of area ratio $A/A_C = 12$ (nozzle A), inserted in a shock tube,[190] and then in a shock tunnel nozzle (nozzle B) of larger dimensions[191] ($A/A_C = 164$). Experimental devices, though very different in detail, were in overall conformity with the diagram of Fig. 132.

For the first example quoted here, the recorded Raman spectrum is shown in Fig. 136 (a), and the corresponding vibrational distribution (up to the fifth level) in (b). It is thus clear that this distribution is close to a Boltzmann distribution at the frozen vibrational temperature 2230 K found from a classical LT calculation (Chapter 10), including the values of relaxation times determined in shock tubes, which is confirmed in Fig. 137, where the evolution of the temperature along the nozzle is shown. Therefore, independently of the experimental uncertainties (relatively important), no significant difference appears between the relaxation times determined in compression and those determined in expansion regimes.

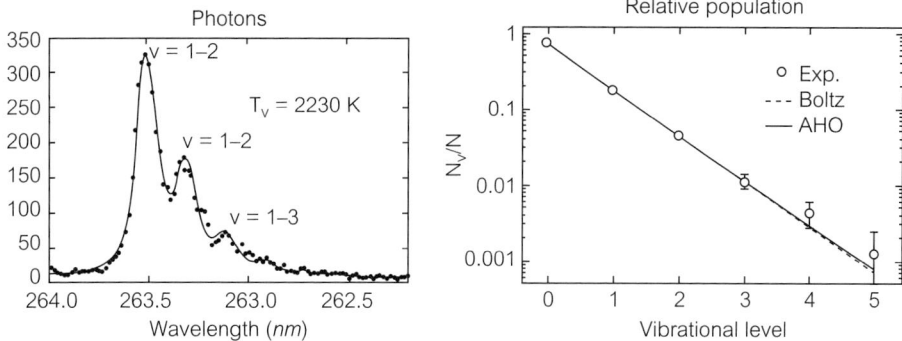

Figure 136. (a) Raman spectrum at 6 cm from the throat; (b) Vibrational population ($A/A_C = 12$, $L = 8$ cm), ($T_0 = 2800$ K, $p_0 = 1.02 \times 10^7$ Pa), (AHO: Anharmonic oscillator) Nozzle A. Excitation by KrF excimer laser (248 nm).

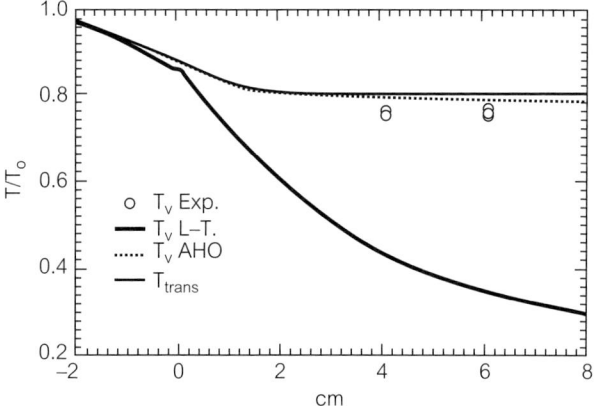

Figure 137. Measured and computed temperatures along the nozzle (nozzle A). (Conditions of Fig. 136).

For direct measurements in a shock tunnel (nozzle B), the degree of purity for the gases employed is not too high, and vibrational 'overpopulations' are observed for levels higher than 3, which may be attributed to an excitation caused by the presence of impurities that are inevitably present in large installations. However, as observed in Fig. 138(a), the distribution for the first three levels is close to a Boltzmann distribution at the frozen vibrational temperature 2153 K, corresponding to the temperature given by a NS–LT calculation with values for relaxation times measured in shock tubes, which confirms the results obtained with nozzle A.

Moreover, a similar measurement carried out at the exit of the nozzle B in front of a model finds that the vibrational temperature calculated as above is about 600 K[191] (Fig. 138(b)).

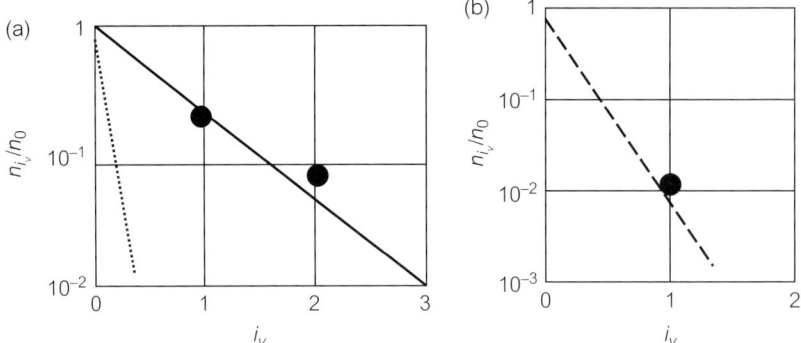

Figure 138. (a) Relative vibrational populations at the exit of nozzle B. ($A/A_C = 164$, $L = 70$ cm), ($T_0 = 2900$ K, $p_0 = 1.70 \times 10^7$ Pa). •: Exp., —— $T_V = 2153$ K, $\cdots$ T $= 270$ K (equilibrium). Excitation by dye laser (500 nm). (b) Relative vibrational populations at 4 mm in front of a model (Nozzle B, – – – $T_0 = 600$ K, $T_V = 600$ K, •: Exp.).

12.2.3 Vibrational catalycity

As previously discussed (Appendix 10.1), for the transitional regime, it is possible to define two accommodation coefficients α_{TR} and α_V for modes TR and V, respectively, and two exchange coefficients γ_{TR} and γ_V, respectively, for the exchanges $TR \to V$ and $V \to TR$. In theory these coefficients can be deduced from measurements of wall heat flux q_w (Appendix 10.1) and therefore from the variation of wall temperature ΔT_w (Appendix 7.2).

Thus, if we measure $\Delta T_w(t)$ at the end wall of a shock tube behind the reflected shock, we obtain signals similar to those in Fig. 139 (a) and (b). We thus compare these signals with those resulting from a Navier–Stokes calculation that includes boundary conditions (Eqns (10.22) and (10.23)). However, in this case, we can neglect the exchanges $V \to TR$, so that we assume $\gamma_V \simeq 0$. Then, iteratively varying the values of the three other coefficients α_{TR}, α_V, and γ_{TR}, we finally calculate theoretical curves that within a 'confidence interval' fit the experimental curves. Obviously, the 'response time' of the temperature probes[192] (Chapter 11) must be much shorter than the timescale of the accommodation and exchange phenomena.

Such measurements have been made[193] with test gases N_2 and CO_2. As a first observation, we may point out that no theoretical curve can coincide with the experimental curve for non-negligible values of α_V, which leads to the conclusion that the wall (here made up by the thermometric probes in glass or ceramics) is non-catalytic for the vibration, i.e. $\alpha_V \simeq 0$.

For nitrogen, it is easy to conduct measurements for conditions where the vibrational excitation is negligible (low shock Mach number); then there is only one coefficient to be determined: α_{TR}. Thus, a value of 0.50 is found for α_{TR}.

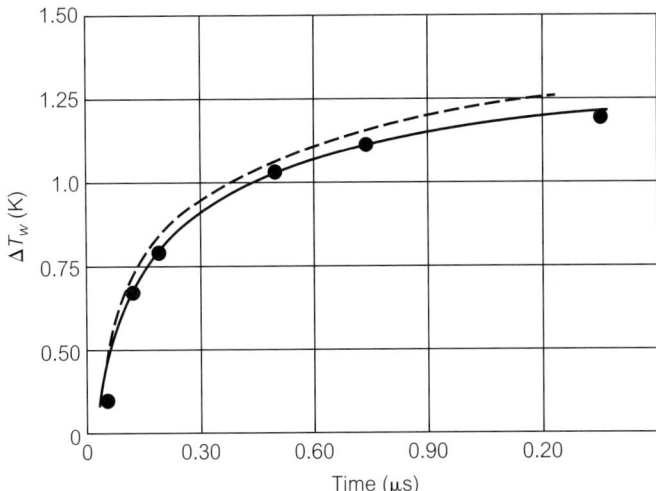

Figure 139. (a) Wall temperature evolution at the end wall of a shock tube (N_2, $M_S = 3.94$, $p_1 = 1.25 \times 10^4$ Pa). •: Experiments, ———: Calculation with $\alpha_{TR} = 0.70$, $\gamma_{TR} = 0.80$, $\alpha_V = 0$; ---: Calculation with $\alpha_{TR} = 0.50$, $\gamma_{TR} = 0.70$, $\alpha_V = 0$.

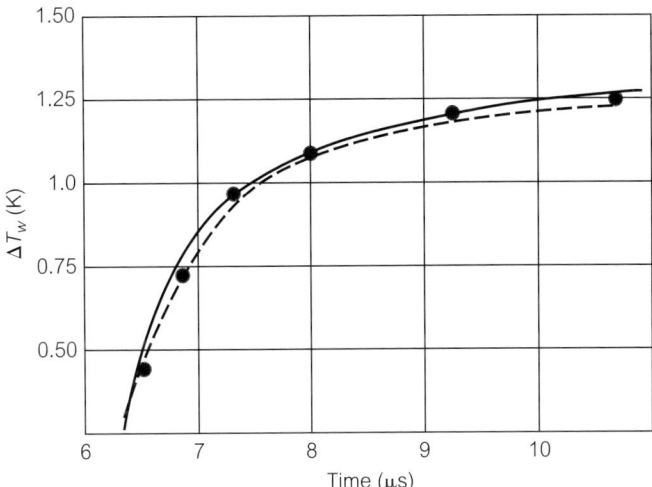

Figure 139. (b) Wall temperature evolution at the end wall of a shock tube (CO_2, $M_S = 3.94$, $p_1 = 300$ Pa). •: Experiments, ---: Calculation with $\alpha_{TR} = 0.66$, $\gamma_{TR} = 0.80$, $\alpha_V = 0$; ———: Calculation with $\alpha_{TR} = 0.70$, $\gamma_{TR} = 0.85$, $\alpha_V = 0$.

Incidentally, we may point out that the value found for argon is $\alpha_{TR} = 0.75$, in conformity with older measurements.[192] Now, when nitrogen is vibrationally excited, the best agreement is obtained with $\alpha_{TR} = 0.70$ and $\gamma_{TR} = 0.80$ (Fig. 139(a)), which shows a significant TR–V coupling, which can modify the value of α_{TR}.

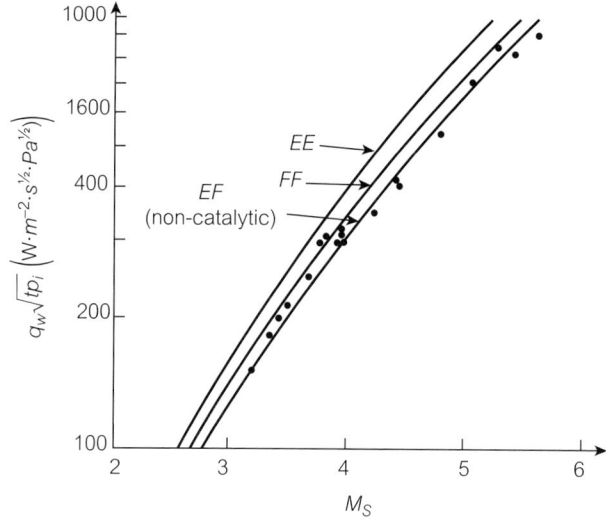

Figure 140. Wall heat flux behind an incident shock wave in CO_2. •: Experimental points.

For carbon dioxide, the best agreement is obtained for the following 'confidence intervals' (Fig. 139(b)):

$$\gamma_V \simeq 0, \quad 0.66 < \alpha_{TR} < 0.70, \quad \text{and} \quad 0.80 < \gamma_{TR} < 0.85$$

TR–V coupling is thus also significant for CO_2.

In order to confirm these results in the continuum regime (essentially the vibrational non-catalycity), measurements of wall heat flux are carried out downstream from an incident shock wave in CO_2, and the values obtained (corresponding to the plateau of ΔT_w) are compared with those deduced from calculations of simplified flows: flow assumed in equilibrium (denoted EE), flow assumed frozen (denoted FF), and flow assumed frozen in the boundary layer with non-catalytic wall for the vibration and in equilibrium outside (denoted EF) (see Fig. 60). Thus, we observe (Fig. 140) that the experimental points are in reasonable agreement with the results of this last calculation, thus confirming the wall non-catalycity for the vibration[194] (Chapter 10).

12.3 Chemical kinetics

12.3.1 Dissociation-rate constants

More than in other fields, the results of chemical kinetics obtained in shock tubes have in the past suffered from measurement techniques that were often indirect (for example, interferometry) or incomplete (emission, absorption, or the like).

The consequence was a large variation in such results, which also arose owing to the methods used for data reduction.

However, there exist a large number of reaction-rate measurements,[100,140,168] in particular for dissociation reactions carried out behind incident and reflected shock waves.

A few results of measurements of dissociation-rate constants k_f in standard gases such as O_2 and N_2 are provided in Figs 141 and 142. These have been obtained by interferometry and UV absorption. They approximately obey the general relation $k_f = C_f T^n \exp(-E_D/kT)$ (Chapter 9).

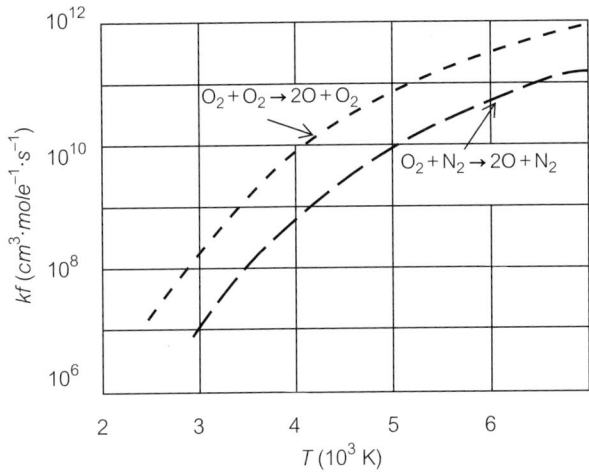

Figure 141. Measured values of dissociation-rate constant (oxygen).[140]

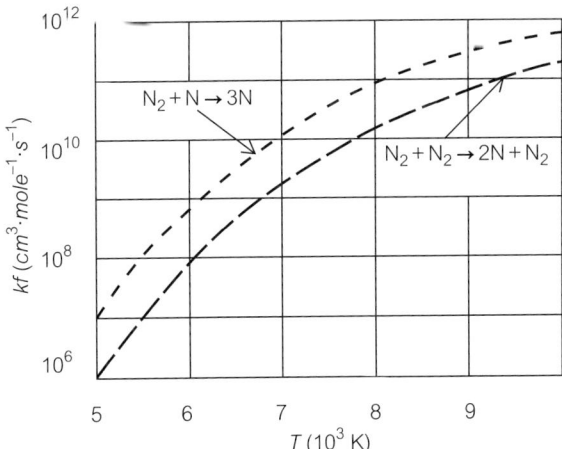

Figure 142. Measured values of dissociation-rate constant (nitrogen).[140]

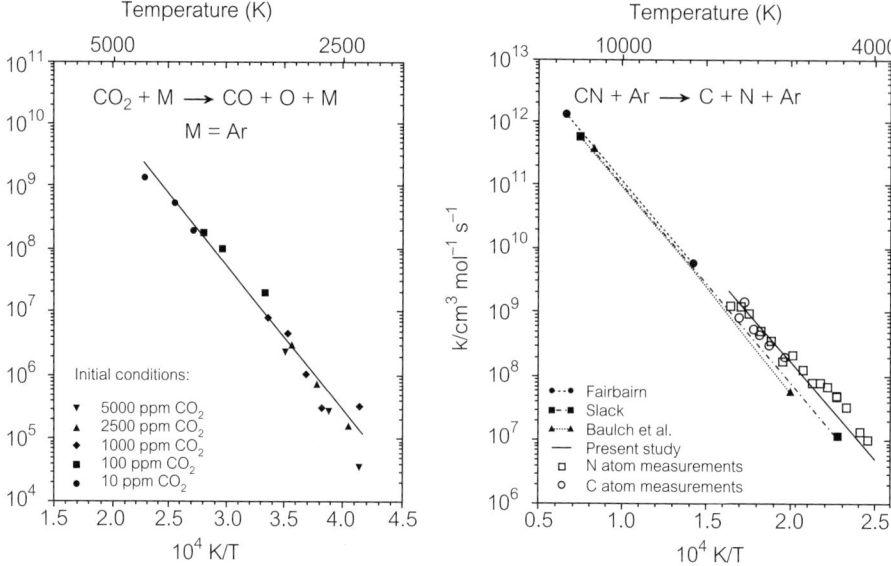

Figure 143. Measured values of dissociation-rate constant (CO_2 and CN).[196]

In fact, at the lowest temperatures, an 'incubation time' is observed, which is due to vibrational relaxation (Chapter 9), since in this case the vibrational relaxation and dissociation proceed on different timescales. For 'moderate' temperatures, the data giving k_f must be interpreted by taking into account the interaction of the dissociation and vibration that proceed on the same timescale. At still higher temperatures, the vibration can be assumed to be in equilibrium, and a 'one temperature' physical model can again be used.

More accurate measurements of dissociation-rate constants for O_2 and N_2 in an excess of argon (thus at quasi-constant temperature) have been obtained from absorption measurements of the characteristic spectral lines (130.5 nm for O and 156.1 nm for N) of oxygen or nitrogen atoms.[196] The same technique can be used to determine the rate constant of any reaction releasing O, N, or C atoms. Thus, in Fig. 143, the dissociation-rate constants of CO_2 and CN are shown.

12.3.2 Time-resolved spectroscopic methods

The time-resolved emission spectra of particular species downstream from a shock wave can allow us not only to ascertain the kinetics of the formation or destruction of these species but also to validate (or not validate) complex calculations of chemical kinetics in which these species are involved. The recording of these spectra (or parts of spectra) during one single experiment clearly

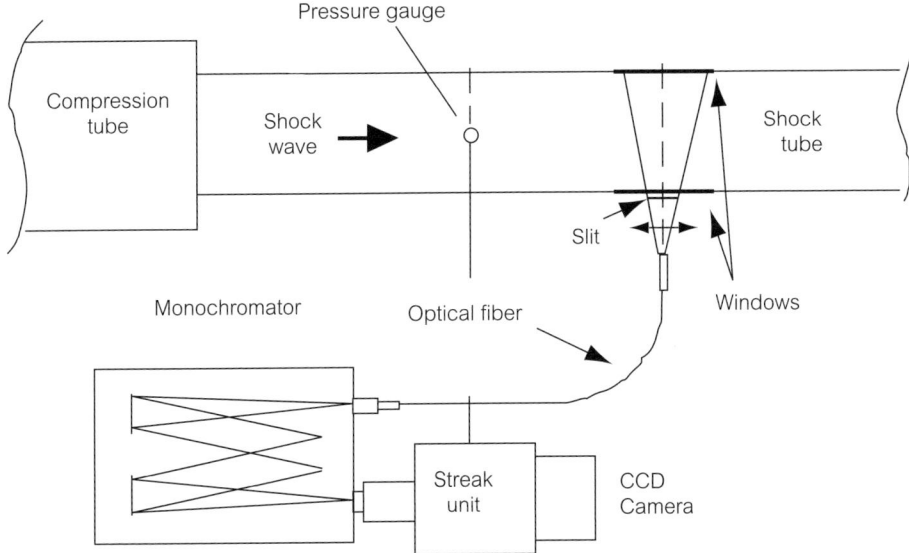

Figure 144. Scheme of optical arrangement for time-resolved emission measurements.

constitutes significant progress compared to the recordings of one or a few particular lines, as were often made in the past.[140]

An example of such an experiment[197] is represented in Fig. 144. The radiation emitted by the shocked gas is focused to a monochromator equipped with a high-resolution holographic grating. At the exit, a streak unit equipped with a CCD camera enables the recording of a significant part of the analysed spectrum. An example of such a recording is shown in Fig. 145. From these pictures, it is possible to extract the variation of the line intensities as functions of time or wavelength.

Three examples of complex chemical kinetics are now presented.

N_2/CH_4/Ar mixture

The study of this mixture (92%N_2, 3%CH_4, 5%Ar) has two objectives. It is of interest theoretically since the reactions developing downstream from a strong shock wave include the production of the radical CN, with the emission of an intense violet band spectrum (Appendix 12.3), so that it is possible to attempt to validate kinetic models for this mixture (Appendix 12.5). The second interest is practical since this mixture constitutes the atmosphere of Titan, which is a satellite of Saturn and was the ultimate target of the Cassini–Huygens probe. The entry of the probe into this atmosphere required knowledge of the radiating heat flux, which is essentially due to CN.[198,199]

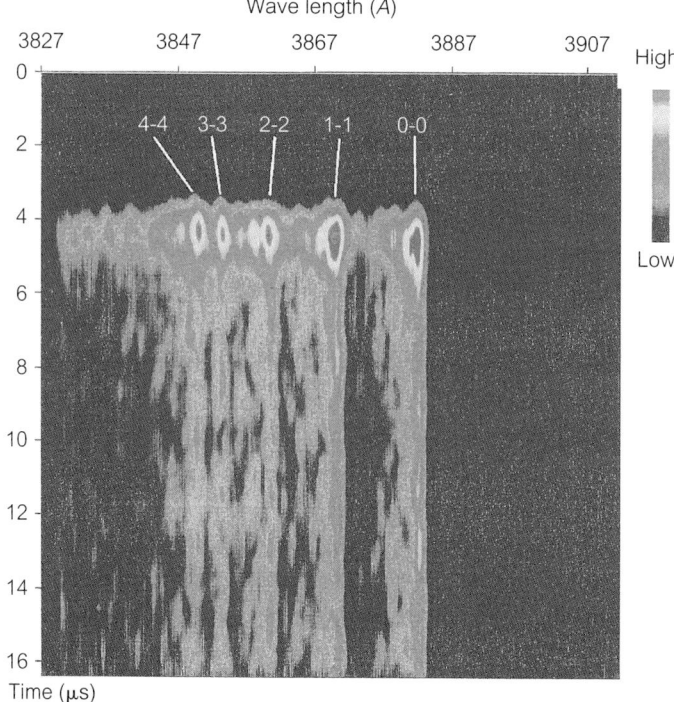

Figure 145. Example of streak image ($\Delta v = 0$ CN band of the mixture $CH_4/N_2/Ar$) ($U_s = 5560$ m/s, $p_1 = 220$ Pa).

Experiments are based on the recording of the spontaneous emission related to the electronic transition $B^2\Sigma^+ \leftrightarrow X^2\Sigma^+$ of CN and more precisely to the $\Delta v = 0$ sequence of this transition, which is known to have the strongest intensity (streak image, Fig. 145).

It is thus possible to deduce from this image the intensity profiles of the lines as functions of the wavelength (Fig. 146) or of the distance to the shock (Fig. 147).[197,200]

In the case considered here, we can observe an 'abnormal' behaviour of the spectrum at its maximum intensity, the 1-1 band having a high intensity compared with the 0-0 band. This tends to disappear, however, with increasing distance from the shock. We also observe an important maximum intensity presented by the various bands close to the shock (overshoot), due to the non-equilibrium.

The experimental relative vibrational populations (up to the level $v = 8$) may also be estimated by using an iterative method (Appendix 12.3). It is then necessary to use a rotational temperature given by the computation (Appendix 12.5), since the intensity of a rotational line corresponding to a given electronic

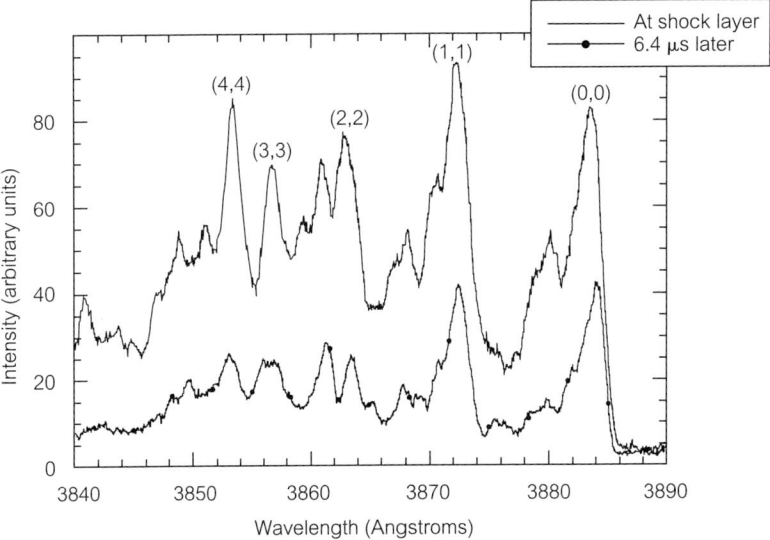

Figure 146. Experimental spectra of the $\Delta v = 0$ band of CN at two instants behind the shock (Conditions of Fig. 145).

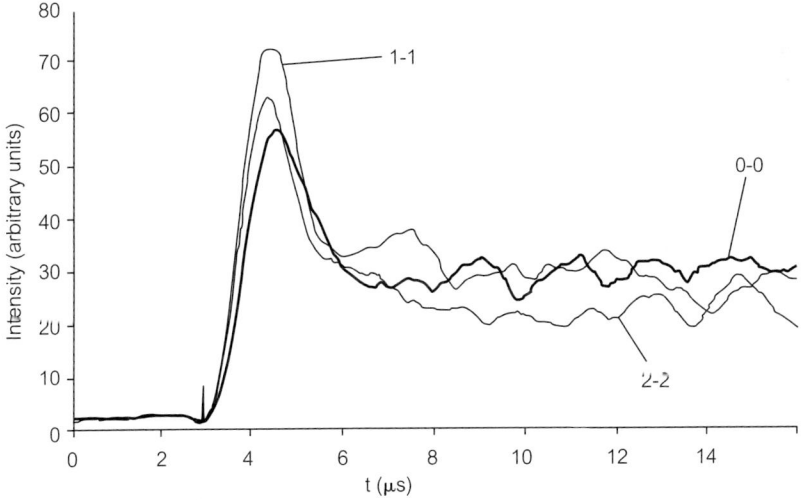

Figure 147. Time evolution of the intensities of the 0-0, 1-1, and 2-2 vibrational bands (Conditions of Fig. 145).

transition is proportional to the population of the excited levels, that is, assuming $T_R = T$:

$$n\left(e', v', j'\right) = K\left(2j' + 1\right) \exp\left(-\frac{\varepsilon_{i_r}}{kT}\right) n\left(v'\right) \qquad (12.5)$$

where no assumption is made a priori on the population of the level v'.

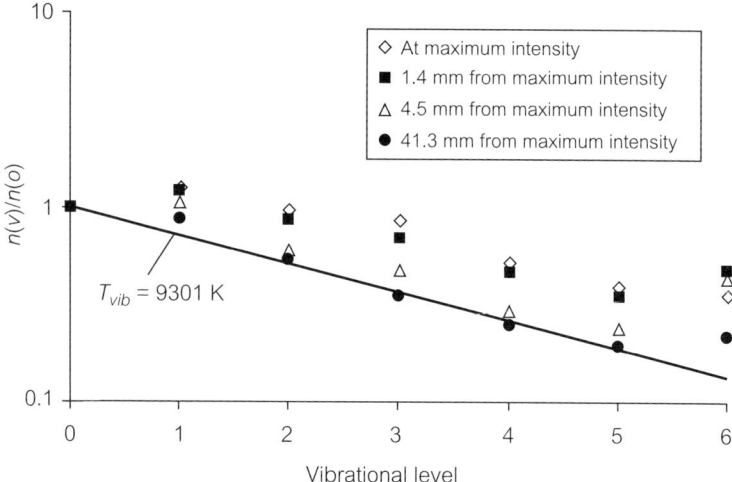

Figure 148. Experimental vibrational population distribution at different distances from the shock wave. (Conditions of Fig. 145).

Thus, the experimental relative vibrational populations corresponding to the case of Fig. 145 may be determined as shown in Fig. 148. This confirms the preceding results showing a non-Boltzmann distribution close to the shock and then a tendency to a Boltzmann distribution.

In addition, a theoretical spectrum can be calculated starting from the kinetic model of Appendix 12.5 by taking into account the broadening of the lines caused by the Doppler effect and by the pressure effect (Voigt profile) as well as by convolution from the 'slit functions' of the measurement system (see example in Fig. 149). By comparing the theoretical and experimental spectra, a very good agreement is obtained in the Boltzmann zone (see example in Fig. 150, where $T_V = 9300\ K \pm 500\ K$).

However, the results of other experiments carried out at higher pressure lead us to consider that the vibrational distribution remains non-Boltzmann along the whole observed distance (a few centimetres) and disagrees with the numerical calculations (Appendix 12.4).

CO_2/N_2 mixture

Similar experiments (spectrum $\Delta v = 0$ of CN) have been carried out in a mixture of 70% CO_2 and 30% N_2. In this case, the vibrational distribution is always a Boltzmann distribution, and the evolution of the vibrational temperature is in relatively good agreement with calculations (Appendix 12.5) (Fig. 151).

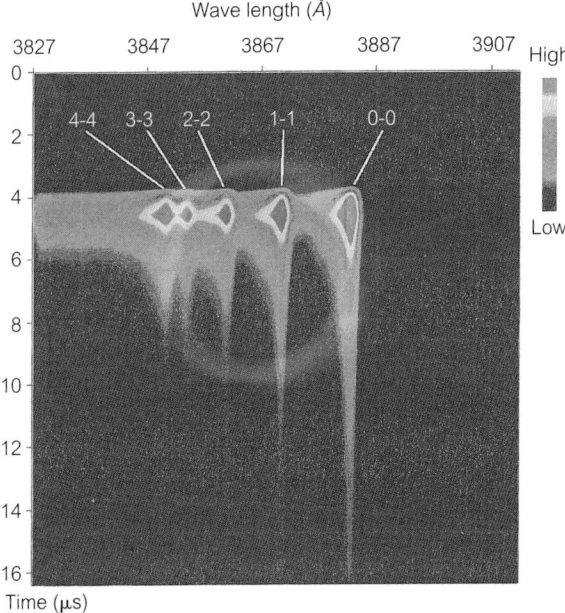

Figure 149. Simulated streak image ($\Delta v = 0$ CN band of the mixture $CH_4/N_2/Ar$). (Conditions of Fig. 145).

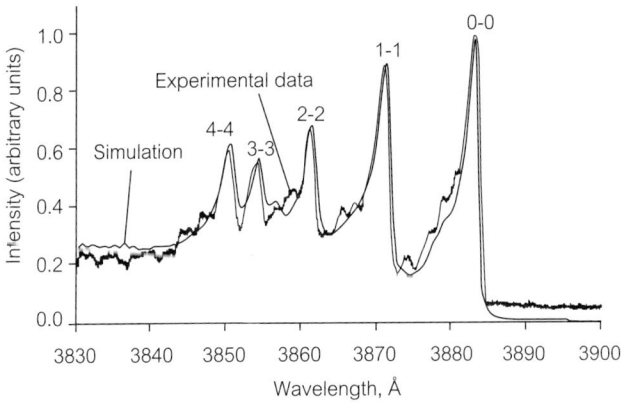

Figure 150. Experimental and calculated spectra of CN in $N_2/CH_4/Ar$. (Conditions of Fig. 145).

Pure CO

The same type of experiment has also been carried out with CO as a test gas to visualize the $\Delta v = 0$ and $\Delta v = +1$ bands of the electronic transition $A^3\Pi_g \leftrightarrow X^3\Pi_u$ of the C_2 molecule (Swan bands), because the intensity of these bands is low in the preceding mixtures. The comparison of measured and calculated spectra shows only global agreement. The differences observed on individual

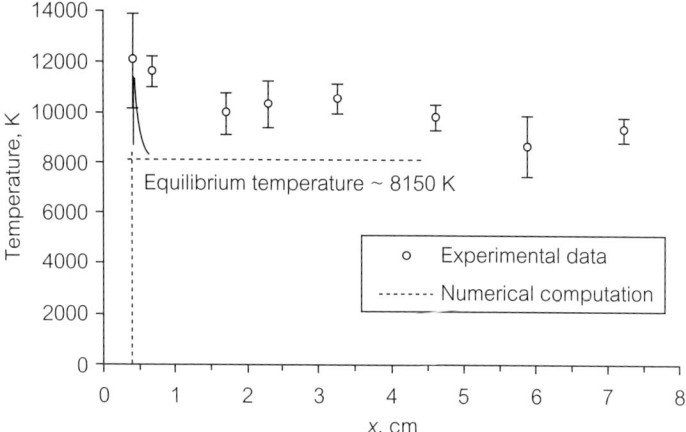

Figure 151. Evolution of vibrational temperature behind a shock in CO_2/N_2 mixture ($U_S = 5800$ m/s, $p_1 = 196$ Pa).

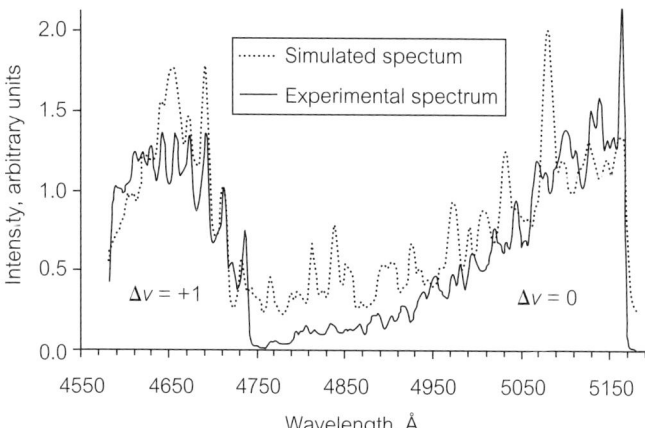

Figure 152. C_2 Swan bands in pure CO ($U_S = 5200$ m/s, $p_1 = 536$ Pa, equilibrium).

lines may be due to the lack of data for spectral parameters of the C_2 molecule at high temperature (Fig. 152). Agreement is much better for the evolution of the intensity integrated over the whole spectrum (Fig. 153).

12.3.3 Chemical catalycity

Similar to the experiments described above for vibrational catalycity, the determination of the recombination coefficient γ and/or the recombination-rate constant at the wall, k_{Rw}, has been obtained starting from measurements of wall heat flux, either in arc tunnels or in real flight.

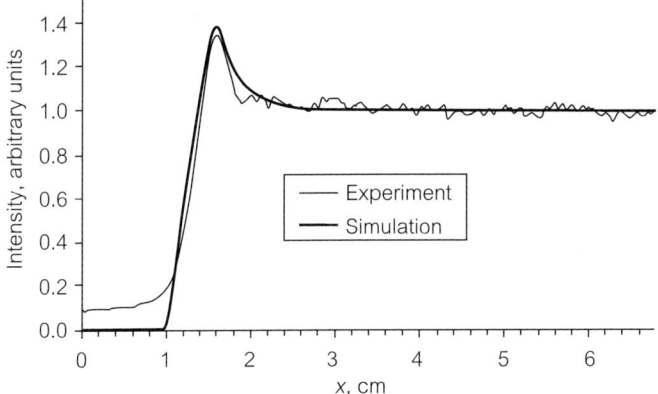

Figure 153. Evolution of the integrated intensity of the C_2 Swan bands behind a shock in CO (Conditions of Fig. 152).

Other types of measurement of γ, as well as of β, the chemical accommodation coefficient (Chapter 10), have been carried out in RF plasma reactors using spectroscopic methods (LIF, actinometry) and calorimetric techniques.

The values found for γ vary greatly depending on the techniques used and the nature and state of the materials tested.[201,204] Thus, for example for atomic oxygen, these values are about 10^{-2} on metallic surfaces (which are quickly oxidized) and about 10^{-3} on non-catalytic materials such as silicon carbide (SiC), for which $\beta \simeq 0.5$.

From these measurements, we can deduce a specific recombination rate k_{Rw} equal to $\gamma\beta(kT/2\pi m)^{1/2}$ (Chapter 10). Thus, for example at 300K, for O_2/Ag we find $k_{Rw} \simeq 220$ cm/s, and for O_2/SiC we find $k_{Rw} \simeq 15$ cm/s. These values increase with temperature, highlighting a progressive change in the mechanism of recombination, of Eley–Rideal type for temperatures less than 900K, and of Langmuir–Hinshelwood type for temperatures in the range 900–1200 K.[205]

12.3.4 Hypersonic flow around bodies

Flows around various bodies (dihedral, cylinder, sphere, and so on) have been studied in shock tunnels, in particular from density measurements by interferometry. Thus, experiments with a sphere[206] have been compared with the computations based on physical and numerical models presented in Chapters 5 and 10.

A parameter relatively simple to measure, and sensitive to non-equilibrium, is the shock detachment distance Δ (or shock stand-off distance, SSD) between the shock and the body along the stagnation line.

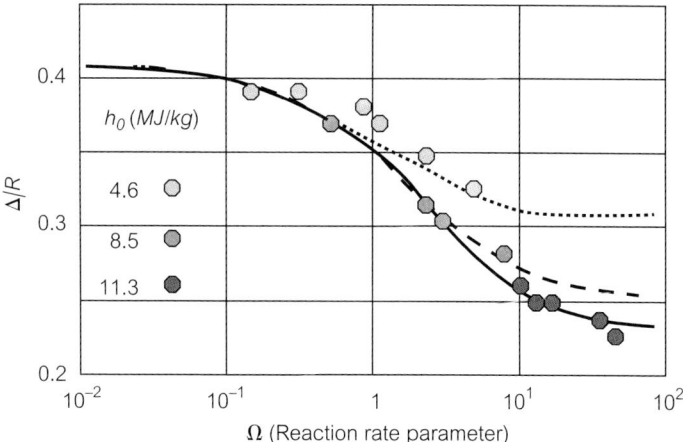

Figure 154. Measured[206] (●) and computed ⋯⋯ stand-off distances in carbon dioxide flows.

Table 7. Shock detachment distance in front of hemisphere-cylinders (airflow, shock tunnel).

Conditions	R (mm)	Δ/R (measured)	Δ/R (computed)
1			
$h_0 = 10.4$ MJ/kg	25	0.129	0.133
$p_0 = 18$ MPa	50	0.121	0.127
2	12.5	0.136	0.132
$h_0 = 4.8$ MJ/kg	25	0.130	0.128
$p_0 = 18$ MPa	50	0.120	0.126

Thus, Fig. 154 shows SSD measurements over a sphere at the exit of a shock tunnel[206] and SSD values computed with a numerical axisymmetric model including the vibration–reaction model developed in Chapter 5.[128] In fact, dimensionless SSD values are reported as a function of a similarity parameter, called the 'reaction-rate parameter' Ω, issued from a one-dimensional flow model[206,207] ($\Omega = 0$: frozen flow; $\Omega \to \infty$: equilibrium flow).

Other SSD measurements have been made in front of hemisphere-cylinders (nose radius R) placed in an airflow at the exit of a shock tunnel nozzle ($M = 7$; see Tables 6 and 7 and Fig. 77 of Chapter 10). Experimental values are deduced from holographic interferometry and schlieren photography,[128] and computed values are obtained from the same physical model as in the preceding example.[108]

A third example of SSD values is given in Fig. 155, where the experimental values are from measurements in a ballistic range,[208] and those labelled 'present model' are from computations similar to the preceding cases.[128] They are

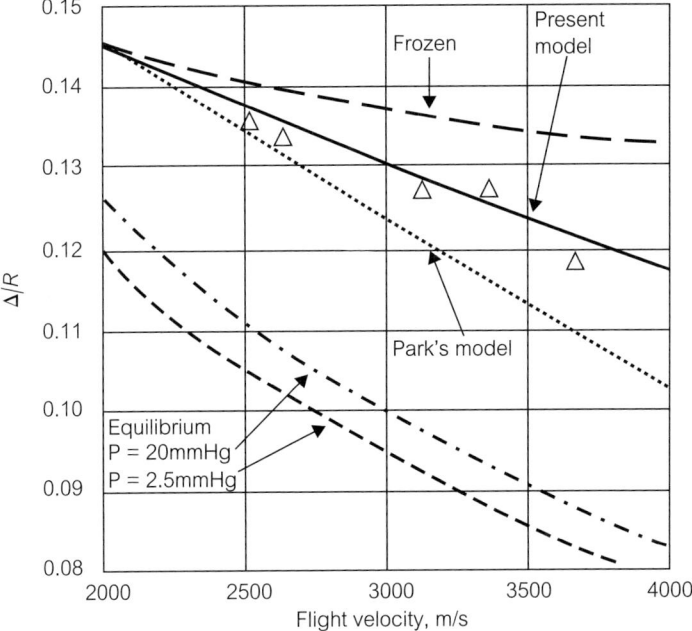

Figure 155. Comparison between measured and computed shock stand-off distance $\rho R = 2 \times 10^{-4}$ kg/m^2, Δ : Experiments (gun tunnel).

also compared to computations obtained from a semi-empirical model (Park's model[106]).

Finally, in the three quoted examples, the agreement between measured and computed SSD values remains within 5%.

Appendix 12.1 Generalities on IR emission

Molecules such as CO that have a permanent dipole moment can spontaneously pass from a vibrational level i to a level j. This spontaneous emission, generally in the infrared spectrum, is characterized by the Einstein coefficients A_i^j related to the probability of this transition. With the harmonic oscillator assumption, only monoquantum transitions occur, and we have:

$$A_i^{i-1} = iA_1^0 \tag{12.6}$$

For two-quantum transitions, related to the anharmonicity of the molecule, we can write:

$$A_i^{i-2} = \frac{i(i-1)}{2} A_2^0 \tag{12.7}$$

If we assume that self-absorption can be neglected, the intensity emitted by the totality of the monoquantum transitions of an 'active' component p, i.e. I_1, is proportional to the number of oscillators, that is:

$$I_1 \sim \sum_i A_i^{i-1} n_i \sim A_1^0 \sum_i i n_i \qquad (12.8)$$

Then, $I_1 \sim n_p E_{Vp}$ (or $\sim \rho c_p e_{Vp}$ per unit volume).

Similarly, for the two-quantum transitions, we have:

$$I_2 \sim \sum_i A_i^{i-2} n_i \sim \sum_i i(i-1) n_i \qquad (12.9)$$

Assuming a Boltzmann distribution for n_i at temperature T_V, we have:

$$I_2 \sim \sum_i i(i-1) \exp\left(-i \frac{\theta_v}{T_v}\right)$$

and applying the results of Appendix 2.3, we find:

$$I_2 \sim (n_p E_{Vp})^2 \qquad (12.10)$$

Appendix 12.2 Models for vibration relaxation times

The (popular) Millikan and White model[11] on TV relaxation times takes into account many experimental results and the SSH theory. Thus, for a relaxation time τ_{pq}^{TV} (where q may be equal to p), we have:

$$p\tau_{pq}^{TV} = \exp\left[0,0016 \tilde{\mu}_{pq}^{1/2} \theta_{vp}^{4/3} \left(T^{-1/3} - 0,015 \tilde{\mu}_{pq}^{1/4}\right) - 18,42\right] \qquad (12.11)$$

where $\tilde{\mu}_{pq}$ is the reduced molecular mass (in g/mole) and p is the pressure (in atmospheres).

For a mixture, $\left(\tau_p^{TV}\right)^{-1} = \sum_q \frac{\xi_q}{\tau_{pq}^{TV}}$ (Eqn. (9.37))

The approximate equation (12.11), valid between 500 and 8000K, may be corrected[182] when the temperature is close to 20 000K by adding to τ_{pq}^{TV} an elastic collision time (Chapter 2), that is $\tau_p^{el} \simeq (n_q \sigma_p \bar{c}_p)^{-1}$, where σ is a cross section equal to $10^{-17} \left(\frac{50000}{T}\right)^2$, and $\bar{c}_p$ is an average characteristic velocity equal to $\left(\frac{8kT}{\pi \mu_{pq}}\right)^{1/2}$. Then, for $T \geq 20\,000$ K, we have:

$$\tau_{pq}^{TV} = \tau_{pq(MW)}^{TV} + \tau_{pq}^{el} \qquad (12.12)$$

Other formulae, more or less empirical, may be found in the literature (Chapter 9), and the same is true of VV relaxation times.

Appendix 12.3 Simulation of emission spectra

The interpretation of experimental spectra corresponding to the emission of electronic transition bands obtained downstream from intense shock waves (CN or C_2, for example) requires comparison with calculated spectra in order to obtain physical parameters (primarily temperatures, possibly concentrations). Here, a few indications about the calculation of these spectra are given.[209–213]

Structure of emission bands

The internal energy of a molecule E is such that

$$E = E_E + E_V + E_R \tag{12.13}$$

By comparison with Eqn. (1.63), Eqn. (12.13) takes into account electronic energy E_E, since in the present chapter we consider molecules in an excited electronic state. In terms of wave number (cm^{-1}), Eqn. (12.13) may be written

$$T = T_e + G + F \tag{12.14}$$

where T_e, G, and F represent the electronic, vibrational, and rotational contributions respectively. The wave numbers of the spectral lines corresponding to a transition between two electronic states and possibly two vibrational and rotational states are such that

$$\nu = (T'_e - T''_e) + (G' - G'') + (F' - F'') \tag{12.15}$$

for a transition between an upper state (denoted by ′) and a lower state (denoted by ″).

Given an electronic transition, $\nu_e = T'_e - T''_e$ is constant, as given a vibrational transition, the following quantity is also constant:

$$\nu_v = \left[\omega'_e\left(v' + \frac{1}{2}\right) - \omega'_e x'_e\left(v' + \frac{1}{2}\right)^2 + \cdots\right]$$

$$- \left[\omega''_e\left(v'' + \frac{1}{2}\right) - \omega''_e x''_e\left(v'' + \frac{1}{2}\right)^2 + \cdots\right] \tag{12.16}$$

(See Eqn. (1.75) for an anharmonic oscillator.)

Therefore, for a given value of $\nu_e + \nu_v = \nu_0$, all possible rotational transitions constitute a single band, for which we have

$$\nu = \nu_0 + F(J') - F(J'') \qquad (12.17)$$

For a vibrating rotator, the term $F(J)$ is written in the form

$$F(J) = B_v J(J+1) - D_v J^2 (J+1)^2 + \cdots \qquad (12.18)$$

(See Appendix 1.2.)

Here, B_v and D_v depend on the vibrational and electronic state, that is:

$$B_v = B_e - \alpha_e \left(v + \frac{1}{2}\right) + \cdots$$

$$D_v = D_e - \beta_e \left(v + \frac{1}{2}\right) + \cdots \qquad (12.19)$$

We can therefore develop F and determine the wave numbers of the rotational lines for a vibrational and electronic transition.

Taking into account forbidden transitions, we eventually obtain a P branch ($\Delta J = -1$), a Q branch ($\Delta J = 0$), and an R branch ($\Delta J = +1$). Spin splitting may also occur, so that we define a new rotational quantum number K, with $J = K + S$ and $J = K - S$.

An example is represented in Figs 156 and 157 for the molecule CN.

Intensity of emission lines

The intensity of a spectral line (energy emitted per second), I_n^m, coming from N_n molecules in the initial state, is defined as

$$I_n^m = N_n h \nu_{nm} A_n^m \qquad (12.20)$$

The transition probability A_n^m (Einstein coefficient) may be expressed as a function of the matrix element of the dipole moment, and for an electronic transition $n' \to n''$ we have

$$I_{n'v'j'}^{n''v''j''} = \frac{64\pi^4 c \nu^4}{3} |Re|^2 q_{v'v''} \frac{S_{j'}}{2j'+1} N(n', v', j') \qquad (12.21)$$

expressed in 10^7 W/(cm^{3}sr), and where $|Re|^2$ is the matrix element of the moment of the electronic transition, $q_{v'v''}$ is the Franck–Condon factor, and $S_{j'}$ is the Hönl–London factor. These spectral parameters are generally well known.

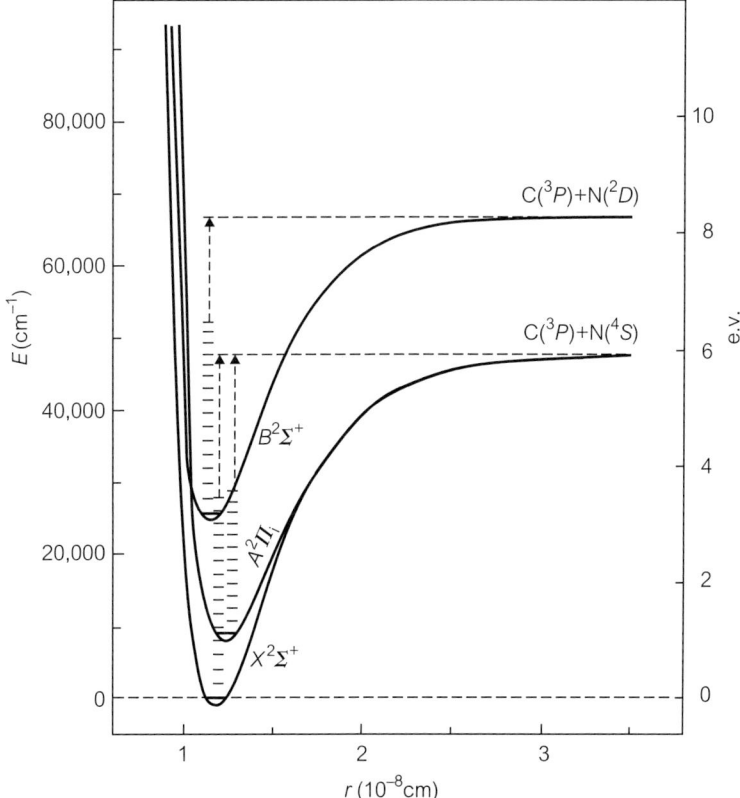

Figure 156. Potential curves for CN.

If we have a Boltzmann distribution, we can write

$$N(n', v', J') = N_0 \frac{(2J'+1)}{Q_{tot}} \exp\left[-\frac{hc}{k}\left(\frac{F(n', v', J')}{T_R} + \frac{G(n', v')}{T_V} + \frac{v'_e}{T_e}\right)\right]$$

(12.22)

Broadening of lines

The 'ideal' spectra obtained using the preceding expressions are composed of monochromatic lines. In reality, however, these lines are broadened depending on the conditions prevailing in the gas: molecular velocities, collisions, lifetimes, pressure effects, electric and magnetic fields, and so on. The majority of these broadenings are of Gaussian or Lorentzian type and can be generally taken into account by a single Voigt profile.

In addition to the broadening due to the gas itself, it is necessary to take into account a global 'apparatus function' in the calculation of the spectrum, for

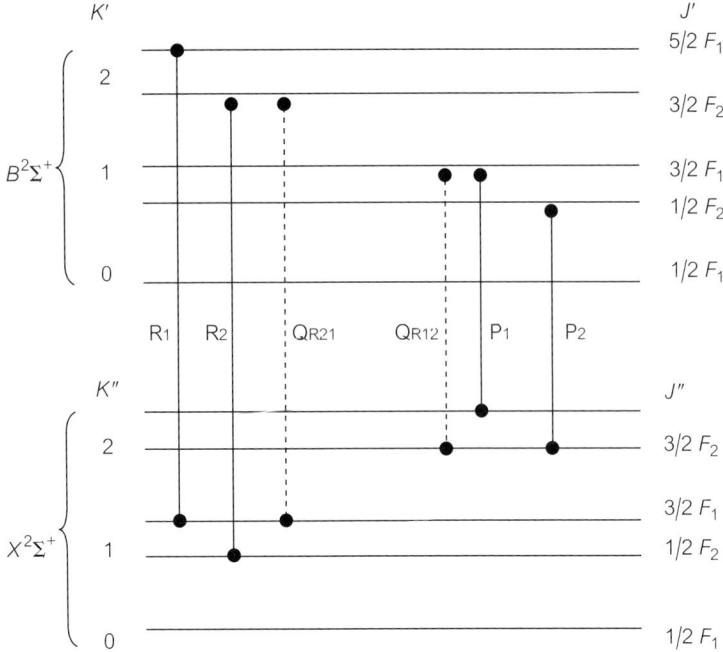

Figure 157. Transitions of the $X^2\Sigma^+ \leftrightarrow B^2\Sigma^+$ band (CN, $S = 1/2$, $\Delta K = \pm 1$, Q_R, Q_P: satellite branches).

comparison with the experimental spectrum. This function, generally because of the non-negligible width of the slits, introduces a convolution of the spectrum, but the function can be experimentally determined.[200]

An example of a calculated spectrum for CN is shown in Fig. 149.

Comparison of measured and computed spectra

The iterative method is summarized in the chart of Fig. 158.

It is also possible to determine the temperature T_R from the ratio of the intensities of two rotational lines of the same vibrational transition (Eqns (12.21) and (12.22)). Similarly, the temperature T_V can be determined from the ratio of the intensities of two rotational lines of two different vibrational bands. Therefore, from the measurement of the intensity of three suitably selected lines, these temperatures (insofar as they can be defined) can be determined, as well as their evolution.[214]

However, because of the monochromator dispersion and the overlapping of the lines, the inaccuracy of T_R and T_V values is significant (about 30%). Comparisons with computed values,[215] however, given in Fig. 159, show qualitative agreement.

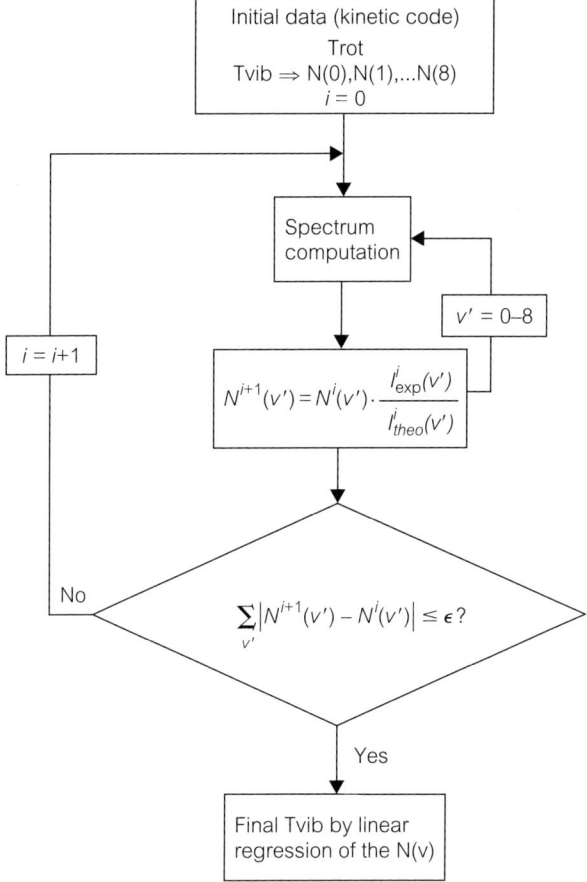

Figure 158. Determination of the vibrational populations of CN from a $\Delta v = 0$ band spectrum.

Appendix 12.4 Precursor radiation in shock tubes

The strong luminosity of the hot gas behind the shock wave may induce an important radiation on the gas ahead of the shock, so that the gas molecules may absorb this radiation and re-emit their own radiation.[100]

An example is given in Fig. 160, which shows the signal given by a photomultiplier sensitive to near ultraviolet and the visible spectrum (2500–6500Å) and registering the passage of all luminous phenomena during an experiment in a shock tube.[200]

This experiment corresponds to that of Fig. 145 with the $N_2/CH_4/Ar$ mixture as a test gas. Thus, we observe a strong precursor signal before the arrival of the

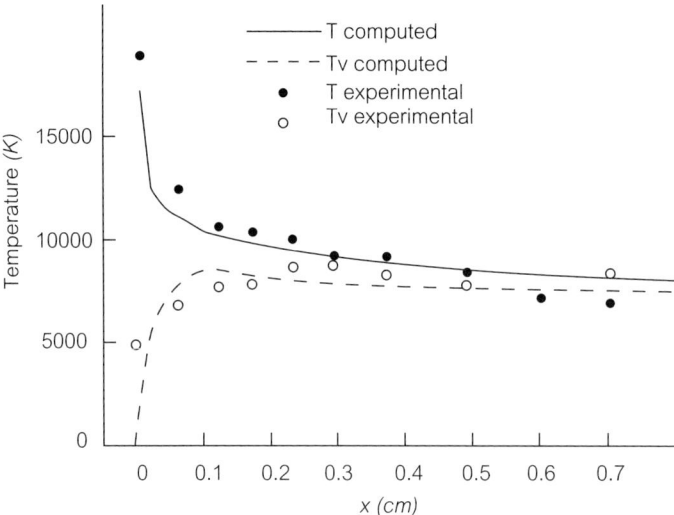

Figure 159. Evolution of CN temperatures behind a shock wave in $N_2/CH_4/Ar$.[214,215] ($U_S = 5680$ m/s, $p_1 = 200$ Pa, $\lambda_1 = 3872$ Å, $\lambda_2 = 3878$ Å, $\lambda_3 = 3882$ Å).

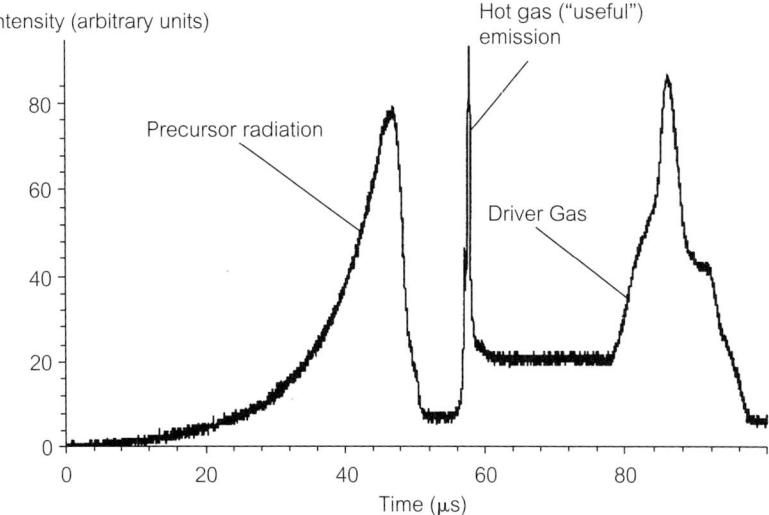

Figure 160. P.M. recording of luminous phenomena in a shock tube (Conditions of Fig. 145).

shock wave. This radiation is essentially composed of ultraviolet light and is not registered by the monochromator–streak unit ensemble (Section 12.3.2). In the same figure, we also observe the radiation due to the mixing zone at the arrival of the driver gas.

Despite the precursor radiation strongly decreasing just before the passage of the shock wave, it is probable that the electronic state of the molecules upstream from the shock is modified and therefore also the vibrational distribution of each state. The macroscopic parameters, however, may not be affected. Further studies on the subject are desirable.

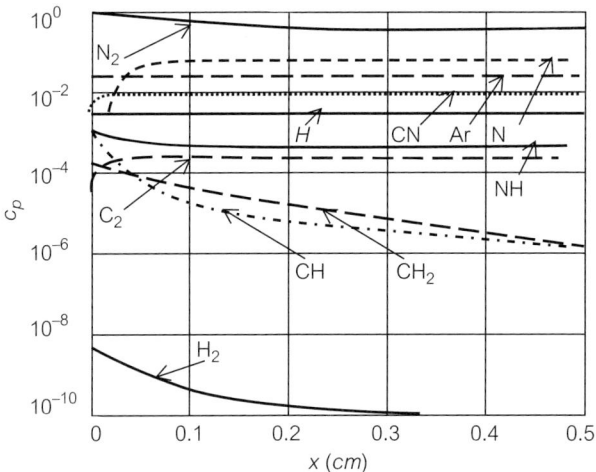

Figure 161. Evolution of species concentrations behind a shock wave in $N_2/CH_4/Ar$ ($U_s = 5600$ m/s, $p_1 = 220$ Pa).

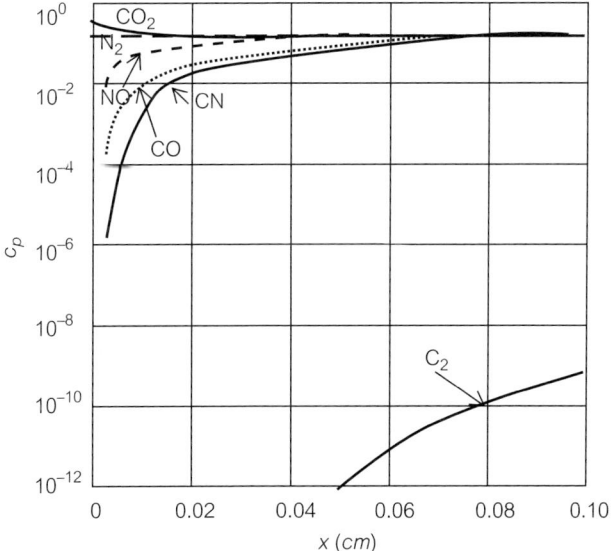

Figure 162. Evolution of species concentrations behind a shock wave in CO_2/N_2 ($U_s = 5800$ m/s, $p_1 = 200$ Pa).

Appendix 12.5 Examples of kinetic models

Calculation of the flow downstream from a shock wave, presented in Chapter 9, is applied here to various mixtures by taking into account the physical processes involved, that is, chemical reactions, vibrational relaxation, interactions, and so on.

The results are used to calculate the emission spectra of CN and C_2 and for comparison with experimental spectra.

$N_2/CH_4/Ar$ mixture

For this mixture, 24 reactions are taken into account,[215] with the following 20 species:

$$H_2, N_2, C_2, CN, NH, CH, CH_2, CH_3, CH_4, H, N, C, Ar,$$
$$H^+, N^+, C^+, Ar^+, CN^+, N_2^+, \text{ and e}$$

It is possible, however, that other species such as C_3 can also be taken into account.

An example of the evolution of the mass fractions of neutral species is shown in Fig. 161 (initial concentrations: N_2: 92% ; CH_4: 3% ; Ar: 5%).

CO_2/N_2 mixture

For this, 19 reactions are taken into account,[216] with the following ten species:

$$CO_2, CO, C, C_2, CN, O_2, O, N_2, N, \text{ and NO}$$

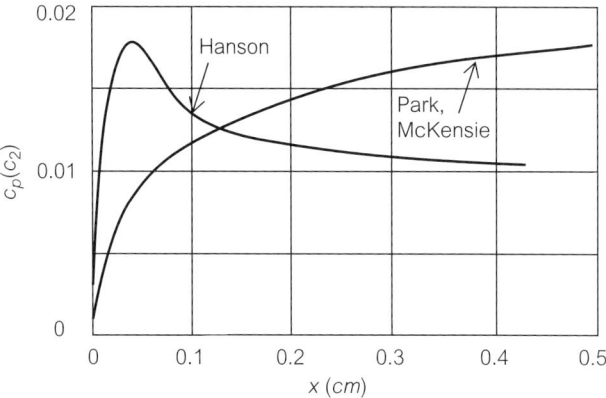

Figure 163. Influence of values of the rate constants on the evolution of the C_2 concentration.

An example of the evolution of the mass fractions is given in Fig. 162 (CO_2: 70% ; N_2: 30%).

Pure CO

Five reactions are taken into account[200] for pure CO, with the following four species:

$$CO, C_2, C, \text{ and } O$$

The influence of the chosen values of the rate constants,[216–218] particularly on the concentration of C_2, is important (Fig. 163).

References

1. S. Chapman, T.G. Cowling, *The Mathematical Theory of Non Uniform Gases*, Cambridge University Press, Cambridge (1970).
2. J.H. Ferziger, H.G. Kaper, *Mathematical Theory of Transport Processes in Gases*, North Holland, New York (1972).
3. J.O. Hirschfelder, C.F. Curtiss, R.B. Bird, *Molecular Theory of Gases and Liquids*, J. Wiley, New York (1959).
4. M.N. Kogan, *Rarefied Gas Dynamics*, Plenum Press, New York (1969).
5. C. Cercignani, *Mathematical Methods in Kinetic Theory*, Macmillan, London (1975).
6. J.L. Waldmann, *Handbuch der Physik*, Vol.12, Springer-Verlag, Berlin (1958).
7. W.G. Vincenti, C.H. Krüger, *Introduction to Physical Gas Dynamics*, R.E. Krieger, Florida (1965).
8. H. Grad, *Comm. Pure Appl. Math.*, 2, 4 (1949).
9. R.M. Velasco, F.G. Uribe, *Physica A*, 134, 339 (1986).
10. J.G. Parker, *Phys. Fluids*, 2, 449 (1959).
11. R.C. Millikan, D.R. White, *J. Chem. Phys.*, 39, 3209 (1963).
12. N. Belouaggadia, R. Brun, *J. Therm. Heat Transf.*, 12, 4, 482 (1998).
13. R. Brun, *Transport et relaxation dans les écoulements gazeux*, Masson, Paris (1986).
14. R. Brun, B. Zappoli, *Phys. Fluids*, 20, 9, 1441 (1977).
15. R.N. Schwartz, Z.I. Slavsky, K.F. Herzfeld, *J. Chem. Phys.*, 20, 1591 (1952).
16. R.N. Schwartz, K.F. Herzfeld, *J. Chem. Phys.*, 22, 767 (1954).
17. C.E. Treanor, J.W. Rich, R.G. Rehm, *J. Chem. Phys.*, 48, 1798 (1967).
18. T.I. MacLaren, J.P. Appleton, *Shock Tube Research*, 27, Chapman and Hall, London (1971).
19. R. Brun, *AIAA Paper* 88-2655 (1988).
20. C.S. Wang Chang, G.E. Uhlenbeck, *Univ. Michigan Report*, CM 681 (1951).
21. C.S. Wang-Chang, G.E. Uhlenbeck, J.De Boer, *Studies in Statistical Mechanics*, II, 242, North Holland Publishing Company, Amsterdam (1964).
22. E.A. Mason, L. Monchick, *J. Chem. Phys.*, 36, 1622 (1962).
23. A. Eucken, *Phys. Z.*, 14, 324 (1913).
24. L. Monchick, A.N. Pereira, E.A. Mason, *J. Chem. Phys.*, 42, 3241 (1965).
25. S.C. Saxena, E.A. Mason, *Molec. Phys.*, 2, 379 (1959).
26. L. Monchick, K.S. Yun, E.A. Mason, *J. Chem. Phys.*, 39, 654 (1963).
27. R. Brun, G. Duran, P.C. Philippi, M.F. Dourrieu, R. Tosello, *Progress in Astro. and Aero.*, 88, 299 (1983).
28. S. Pascal, R. Brun, *Phys. Rev. E*, 47, 5, 3251 (1993).
29. E.P. Gross, E.A. Jackson, *Phys. Fluids*, 12, 432 (1959).
30. F.B. Hanson, T.F. Morse, *Phys. Fluids*, 10, 345 (1967).

31. J. Mac Cormack, *Phys. Fluids*, 16, 2095 (1973).
32. P.C. Philippi, R. Brun, *Physica*, 105A, 147 (1981).
33. C.D. Levermore, *J. Stat. Phys.*, 83, 5–6, 1021 (1996).
34. C.E. Treanor, P.V. Marrone, *Phys. Fluids*, 5, 9, 1022 (1962).
35. P.V. Marrone, C.E. Treanor, Phys. Fluids, 6, 9, 1215 (1963).
36. C. Park, *J. Therm. Heat Transf.*, 2, 1, 8 (1988).
37. C. Park, *Molecular Physics and Hypersonic Flow*, NATO-ASI Series C, 482, 581, Kluwer Acad. Pub., Dordrecht (1996).
38. S.O. Macheret, J.W. Rich, *AIAA Paper* 93–2860 (1993).
39. R. Brun, A. Chikaoui, *Rarefied Gas Dynamics*, 2, 743, CEA, Paris (1979).
40. M.N. Kogan, V.S. Galkin, N.K. Makashev, *Rarefied Gas Dynamics*, 2, 653, CEA, Paris (1979).
41. R. Brun, M.P. Villa, J.G. Meolans, *Rarefied Gas Dynamics*, 2, 593, Univ. Tokyo Press, Tokyo (1984).
42. R. Brun, N. Belouaggadia, *3rd Eur. Symp. on Aero. for Space Vehicles*, 266, ESTEC, Nordwijk (1998).
43. C.R. Wilke, *J. Chem. Phys.*, 18, 4 (1949).
44. J.M. Yos, *Tech. Memor.*, TM63.7 AVCO-RAD, Wilmington (1963).
45. P.D. Neufeld, A.R. Janzon, A.R. Aziz, *J. Chem. Phys.*, 57, 3 (1972).
46. R. Brun, J.P. El Haouari, *Rarefied Gas Dynamics*, 185, Plenum Press, New York (1985).
47. P. Bhatnagar, E.P. Gross, M.K. Krook, *Phys. Rev.*, 94, 511 (1954).
48. T.F. Morse, *Phys. Fluids*, 7, 159 (1964).
49. L.B. Thomas, *Rarefied Gas Dynamics*, 1, 155, Academic Press, New York (1967).
50. P. Welander, *Ark. For Fysik*, 7, 507, (1954).
51. R. Goniak, *Thèse de Doctorat*, Université de Provence, Marseille (1994).
52. F. Nasuti, M. Barbato, C. Bruno, *J. Therm. Heat Transf.*, 10, 131 (1996).
53. E.J. Jumper, *Molecular Physics and Hypersonic Flow*, NATO-ASI Series C, 482, 181, Kluwer Acad. Pub., Dordrecht (1996).
54. M. Larini, R. Brun, *Int. J. Heat Mass Transf.*, 16, 2189 (1973).
55. C.D. Scott, *NASA* TM X. 58111 (1973).
56. R.N. Gupta, C.D. Scott, J.N. Moss, *AIAA Paper* 84–1732 (1984).
57. G.N. Patterson, *Molecular Flow of Gases*, J. Wiley and Sons Inc., New York (1956).
58. G.A. Bird, *Molecular Gas Dynamics*, Oxford Univ. Press (1976).
59. K. Nanbu, *J. Phys. Soc. Jap.*, 52, 2654 (1983).
60. K. Koura, *Phys. Fluids*, 9, 11, 3543 (1997).
61. C. Borgnake, P.S. Larsen, *J. Comput. Phys.*, 18, 405 (1975).
62. G.A. Bird, *Rarefied Gas Dynamics*, 175, Univ. Tokyo Press, Tokyo (1984).
63. A. Jarrar, *Thèse de Doctorat*, Université de Provence, Marseille (1987).
64. R. Brun, *Entropie*, 164 (1992).
65. R.A. Hartunian, R.L. Warvig, *Phys. Fluids*, 5, 2 (1952).
66. A.J. Chabai, R.J. Emrich, *J. Appl. Phys.*, 26, 779 (1955).

67. M.E. Ben Salah, R. Brun, *J. Mec. Theo. Appl.* 5, 2, 217 (1986).
68. S. Chandrasekhar, *Hydodynamic and Hydromagnetic Stability*, Dover Pub., Oxford Univ. Press, Oxford (1961).
69. J. Galan, R. Brun, M. Fortunato, J. Papin, *Int. J. Heat Mass Transf.*, 26, 10, 1453 (1983).
70. W.H. Dorrance, *Viscous Hypersonic Flow*, McGraw Hill, New York (1962).
71. J.C. Slattery, *Momentum, Energy and Mass Transfer in Continua*, McGraw Hill, New York (1972).
72. F.F. Chen, *Introduction to Plasma Physics and Controlled Fusion*, Plenum Press, New York (1984).
73. D.R. Nicholson, *Introduction to Plasma Physics*, Krieger Publishing, Malabar FL (1992).
74. T.J.M. Boyd, J.J. Sanderson, *The Physics of Plasmas*, Cambridge Univ. Press, Cambridge (2003).
75. D.A. Gurnett, A. Bhattacharjee, *Introduction to Plasma Physics*, Cambridge Univ. Press, Cambridge (2005).
76. T.G. Cowling, *Magnetohydrodynamics*, Interscience, New York (1957).
77. Shih.I. Pai, *Magnetogasdynamics and Plasma Dynamics*, Springer-Verlag, Vienna (1962).
78. J.A. Bittencourt, *Fundamentals of Plasma Physics*, Springer, New York (2004).
79. T.K. Bose, *High Temperature Gas Dynamics*, Springer, Berlin (1979).
80. A.H. Shapiro, *The Dynamics and Thermodynamics of Compressible Fluid Flow*, Ronald Press Company, New York (1948).
81. R. Courant, K.O. Friedrichs, *Supersonic Flow and Shock Waves*, Interscience, New York (1948).
82. W.D. Hayes, R.F. Probstein, *Hypersonic Flow Theory*, Academic Press, New York (1966).
83. A.H.W. Liepmann, A. Roshko, *Elements of Gas Dynamics*, J. Wiley and Sons, New York (1957).
84. G. Rudinger, *Nonsteady Duct Flow, Wave Diagram Analysis*, Dover Pub. Inc., New York (1969).
85. H.M. Mott-Smith, *Phys. Rev.*, 82, 885 (1951).
86. D.R. Hartree, *Numerical Analysis*, Oxford Univ. Press, Oxford (1958).
87. G. Ben-Dor, *Phenomena of Shock Reflections*, Springer-Verlag, Berlin (1992).
88. Z. Han, X. Yin, *Shock Dynamics*, Kluwer Acad. Pub., Dordrecht (1993).
89. M. Van Dyke, *Perturbation Methods in Fluid Mechanics*, Academic Press, New York (1975).
90. F.K. Moore, *Theory of Laminar Flow*, IV, Princeton Univ. Press (1964).
91. J.P. Brazier, B. Aupoix, J. Cousteix, *C.R. Acad. Sci.*, 310, II, 1583, Paris (1990).
92. C.B. Cohen, E. Reshotko, *NACA Rep.* 1293 (1956).
93. M. Lesieur, *Turbulence in Fluids*, Kluwer Acad. Pub., Dordrecht (1990).
94. A. Favre, *J. Mec.*, 4, 361 & 391 (1969).
95. T. Cebeci, A.M.O. Smith, *Analysis of Turbulent Boundary Layer*, Academic Press, New York (1974).
96. J.S. Shang, S.T. Suzhikov, *AIAA J.*, 43, 8, 1633 (2005).
97. R. Meyer, N. Nishihara, A. Hicks, N. Chintola, M. Cundy, W.R. Lempert, I. Adamovich, S. Gogimeni, *AIAA J.*, 43, 9, 1923 (2005).
98. J.P. Leoni, *Thèse de Doctorat*, Université de Provence, Marseille (1984).
99. J.F. Clarke, M. McChesney, *Dynamics of Relaxing Gases*, Butterworths, London (1976).

100. A.G. Gaydon, I.R. Hurle, *The Shock Tube in High Temperature Chemical Research*, Chapman and Hall, London (1963).
101. J. William, *Thèse de Doctorat*, Université de Provence, Marseille (1999).
102. I. Armenise, M. Capitelli, G. Colonna, C. Gorce, *J. Therm. Heat Transf.*, 10, 3 (1996).
103. M. Capitelli, I. Armenise, C. Gorce, *AIAA Paper* 96-1985 (1996).
104. M. Billiotte, *Thèse de Doctorat és Sciences*, Université de Provence, Marseille (1972).
105. C. Park, *AIAA Paper* 85–0247 (1985).
106. C. Park, *AIAA Paper* 89–1740 (1989).
107. N. Belouaggadia, T. Saito, R. Brun, K. Takayama, *Interdisciplinary Shock Wave Research*, 79, ISISW, Sendai (2004).
108. N. Belouaggadia, R. Brun, *J. Therm. Heat Transf.*, 20, 1, 148 (2006).
109. J.G. Meolans, C. Nicoli, R. Brun, *Progress in Astro. and Aero.*, 76, 106 (1981).
110. M.C. Druguet, D. Zeitoun, R. Brun, *High Performance Computing*, 479, Elsevier Science Publishers (1991).
111. P. Colas, *Thèse de Doctorat*, Université de Provence, Marseille (1989).
112. D. Zeitoun, E. Bocaccio, M.C. Druguet, M. Imbert, R. Brun, *AGARD Conf. Proc.* 514, 34, 1 (1993).
113. H.G. Hornung, *3rd Joint Europe/US Course in Hypersonics*, RWTH Aachen (1990).
114. F. Lordet, J.G. Meolans, A. Chauvin, R. Brun, *Shock Waves*, 4, 299 (1995).
115. J.G. Meolans, A. Chauvin, *AIAA Paper* 91–1340 (1991).
116. P. Gubernatis, *Thèse de Doctorat*, Université de Provence, Marseille (1989).
117. G.S.R. Sarma, *Progress in Aerospace Sciences*, 36, 281 (2000).
118. J.H. Lee, *AIAA Paper* 84-1729 (1984).
119. R. Brun, B. Zappoli, D. Zeitoun, *Phys. Fluids*, 22, 4, 786 (1979).
120. J.A. Fay, F.A. Riddel, *J. Aero. Sci.*, 73 (1958).
121. P.H. Rose, N.L. Stark, *J. Aero. Sci.*, (1957).
122. R. Goulard, *Jet Propulsion*, 737 (1958).
123. L. Lees, *Jet Propulsion*, 26, 259 (1956).
124. D. Zeitoun, R. Brun, *Int. J. Heat Mass Transf.*, 16, 427 (1973).
125. R. Brun, *Hypersonics I, Defining the Hypersonic Environment*, 81, Birkäuser, Boston (1989).
126. M.F. Dourrieu, *Thèse de Doctorat*, Université de Provence, Marseille (1980).
127. N. Belouaggadia, R. Brun, K. Takayama, *Shock Waves*, 16, 1, 16 (2006).
128. N. Belouaggadia, T. Hashimoto, S. Nonaka, K. Takayama, R. Brun, *AIAA J.*, 45, 6, 1420 (2007).
129. J. Anderson, *Gasdynamic Laser: An Introduction*, Academic Press, New York (1976).
130. D. Zeitoun, M. Maurel, M. Imbert, R. Brun, *AIAA J.*, 29, 3, 425 (1991).
131. R. Brun, S. Elkesslassy, I. Chemouni, *Progress in Astro. and Aero.*, 116, 542 (1989).
132. R. Brun, C. Chauvin, D. Zeitoun, *Mech. Res. Com.*, 8, 3, 187 (1981).
133. R.W. MacCormack, *AIAA Paper* 85–0032 (1985).
134. G.V. Candler, R.W. MacCormack, *AIAA Paper* 88–0511 (1988).
135. J.L. Steger, R.F. Warming, *J. Comput. Phys.*, 40, 263 (1981).

136. I.V. Rakich, D.A. Stewart, M.J. Lanfranco, *AIAA Therm. Fl. Plasma Heat Transf. Conf.*, St Louis (1982).

137. A. Ferri *(Ed)*, *Fundamental Data Obtained from Shock Tube Experiments*, Pergamon Press (1961).

138. G.N. Bradley, *Shock Waves in Chemistry and Physics*, London (1962).

139. H. Oertel, *Stossrohre*, Springer-Verlag, Wien (1966).

140. Y.V. Stupochenko, S.A. Losev, A.I. Osipov, *Relaxation in Shock Waves*, Springer-Verlag, Berlin (1967).

141. L.Z. Dumitrescu, *Cercetari in Tuburile de Soc*, Ed. Acad. Rep. Soc. Rom. Bucuresti (1969).

142. H. Mirels, *Phys. Fluids*, 6, 1201 (1963).

143. J. Sidés, R. Brun, *J. Mec.*, 14, 3, 387 (1975).

144. D. Zeitoun, M. Imbert, *AIAA J.*, 17, 8, 821 (1979).

145. H. Mirels, *AIAA J.*, 2, 1, 84 (1964).

146. C.J. Doolan, P.A. Jacobs, *AIAA J.*, 34, 6, 1291 (1996).

147. D.R. White, *J. Fl. Mech.*, 4, 585 (1958).

148. C.J.S. Simpson, T.R. Chandler, K.B. Bridgman, *Phys. Fluids*, 10, 1894 (1967).

149. R.S. Hickman, L.C. Farrar, J.B. Kyser, *Phys. Fluids*, 18, 1249 (1970).

150. R. Brun, R. Reboh, *AIAA J.*, 15, 9, 1344 (1977).

151. D. Zeitoun, R. Brun, M.J. Valetta, *Shock Tubes and Waves*, 180, The Magness Press, Jerusalem (1980).

152. K. Tajima, E. Outa, G. Nakada, *Bull. JSME*, 11, 43 (1968).

153. R. Brun, G. Muret, *C.R. Acad. Sci.*, Paris, 268, 902 A (1969).

154. R.A. Hartunian, A.L. Russo, P.V. Marrone, *J. Aerosp. Sci.*, 27, 8, 587 (1960).

155. J.L. Hall, *AIAA J.*, 13, 4, 519 (1975).

156. R. Brun, P. Auberger, N.G. Van Que, *Acta Astronautica*, 5, 1145 (1978).

157. A. Herzberg, W.E. Smith, M.S. Glick, W. Squire, *Cornell Aero. Lab. Rep.* AD-789 A.2 (1955).

158. H. Mark, *J. Aero. Sci.*, 24, 304 (1957).

159. M.P. Dumitrescu, *Shock Waves*, 1581, World Scientific, Singapore (1995).

160. A.J. Neely, R.G. Morgan, *Aeronaut. J.*, 946, 175 (1994).

161. H.R. Yu, B. Esser, M. Lennartz, H. Grönig, *Shock Waves*, 245, Springer-Verlag, Berlin, (1992).

162. R.J. Stalker, *AIAA J.*, 5, 12, 2160 (1967).

163. H. Hornung, *GALCIT Rep.* FM 88-1 (1988).

164. M.P. Dumitrescu, *Shock Waves*, 1487, World Scientific, Singapore (1995).

165. K. Itoh, *Shock Waves*, 43, World Scientific, Singapore (1995).

166. Y. Burtschell, R. Brun, D. Zeitoun, *Shock Waves*, 583, Springer-Verlag, Berlin (1992).

167. R. Brun, P. Humeau, J.C. Thomas, *J. Phys. Appl.*, 26, 15 (1965).

168. R. Marmey, J.P. Guibergia, R. Brun, *Rom. J. Tech. Sci. Appl. Mech.*, 16, 3, 645 (1971).

169. C. Chartier, R. Brun, Y.M. Grellier, *Les tubes à choc: Conception et recherches*, Pub. Sci Tech. Ministére de l'Air, N.T. 130 (1963).

170. P. Sagnier, G. François, *Shock Waves Marseille*, 233, Springer, Berlin (1995).

171. H. Grönig, *Shock Waves*, 3, Springer-Verlag, Berlin (1992).

172. A. Chaix, M.P. Dumitrescu, L.Z. Dumitrescu, R. Brun, *Shock Waves*, 21st ISSW, Paper 2061, Great Keppel (1997).
173. R. Brun, J.C. Clebant, H. Dahel, *Rev. Phys. Appl.*, 1, 2, 81 (1966).
174. R.D. Richtmeyer, *Com. Pure Appl. Math.*, 13, 297 (1960).
175. M.A. Levin, *Phys. Fluids*, 13, 5, 1166 (1970).
176. V.A. Andronov, S.M. Bakhrakh, E.E. Meshkov, V.V. Nikiforov, A.V. Pernitokii, A.I. Tolshmyakov, *Sov. Phys. Dokl.*, 27, 393 (1962).
177. L. Houas, R. Brun, M. Hanana, *AIAA J.*, 24, 8, 1254 (1986).
178. R. Brun, J.P. Guibergia, *J. Mec.*, 6, 1, 79, Paris (1967).
179. P. Dumas, *Thèse de Doctorat*, Université de Provence, Marseille (1979).
180. T.O. Lambert, *Vibrational and Rotational Relaxation in Gases*, Clarendon Press, Oxford (1977).
181. N.H. Johannesen, H.K. Zienziewitz, P.A. Blythe, J.H. Gerrard, *J. Fl. Mech.*, 13, 213 (1962).
182. C. Park, *Nonequilibrium Hypersonic Aerothermodynamics*, J. Wiley, New York (1990).
183. S. Sato, C. Tsuchiya, A. Kuratami, *J. Chem. Phys.*, 54, 3049 (1969).
184. C.W. Von Rosenberg, K.N.C. Bray, J. Pratt, *J. Chem. Phys.*, 56, 3230 (1971).
185. J.P. Martin, M.R. Buckingham, J.A. Chenery, C.J.S.M. Simpson, *J. Chem. Phys.*, 74, 15 (1983).
186. J.G. Meolans, R. Brun, *Rarefied Gas Dynamics*, 345, Teubner, Stuttgart (1986).
187. I.R. Hurle, A.L. Russo, J.G. Hall, *J. Chem. Phys.*, 40, 2076 (1964).
188. C.W. Von Rosenberg, R.L. Taylor, J.D. Teare, *J. Chem. Phys.*, 54, 5 (1974).
189. D.I. Sebacher, *Heat Transf. Fl. Mech. Institute*, 240, Stanford University Press (1966).
190. S. Sharma, S.M. Ruffin, W.D. Gillespie, S.A. Meyer, *J. Therm. Heat Transf.*, 7, 4, 697 (1993).
191. H. Pilverdier, R. Brun, M.P. Dumitrescu, *J. Therm. Heat Transf.*, 15, 4, (2001).
192. J.R. Busing, J.F. Clarke, *7th AGARD Coll.*, Oslo (1966).
193. R. Brun, A.K. Moustaghfir, J.G. Meolans, *Shock Waves*, 21st ISSW, Paper 1611 Great Keppel (1997).
194. R. Brun, J.P. Guibergia, *C.R. Acad. Sci.*, Paris, 268, 166A (1969).
195. R.K. Hanson, *Shock Waves @ Marseille*, II, 7, Springer, Berlin (1995).
196. P. Roth, *Aerothermochemistry of Spacecraft and Associated Hypersonic Flows*, 267, Jouve, Paris (1999).
197. D. Ramjaun, M.P. Dumitrescu, R. Brun, *J. Therm. Heat Transf.*, 13, 2, 219 (1999).
198. C.S. Park, D. Bershader, C. Park, *J. Therm. Heat Transf.*, 10, 4, 563 (1996).
199. M. Baillon, C. Taquin, J. Soler, *Shock Waves @ Marseille*, 2, 339, Springer, Berlin (1995).
200. D. Ramjaun, *Thèse de Doctorat*, Université de Provence, Marseille (1998).
201. C.D. Scott, *Molecular Physics and Hypersonic Flow*, NATO-ASI Series C, 482, 161, Kluwer Acad. Pub. (1996).
202. G. Eitelberg, *Int. Rep.*, DLR IB-223–96 A54 (1997).
203. P. Kolodziej, D.A. Stewart, *AIAA Paper* 87-1637 (1987).
204. P. Cauquot, S. Cavadias, J. Amouroux, *J. Therm. Heat Transf.*, 12, 2, 206 (1998).
205. Z.A. Zanguill, *Physics at Surfaces*, Cambridge Univ. Press, Cambridge (1988).
206. C.Y. Wen, H.G. Hornung, *J. Fl. Mech.*, 299, 389 (1996).
207. H. Olivier, *J. Fluid Mech.*, 413, 345 (2000).

208. S. Nonaka, K. Takayama, *AIAA Paper* 99–1025 (1999).
209. G. Herzberg, *Molecular Spectra and Molecular Structure*, I, Van Nostrand, New York (1966).
210. R. Engleman, *J. Mol. Spect.*, 49, 1, 106 (1974).
211. R.J. Spindler, *J. Quant. Spect. Rad. Transf.*, 5, 1, 165 (1965).
212. H.F. Nelson, C. Park, E.E. Whiting, *J. Therm. Heat Transf.*, 5, 2, 157 (1991).
213. J.O. Arnold, E.E. Whiting, G.L. Lyle, *J. Quant. Spect. Rad. Transf.*, 9, 4, 775 (1969).
214. L. Labracherie, *Thèse de Doctorat*, Université de Provence, Marseille (1994).
215. K. Koffi-Kpante, *Thèse de Doctorat*, Université de Provence, Marseille (1996).
216. F. Mazoué, *Thèse de Doctorat*, Université de Provence, Marseille (1994).
217. R.K. Hanson, *J. Chem. Phys.*, 60, 12 (1970).
218. C.S. Park, J.T. Howe, R.L. Jaffe, *J. Therm. Heat Transf.*, 8, 1, 9 (1994).

Index

Absorption
  coefficients, 319
  measurements, 366
Accommodation
  coefficients, 179
  measurements, 372
Air
  chemical kinetics, 274
  equilibrium chemistry, 263, 286
  ionization, 287

Blow-down tunnel
  arc, 326, 347
Boundary conditions
  Boltzmann equation, 178
  boundary layer, 246
  Navier–Stokes equations, 200
Boundary layer
  behind an incident shock, 300
  behind a reflected shock, 303
  equilibrium (in), 295
  general equations, 233
  nozzle (in), 301
  transformations, 244
  turbulent, 252
  typical cases, 247
  unsteady, 300, 316
  vibrational non-equilibrium, 300

Catalycity
  chemical, 298, 382
  vibrational, 313, 372
Chemical reactions, 259
Collision cross-section
  differential, 20
  total, 22
Collision frequency
  definition, 19
  elastic, 19
  inelastic, 20, 62
  Maxwellian regime, 58
Collision rate, 44, 270
Collisional integrals, 72, 78
  mixture, 117, 153
  pure gas, 92, 149, 163, 170
Collisional invariants, 15, 17, 47

Collisions
  elastic, 14, 19, 33
  generality, 14
  inelastic, 17
  models, 31
  reactive, 18
  resonant, 49
  TV, VV, 47
Concentrations
  mass, 105
  molar, 16, 260
  number density, 8
Conductivity
  approximate expressions, 86, 123, 137, 165
  pure gases, 74, 134
  mixtures, 105, 136
Control
  mass, volume, 220
Cramer systems
  mixtures, 113, 150
  pure gases, 91, 147

Density, 7
Diaphragm
  opening time, 351
Diffusion
  binary, 106
  Raman, 366, 370
  self-diffusion, 87
Dimensionless numbers
  Brinkmann, 205
  Damköhler, 204
  Eckert, 203
  Knudsen, 37
  Lewis, 205
  Mach, 206
  Nusselt, 207
  Peclet, 205
  Prandtl, 203
  Reynolds, 202
  Schmidt, 203
  Stanton, 207
  vibrational number, 214
Direct simulation
  Monte-Carlo methods, 183

Dissociation
  energy (of), 30
  rate constants, 140, 267, 374
Distribution function
  Boltzmann, 41
  Maxwell, 39
  Maxwell–Boltzmann, 41
  Treanor, 47
  Vibrational non-equilibrium, 43

Eigenfunctions, eigenvalues, 87
Emission
  infrared, 385
  line intensity, 388
  measurements, 361
  simulation, 387
  spectra, 376
Energy
  definitions, 8, 26
  electronic, 387
  flux, 10
  rotation, 27
  translation, 26
  vibration, 28
Enthalpy
  component, 105
  mixture, 197
  stagnation, total, 198
Entropy
  definition, balance, 199, 214
Equilibrium
  equilibrium rate constant, 55, 260
  regimes, 68, 295
Equations
  Boltzmann, 11
  conservation, 12, 212
  dimensionless, 202, 213
  Euler, 40
  Landau–Teller, 45, 361
  Navier–Stokes, 74, 80, 212
  particular forms, 197
  relaxation, 44, 319
  total balances, 220
Exchange
  coefficients, 313
  measurement, 372

Flow regime
  collisional, 37
  equilibrium (in), 38
  free molecular, 181
  hypersonic, 187
  non-equilibrium (in), 131
  SNE, WNE, 69
Flows
  around a body, 308
  dissipative, 231
  flat plate, 248
  isentropic, 199, 226
  nozzle (in), 292, 305
  one-dimensional, 226, 227
  reactive, 259
  stagnation point, 296
  two-dimensional, 210

Gas-dynamic laser, 311, 317

H theorem, 56
Heat flux
  interface, 216
  measurement, 341, 353
  translation, rotation, vibration, 10
  wall, 249

Ideal gas, 224
Interaction
  shock–shock, 241
  shock–interface, 242, 355
  vibration–dissociation, 141
  vibration–reaction, 275
  vibration–recombination, 142
Interaction potential
  Lennard–Jones, 32
  Morse, 29
  repulsive centre, 32
  rigid elastic sphere, 31
Interface
  gas–gas, 241, 327
  gas–liquid, 217
  gas–solid, 216
Interferometry, 342

Kinetics
  air, 263, 274
  chemical, 266, 374
  models, 394
  vibrational, 271, 300

Line broadening, 389

Magnetohydrodynamics
  equations, 221
  example of flow, 255
Mass action (law of), 260
Mass flux, 104, 196

Mean free path, 58
Methods
   BGK, 175
   Chapman–Enskog (CE), 70, 100
   Generalized Chapman–Enskog (GCE), 160
   Gross–Jackson, 124
   moments (of), 128
Mixtures
   binary, 100
   CO_2/N_2, 380, 394
   diatomic gases, 106
   monatomic gases, 101
   $N_2/CH_4/Ar$, 377, 394
   reactive gas, 53, 112
Models
   harmonic oscillator, 28
   kinetic, 394
   rigid rotator, 27
   turbulent boundary layer, 254
Momentum flux, 9

Non-equilibrium
   rotational, 80
   vibrational, 84
Nozzle
   calibration, 349
   hypersonic, 305
   ideal gas, 237
   source effect, 358
   supersonic, 237, 277
Numerical methods
   direct simulation Monte-Carlo, 183
   method of characteristics, 236
   unsteady methods, 323

Polynomials
   Hermite, 128
   Sonine–Laguerre, 88
   Wang-Chang–Uhlenbeck, 89
Pressure
   hydrostatic, 40
   measurements, 341
   relaxation, 134, 320

Quantities
   characteristic, 201
   state, 7
   transport, 9

Rate constants
   dissociation, 140
   reaction, 54
   recombination, 142

Real flight, 350
Relaxation time
   characteristic, 21
   rotation, 81
   vibration, 46, 361

Shock tube
   boundary layer, 330
   configurations, 338
   free piston, 340
   measurements, 341
   perturbations, 330
   precursor radiation, 391
   reflected shock, 334
   simple, 327
   test time, 352
   visualizations, 343
Shock tunnel
   calibration, 348
   free piston, 348
   reflected shock, 347
   simulation, 356
Shock wave
   generalized Rankine–Hugoniot, 315
   ideal gas, 230
   interactions, 241
   oblique, 329
   Rankine–Hugoniot relations, 229
   reflection, 240
   straight, 229
Similarity solutions, 247
Specific heat
   rotation, vibration, 59
   translation, 74
Spectroscopy
   absorption, 366
   emission, 377
   Raman, 366
Stress tensor, 196
Supersonic jets, 311

Temperature
   jump, 180
   measurement, 392
   rotation, 80
   translation, 9
   vibration, 48
Test time
   shock tube, 352
   shock tunnel, 348
Total balances, 208
Transition probability, 19

Transport
　quantities, 9
Turbulence
　boundary layer, 262

Velocity
　diffusion, 150
　macroscopic, 7
　mean quadratic, 58
　slip, 180
　thermal, 8

Vibrational populations
　determination, 366
　evolution, 44, 283
Vibrational relaxation
　general equations, 45
　linearization, 96
　mixtures, 52
　particular equations, 46
Viscosity
　approximate expressions, 122
　bulk, 79
　dynamic, 73